Springer Texts in Business and Economics

More information about this series at http://www.springer.com/series/10099

Felix Munoz-Garcia · Daniel Toro-Gonzalez

Strategy and Game Theory

Practice Exercises with Answers

 Springer

Felix Munoz-Garcia
School of Economic Sciences
Washington State University
Pullman, WA
USA

Daniel Toro-Gonzalez
School of Economics and Business
Universidad Tecnológica de Bolívar
Cartagena, Bolivar
Colombia

ISSN 2192-4333   ISSN 2192-4341   (electronic)
Springer Texts in Business and Economics
ISBN 978-3-319-81410-0   ISBN 978-3-319-32963-5   (eBook)
DOI 10.1007/978-3-319-32963-5

Printed on acid-free paper

This Springer imprint is published by Springer Nature
The registered company is Springer International Publishing AG Switzerland

The original version of the book was revised:
For detailed information please see erratum.
The erratum to this book is available at
10.1007/978-3-319-32963-5_11

Preface

This textbook presents worked-out exercises on Game Theory, with detailed step-by-step explanations, which both undergraduate and master's students can use to further understand equilibrium behavior in strategic settings. While most textbooks on Game Theory focus on theoretical results; see, for instance, Tirole (1991), Gibbons (1992) and Osborne (2004), they offer few practice exercises. Our goal is, hence, to complement the theoretical tools in current textbooks by providing practice exercises in which students can learn to systematically apply theoretical solution concepts to different fields of Economics and Business, such as industrial economics, public policy and regulation.

The textbook provides many exercises with detailed verbal explanations (97 exercises in total), which cover the topics required by Game Theory courses at the undergraduate level, and by most courses at the Masters level. Importantly, our textbook emphasizes the economic intuition behind the main results, and avoids unnecessary notation when possible, and thus is useful as a reference book regardless of the Game Theory textbook adopted by each instructor. Importantly, these points differentiate our presentation from that found in solutions manuals. Unlike these manuals, which can be rarely read in isolation, our textbook allows students to essentially read each exercise without difficulties, thanks to the detailed verbal explanations, figures, and intuitions. Furthermore, for presentation purposes, each chapter ranks exercises according to their difficulty (with a letter A to C next to the exercise number), allowing students to first set their foundations using easy exercises (type-A), and then move on to harder applications (type-B and C exercises).

Organization of the Book

We first examine games that are required in most courses at the undergraduate level, and then advance to more challenging games (which are often the content of master's courses), both in Economics and Business programs. Specifically, Chaps. 1–6 cover complete-information games, separately analyzing simultaneous-move and sequential-move games, with applications from industrial economics and regulation; thus helping students apply Game Theory to other fields of research.

Chapters 7–9 pay special attention to incomplete information games, such as signaling games, cheap talk games, and equilibrium refinements. These topics have experienced a significant expansion in the last two decades, both in the theoretical and applied literature. Yet to this day most textbooks lack detailed worked-out examples that students can use as a guideline, leading them to especially struggle with this topic, which often becomes the most challenging for both undergraduate and graduate students. In contrast, our presentation emphasizes the common steps to follow when solving these types of incomplete information games, and includes graphical illustrations to focus students' attention to the most relevant payoff comparisons at each point of the analysis.

How to Use This Textbook

Some instructors may use parts of the textbook in class in order to clarify how to apply certain solution concepts that are only theoretically covered in standard textbooks. Alternatively, other instructors may prefer to assign certain exercises as a required reading, since these exercises closely complement the material covered in class. This strategy could prepare students for the homework assignment on a similar topic, since our practice exercises emphasize the approach students need to follow in each class of games, and the main intuition behind each step. This strategy might be especially attractive for instructors at the graduate level, who could spend more time covering the theoretical foundations in class, asking students to go over our worked-out applications of each solution concept on their own. In addition, since exercises are ranked according to their difficulty, instructors at the undergraduate level can assign the reading of relatively easy exercises (type-A) and spend more time explaining the intermediate level exercises in class (type-B questions), whereas instructors teaching a graduate-level course can assume that students are reading most type-A exercises on their own, and only use class time to explain type-C (and some type-B) exercises.

Acknowledgments

We would first like to thank several colleagues who encouraged us in the preparation of this manuscript: Ron Mittlehammer, Jill McCluskey, and Alan Love. Ana Espinola-Arredondo reviewed several chapters on a short deadline, and provided extremely valuable feedback, both in content and presentation; and we extremely thankful for her insights. Felix is especially grateful to his teachers and advisors at the University of Pittsburgh (Andreas Blume, Esther Gal-Or, John Duffy, Oliver Board, In-Uck Park, and Alexandre Matros), and at the University of Barcelona (Carles Rafels, Marina Nunez, and Francisco Javier Martinez) who taught him Game Theory and Industrial Organization, instilling a passion for the use of these topics in applied settings which hopefully transpires in the following pages. We are also thankful to the "team" of teaching and research assistants, both at Washington State University and at Universidad Tecnologica de Bolivar, who helped us with this project over several years: Diem Nguyen, Gulnara Zaynutdinova, Donald Petersen, Qingqing Wang, Jeremy Knowles, Xiaonan Liu, Ryan Bain, Eric Dunaway, Tongzhe Li, Wenxing Song, Pitchayaporn Tantihkarnchana, Roberto Fortich, Jhon Francisco Cossio Cardenas, Luis Carlos Díaz Canedo, Pablo Abitbol, and Kevin David Gomez Perez. We also appreciate the support of the editors at Springer-Verlag, Rebekah McClure, Lorraine Klimowich, and Dhivya Prabha. Importantly, we would like to thank our wives, Ana Espinola-Arredondo and Ericka Duncan, for supporting and inspiring us during the (long!) preparation of the manuscript. We would not have been able to do it without your encouragement and motivation.

Felix Munoz-Garcia
Daniel Toro-Gonzalez

Contents

Dominance Solvable Games

Introduction

This chapter first analyzes how to represent games in normal form (using matrices) and in extensive form (using game trees). We afterwards describe how to systematically detect strictly dominated strategies, i.e., strategies that a player would not use regardless of the action chosen by his opponents.

Strictly dominated strategies. Player i finds strategy s_i^* as strictly dominated by another strategy s_i' if

$$u_i\left(s_i', s_{-i}\right) > u_i(s_i^*, s_{-i}) \text{ for every strategy profile } s_{-i} \in S_{-i}$$

where $s_{-i} = (s_1, s_2, .., s_{i-1}, s_{i+1}, \ldots, s_N)$ represents the profile of strategies selected by player i's opponents, i.e., a vector with $N - 1$ components. In words, strategy s_i' strictly dominates s_i^* if it yields a strictly higher utility than strategy s_i^* *regardless* of the strategy profile that player i's rival choose.

Since we can anticipate that strictly dominated strategies will not be selected by rational players, we apply the Iterative Deletion of Strictly Dominated Strategies (IDSDS) to predict players' behavior. We elaborate on the application of IDSDS in games with two and more players, and in games where players are allowed to choose between two strategies, between more than two strategies, or a continuum of strategies. In some games, we will show that the application of IDSDS is powerful, as it rules out dominated strategies and leaves us with a relatively precise equilibrium prediction, i.e., only one or two strategy profiles surviving the application of IDSDS. In other games, however, we will see that IDSDS "does not have a bite" because no strategies are dominated; that is, a strategy does not provide a strictly lower payoff to

The original version of the chapter was revised: The erratum to the chapter is available at: 10.1007/978-3-319-32963-5_11

player *i* regardless of the strategy profile selected by his opponents (it can provide a higher payoff under some of his opponents' strategies). In this case, we won't be able to offer an equilibrium prediction, other than to say that the entire game is our most precise equilibrium prediction! In subsequent chapters, however, we explore other solution concepts that provide more precise predictions that IDSDS.

Finally, we study the deletion of *weakly* dominated strategies does not necessarily lead to the same equilibrium outcomes as IDSDS, and its application is in fact sensible to deletion order. We can apply the above definition of strictly dominated strategies to define weakly dominated strategies. Specifically, we say that strategy s_i' weakly dominates s_i^* if

$$u_i(s_i', s_{-i}) \geq u_i(s_i^*, s_{-i}) \text{ for every strategy profile } s_{-i} \in S_{-i}, \text{ and}$$

$$u_i(s_i', s_{-i}) \geq u_i(s_i^*, s_{-i}) \text{ for at least one strategy profile } s_{-i} \in S_{-i}.$$

Exercise 1—From Extensive Form to Normal form Representation-I[A]

Represent the extensive-form game depicted in Fig. 1.1 using its normal-form (matrix) representation.

Answer
We start identifying the strategy sets of all players in the game. The cardinality of these sets (number of elements in each set) will determine the number of rows and columns in the normal-form representation of the game.

Starting from the initial node (in the root of the game tree located on the left-hand side of the figure), Player 1 must select either strategy A or B, thus implying that the strategy space for player 1, S_1, is:

$$S_1 = \{A, B\}$$

In the next stage of the game, Player 2 conditions his strategy on player 1's choice, since player 2 observes such a choice before selecting his own. We need to consider that the strategy profile of player 2 (S_2) must be a complete plan of action (complete contingent plan). Therefore, his strategy space becomes:

Fig. 1.1 Extensive-form game

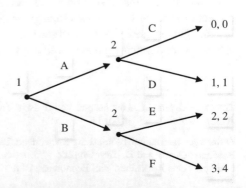

$$S_2 = \{CE, CF, DE, DF\}$$

where the first component of every strategy describes how player 2 responds upon observing that player 1 chose A, while the second component represents player 2's response after observing that player 1 selected B. For example, strategy CE describes that player 2 responds with C after player 1 chooses A, but with E after player 1 chooses B.

Using the strategy space of player 1, with only two available strategies $S_1 = \{A, B\}$, and that of player 2, with four available strategies $S_2 = \{CE, CF, DE, DF\}$, we obtain the 2×4 payoff matrix represented in Fig. 1.2. For instance, the payoffs associated with the strategy profile where player 1 chooses A and player 2 chooses C if A and E if B, $\{A, CE\}$, is (0,0).

Remark: Note that, if player 2 could not observe player 1's action before selecting his own (either C or D), then player 2's strategy space would be $S_2 = \{C, D\}$, implying that the normal form representation of the game would be a 2×2 matrix with A and B in rows for player 1, and C and D in columns for player 2.

Exercise 2—From Extensive Form to Normal Form Representation-II^A

Consider the extensive form game in Fig. 1.3

Part (a) Describe player 1's strategy space.
Part (b) Describe player 2's strategy space.
Part (c) Take your results from parts (*a*) and (*b*) and construct a matrix representing the normal form game of this game tree.

Answer

Part (a) Player 1 has three available strategies, implying a strategy space of $S_1 = \{H, M, L\}$
Part (b) Since in this game players act sequentially, the second mover can condition his move on the first player's action. Hence, player 2's strategy set is

$$S_2 = \{aaa, aar, arr, rrr, rra, raa, ara, rar\}$$

where each of the strategies represents a complete plan of action that specifies player 2's response after player 1 chooses H, after player 1 selects M, and after player 1 chooses L, respectively. For instance *arr* indicates that player 2 responds

	Player 2			
Player 1	CE	CF	DE	DF
A	0, 0	0, 0	1, 1	1, 1
B	2, 2	3, 4	2, 2	3, 4

Fig. 1.2 Normal-form game

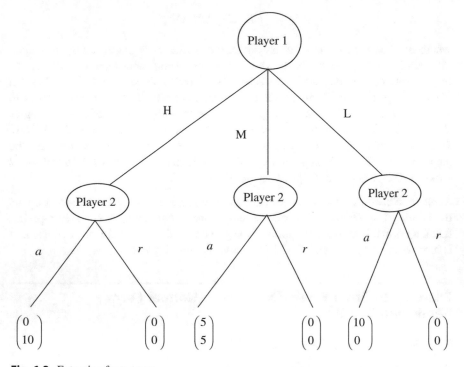

Fig. 1.3 Extensive-form game

with *a* after observing that player 1 chose *H*, with *r* after observing *M*, and with *r* after observing *L*.

Remark: If player 2 could not observe player 1's choice, the extensive-form representation of the game would depict a long dashed line connecting the three nodes at which player 2 is called on to move. (This dashed line is often referred as player 2's "information set") In this case, player 2 would not be able to condition his choice (since he cannot observe which action player 1 chose before him), thus implying that player 2's set of available actions reduces to only two (accept or reject), i.e., $S_2 = \{a, r\}$.

Part (c) If we take the three available strategies for player 1, and the above eight strategies for player 2, we obtain the following normal form game (Fig. 1.4).

Notice that this normal form representation contains the same payoffs as the game tree. For instance, after player 1 chooses *M* (in the second row), payoff pairs

Player 1		Player 2							
		aaa	*aar*	*arr*	*rrr*	*rra*	*raa*	*ara*	*rar*
	H	0,10	0,10	0,10	0,0	0,0	0,0	0,10	0,0
	M	5,5	5,5	0,0	0,0	0,0	5,5	0,0	5,5
	L	10,0	0,0	0,0	0,0	10,0	10,0	10,0	0,0

Fig. 1.4 Normal-form game

only depend on player 2's response after observing M (the second component of every strategy triplet in the columns). Hence, payoff pairs are either (5, 5), which arise when player 2 responds with a after M, or (0, 0), which emerge when player 2 responds instead with r after observing M.

Exercise 3—From Extensive Form to Normal Form Representation-III[B]

Consider the extensive-form game in Fig. 1.5. Provide its normal form (matrix) representation.

Answer
Player 2. From the extensive form game, we know player 2 only plays once and has two available choices, either A or B. The dashed line connecting the two nodes at which player 2 is called on to move indicates that player 2 cannot observe player 1's choice. Hence, he cannot condition his choice on player 1's previous action, ultimately implying that his strategy space reduces to $S_2 = \{A, B\}$.
Player 1. Player 1, however, plays twice (in the root of the game tree, and after player 2 responds) and has multiple choices:

1. First, he must select either U or D, at the initial node of the tree, i.e., left-hand side of the figure;

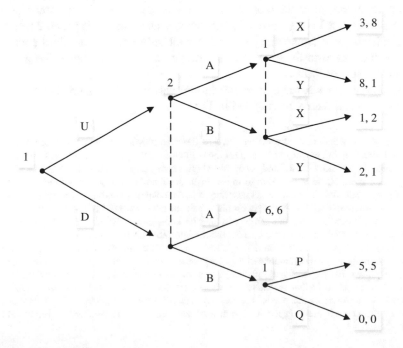

Fig. 1.5 Extensive-form game

Fig. 1.6 Normal-form game

	2	
1	A	B
UXP	3, 8	1, 2
UXQ	3, 8	1, 2
UYP	8, 1	2, 1
UYQ	8, 1	2, 1
DXP	6, 6	5, 5
DXQ	6, 6	0, 0
DYP	6, 6	5, 5
DYQ	6, 6	0, 0

2. Then choose X or Y, in case that he played U at the beginning of the game. (Note that in this event he cannot condition his choice on player 2's choice, since he cannot observe whether player 2 selected A or B); and
3. Then choose P or Q, which only becomes available to player 1 in the event that player 2 responds with B after player 1 chose D.

Therefore, player 1's strategy space is composed of triplets, as follows,

$$S_1 = \{UXP, UXQ, UYP, UYQ, DXP, DXQ, DYP, DYQ\}$$

whereby the first component of every triplet describes player 1's choice at the beginning of the game (the root of the game tree), the second component represents his decision (X or Y) in the event that he chose U and afterwards player 2 responded with either A or B (something player 1 cannot observe), and the third component reflects his choice in the case that he chose D at the beginning of the game and player 2 responds with B.[1]

As a consequence, the normal-form representation of the game is given by the following 8×2 matrix represented in Fig. 1.6.

---

[1]You might be wondering why do we have to describe player 1's choice between X and Y in triplets indicating that player 1 selected D at the beginning of the game. The reason for this detailed description is twofold: on one hand, a complete contingent plan must indicate a player's choices at every node at which he is called on to move, even those nodes that would not emerge along the equilibrium path. This is an important description in case player 1 submits his complete contingent plan to a representative who will play on his behalf. In this context, if the representative makes a mistake and selects U, rather than D, he can later on know how to behave after player 2 responds. If player 1's contingent plan was, instead, incomplete (not describing his choice upon player 2's response), the representative would not know how to react afterwards. On the other hand, a players' contingent plan can induce certain responses from a player's opponents. For instance, if player 2 knows that player 1 will only plays Q in the last node at which he is called on to move, player 2 would have further incentives to play A. Hence, complete contingent plans can induce certain best responses from a player's opponents, which we seek to examine. (We elaborate on this topic in the next chapters, especially when analyzing equilibrium behavior in sequential-move games.).

Exercise 4—Representing Games in Its Extensive Form[A]

Consider the standard rock-paper-scissors game, which you probably played in your childhood. If you did not play this old game before, do not worry, we will explain it next. Two players face each other with both hands on their back. Then each player simultaneously chooses rock (R), paper (P) or scissors (S) by rapidly moving one of his hands to the front, showing his fits (a symbol of a rock), his extended palm (representing a paper), or two of his fingers in form of a V (symbolizing a pair of scissors). Players seek to select an object that is ranked superior to that of his opponent, where the ranking is the following: scissors beat paper (since they cut it), paper beats rock (because it can wrap it over), and rock beats scissors (since it can smash them). For simplicity, consider that a player obtains a payoff of 1 when his object wins, -1 when it losses, and 0 if there is a tie (which only occurs when both players select the same object). Provide a figure with the extensive-form representation of this game.

Answer

Since the game is simultaneous, the extensive-form representation of this game will have three branches in its root (initial node), corresponding to Player 1's choices, as in the game tree depicted in Fig. 1.7. Since Player 2 does not observe Player 1's choice before choosing his own, Player 2 has three available actions (Rock, Paper and Scissors) which cannot be conditioned on Player 1's actual choice. We graphically represent Player 2's lack of information when he is called on to move by connecting Player 2's three nodes with an information set (dashed line in Fig. 1.7).

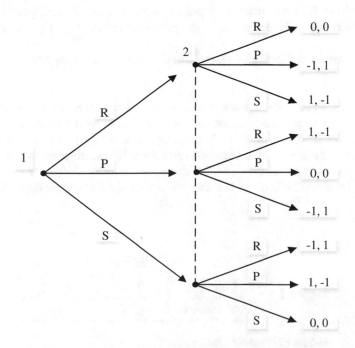

Fig. 1.7 Extensive-form of the Rock, Paper and Scissors game

Finally, to represent the payoffs at the terminal nodes of the tree, we just follow the ranking specified above. For instance, when player 1 chooses rock (R) and player 2 selects scissors (S), player 1 wins, obtaining a payoff of 1, while player 2 losses, accruing a payoff of -1, this set of payoffs entails the payoff pair $(1, -1)$. If, instead, player 2 selected paper, he would become the winner (since paper wraps the rock), entailing a payoff of 1 for player 2 and -1 for player 1, that is $(-1, 1)$. Finally, notice that in those cases in which the objects players display coincide, i.e., $\{R, R\}$, $\{P, P\}$ or $\{S, S\}$, the payoff pair becomes $(0, 0)$.

Exercise 5—Prisoners' Dilemma Game[A]

Two individuals have been detained for a minor offense and confined in separate cells. The investigators suspect that these individuals are involved in a major crime, and separately offer each prisoner the following deal, as depicted in Fig. 1.8: if you confess while your partner doesn't, you will leave today without serving any time in jail; if you confess and your partner also confesses, you will have to serve 5 years in jail (since prosecutors probably can accumulate more evidence against each prisoner when they both confess); if you don't confess and your partner does, you will have to serve 15 years in jail (since you did not cooperate with the prosecution but your partner's confession provides the police with enough evidence against you); finally, if neither of you confess, you will have to serve one year in jail.

Part (a) Draw the prisoners' dilemma game in its extensive form representation.
Part (b) Mark its initial node, its terminal nodes, and its information set. Why do we represent this information set in the prisoners' dilemma game in its extensive form?
Part (c) How many strategies player 1 has? What about player 2?

Answer

Part (a) Since both players must simultaneously choose whether or not to confess, player 2 cannot condition his strategy on player 1's decision (which he cannot observe). We depict this lack of information by connecting both of the nodes at which player 2 is called on to move with an information set (dashed line) in Fig. 1.9.
Part (b) Its initial node is the "root" of the game tree, whereby player 1 is called on to move between Confess and Not confess, the terminal nodes are the nodes where

		Prisoner 2	
		Confess	Not confess
Prisoner 1	Confess	-5,-5	0,-15
	Not confess	-15,0	-1,-1

Fig. 1.8 Normal-form of Prisoners' Dilemma game

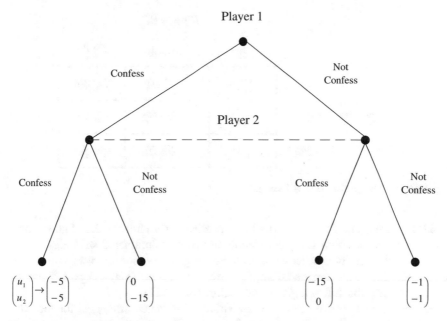

Player 1

Confess

Not
Confess

Player 2

Confess

Not
Confess

Confess

Not
Confess

$$\begin{pmatrix} u_1 \\ u_2 \end{pmatrix} \rightarrow \begin{pmatrix} -5 \\ -5 \end{pmatrix}$$
$$\begin{pmatrix} 0 \\ -15 \end{pmatrix}$$
$$\begin{pmatrix} -15 \\ 0 \end{pmatrix}$$
$$\begin{pmatrix} -1 \\ -1 \end{pmatrix}$$

Fig. 1.9 Prisoners' dilemma game in its extensive-form

the game ends (and where we represent the payoffs that are accrued to every player), and the information set is a dashed line connecting the nodes in which the second mover is called to move. We represent this information set to denote that the second mover is choosing whether to Confess or Not Confess without knowing exactly what player 1 did.
Part (c) Player 1 only has two possible strategies: $S_1 = \{$Confess, Not Confess$\}$. The second player has only two possible strategies $S_2 = \{$Confess, Not Confess$\}$ as well, since he is not able to observe what player 1 did before taking his decision. As a consequence, player 2 cannot condition his strategy on player 1's choice.

Exercise 6—Dominance Solvable Games[A]

Two political parties simultaneously decide how much to spend on advertising, either low, medium or high, yielding the payoffs in the following matrix (Fig. 1.10) in which the Red party chooses rows and the Blue party chooses columns. Find the strategy profile/s that survive IDSDS.

Answer
Let us start by analyzing the Red party (row player). First, note that High is a strictly dominant strategy for the Red party, since it yields a higher payoff than both Low and Middle, regardless of the strategy chosen by the Blue party (i.e., independently of the column the Blue party selects). Indeed, $100 > 80 > 50$ when the Blue party chooses the Low column, $80 > 70 > 0$ when the Blue party selects the

		Blue party		
		Low	Middle	High
	Low	80,80	0,50	0,100
Red party	Middle	50,0	70,70	20,80
	High	100,0	80,20	50,50

Fig. 1.10 Political parties normal-form game

Middle column, and $50 > 20 > 0$ when the Blue party chooses the High column. As a consequence, both Low and Middle are strictly dominated strategies for the Red party (they are both strictly dominated by High in the bottom row), and we can thus delete the rows corresponding to Low and Middle from the payoff matrix, leaving us with the following reduced matrix (Fig. 1.11).

We can now check if there are any strictly dominated strategies for the Blue party (in columns). Similarly as for the Red party, High strictly dominates both Low and Middle since $50 > 20 > 0$; and we can thus delete the columns corresponding to Low and Middle from the payoff matrix, leaving us with a single cell, (High, High). Hence, (High, High) is the unique strategy surviving IDSDS.

Exercise 7—Applying IDSDS (Iterated Deletion of Strictly Dominated Strategies)[A]

Consider the simultaneous-move game depicted in Fig. 1.12., where two players choose between several strategies.

Find which strategies survive the iterative deletion of strictly dominated strategies, IDSDS.

Answer
Let us start by identifying the strategies of player 1 that are strictly dominated by other of his own strategies. When player 1 chooses a, in the first row, his payoff is either 1 (when player 2 chooses x or y) or zero (when player 2 chooses z, in the third

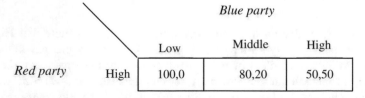

		Blue party		
		Low	Middle	High
Red party	High	100,0	80,20	50,50

Fig. 1.11 Political parties reduced normal-form game

Fig. 1.12 Normal-form
game with four available
strategies

Player 2

Player 1

	x	y	z
a	1,2	1,2	0,3
b	4,0	1,3	0,2
c	3,1	2,1	1,2
d	0,2	0,1	2,4

column). These payoffs are unambiguously lower than those in strategy c in the third row. In particular, when player 2 chooses x (in the first column), player 1 obtains a payoff of 3 with c but only a payoff of 1 with a; when player 2 chooses y, player 1 earns 2 with c but only 1 with a; and when player 2 selects z, player 1 obtains 1 with c but a zero payoff with a. Hence, player 1's strategy a is strictly dominated by c, since the former yields a lower payoff than the latter regardless of the strategy that player 2 selects (i.e., regardless of the column he uses). Thus, the strategies of player 1 that survive one round of the iterative deletion of strictly dominated strategies (IDSDS) are b, c and d, as depicted in the payoff matrix in Fig. 1.13.

Let us now turn to player 2 (by looking at the second payoff within every cell in the matrix). In particular, we can see that strategy z strictly dominates x, since it provides to player 2 a larger payoff than x regardless of the strategy (row) that player 1 uses Specifically, when player 1 chooses b (top row), player 2 obtains a payoff of 2 by selecting z (see the right-hand column) but only a payoff of zero from choosing x (in the left-hand column). Similarly, when player 1 chooses c (in the middle row), player 2 earns a payoff of 2 from z but only a payoff of 1 from x. Finally, when player 1 selects d (in the bottom row), player 2 obtains a payoff of 4 from z but only a payoff of 2 from x. Hence, strategy z yields player 2 a larger payoff independently of the strategy chosen by player 1, i.e., z strictly dominates x, which allows us to delete strategy x from the payoff matrix. Thus, the strategies of player 2 that survive one additional round of the IDSDS are y and z, which helps us further reduce the payoff matrix to that in Fig. 1.14.

We can now move to player 1 again. For him, strategy c strictly dominates b, since it provides an unambiguously larger payoff than b regardless of the strategy selected by player 2 (regardless of the column). In particular, when player 2 chooses

Fig. 1.13 Reduced
normal-form game after
one round of IDSDS

Player 2

Player 1

	x	y	z
b	4,0	1,3	0,2
c	3,1	2,1	1,2
d	0,2	0,1	2,4

Player 2

Player 1

	y	z
b	1,3	0,2
c	2,1	1,2
d	0,1	2,4

Fig. 1.14 Reduced normal-form game after two rounds of IDSDS

Player 2

Player 1

	y	z
c	2,1	1,2
d	0,1	2,4

Fig. 1.15 Reduced normal-form game

y (left-hand column), player 1 obtains a payoff of 2 from selecting strategy c but only one from strategy b. Similarly, if player 2 chooses z (in the right-hand column), player 1 obtains a payoff of one from strategy c but a payoff of zero from strategy b. As a consequence, strategy b is strictly dominated, which allows us to delete strategy b from the above matrix, obtaining the reduced matrix in Fig. 1.15.

At this point, returning to player 2, we note that z strictly dominates y, so we can delete strategy y for player 2. Finally, considering player 2 always chooses z, for player 1 strategy d strictly dominates c, since the payoff of 2 is higher than one unit derived from playing c. Therefore, our most precise equilibrium prediction after using IDSDS is remaining strategy profile (d,z), indicating that player 1 will always choose d, while player 2 will always select z.

Exercise 8—Applying IDSDS When Players Have Five Available Strategies[A]

Two students in the Game Theory course plan to take an exam tomorrow. The professor seeks to create incentives for students to study, so he tells them that the student with the highest score will receive a grade of A and the one with the lower score will receive a B. Student 1's score equals $x_1 + 1.5$, where x_1 denotes the amount of hours studying. (That is, he assume that the greater the effort, the higher her score is.) Student 2's score equals x_2, where x_2 is the hours she studies. Note that these score functions imply that, if both students study the same number of hours, $x_1 = x_2$, student 1 obtains a highest score, i.e., she might be the smarter of the two. Assume, for simplicity, that the hours of studying for the game theory class is an integer number, and that they cannot exceed 5 h, i.e., $x_i \in \{1, 2, \ldots, 5\}$. The payoff to every student i is $10 - x_i$ if she gets an A and $8 - x_i$ if she gets a B.

Part (a) Find which strategies survive the iterative deletion of strictly dominated strategies (IDSDS).

Part (b) Which strategies survive the iterative deletion of weakly dominated strategies (IDWDS).

Answer

Part (a) Let us first show that for either player, exerting a zero effort i.e., $x_i = 0$, strictly dominates effort levels of 3, 4, and 5. If $x_i = 0$ then player i's payoff is at least 8, which occurs when she gets a B. By choosing any other effort level x_i, the highest possible payoff is $10 - x_i$, which occurs when she gets an A. Since $8 > 10 - x_i$ when $x_i > 2$, then zero effort strictly dominates efforts of 3, 4, or 5. Intuitively, we consider which is the lowest payoff that player i can obtain from exerting zero effort, and compare it with the highest payoff he could obtain from deviating to a positive effort level $x_i \neq 0$. If this holds for some effort levels (as it does here for all $x_i > 2$), it means that, regardless of what the other student $j \neq i$ does, student i is strictly better off choosing a zero effort than deviating.

Once we delete effort levels satisfying $x_i > 2$, i.e., $x_i = 3$, 4, and 5 for both player 1 (in rows) and player 2 (in columns), we obtain the 3×3 payoff matrix depicted in Fig. 1.16.

As a practice of how to construct the payoffs in this matrix, note that, for instance, when player 1 chooses $x_1 = 1$ and player 2 selects $x_2 = 2$, player 1 still gets the highest score, i.e., player 1's score is $1 + 1.5 = 2.5$ thus exceeding player 2's score of 2, which implies that player 1's payoff is $10 - 1 = 9$ while player 2's payoff is $8 - 2 = 6$.

At this point, we can easily note that player 2 finds $x_2 = 1$ (in the center column) to be strictly dominated by $x_2 = 0$ (in the left-hand column). Indeed, regardless of which strategy player 1 uses (i.e., regardless of the particular row you look at) player 2 obtains the lowest score on the exam when he chooses an effort level of 0 or 1, since player 1 benefits from a 1.5 score advantage. Indeed, exerting a zero effort yields player 2 a payoff of 8, which is unambiguously higher than the payoff he obtains from exerting an effort of $x_2 = 1$, which is 7, regardless of the particular effort level exerted by player 1. We can thus delete the column referred to $x_2 = 1$ for player 2, which leaves us with the reduced form matrix in Fig. 1.17.

		Player 2		
		0	1	2
Player 1	0	10,8	10,7	8,8
	1	9,8	9,7	9,6
	2	8,8	8,7	8,6

Fig. 1.16 Reduced normal-form game

Now consider player 1. Notice that an effort of $x_1 = 1$ (in the middle row) strictly dominates $x_1 = 2$ (in the bottom row), since the former yields a payoff of 9 regardless of player 2's strategy (i.e., independently on the column), while an effort of $x_1 = 2$ only provides player 1 a payoff of 8. We can, therefore, delete the last row (corresponding to effort $x_1 = 2$) from the above matrix, which helps us further reduce the payoff matrix to the 2×2 normal-form game in Fig. 1.18.

Unfortunately, we can no longer find any strictly dominated strategy for either player. Specifically, neither of the two strategies of player 1 is dominated, since an effort of $x_1 = 0$ does not yield a weakly larger payoff than $x_1 = 1$, i.e., it entails a strictly larger payoff when player 2 chooses $x_2 = 0$ (in the left-hand column) but lower payoff when player 2 selects $x_2 = 2$ (in the right-hand column). Similarly, none of player 2's strategies is strictly dominated either, since $x_2 = 0$ yields the same payoff as $x_2 = 2$ when player 1 selects $x_1 = 0$ (top row), but a larger payoff when player 1 chooses $x_1 = 1$ (in the bottom row).

Therefore, the set of strategies that survive IDSDS for player 1 are $\{0, 1\}$ and for player 2 are $\{0, 2\}$. Thus, the IDSDS predicts that player 1 will exert an effort of 0 or 1, and that player 2 will exert effort of 0 or 2.

Part (b) Now let us repeat the analysis when we instead iteratively delete *weakly* dominated strategies. From part (a), we know that effort levels 3, 4, and 5 are all strictly dominated for both players, and therefore weakly dominated as well. By the same argument, we can delete all strictly dominated strategies (since they are also weakly dominated), leaving us with the reduced normal-form game we examined at the end of our discussion in part (a), which we reproduce below (Fig. 1.19).

Player 2

		0	2
Player 1	0	10,8	8,8
	1	9,8	9,6
	2	8,8	8,6

Fig. 1.17 Reduced normal-form game

Player 2

		0	2
Player 1	0	10,8	8,8
	1	9,8	9,6

Fig. 1.18 Reduced normal-form game

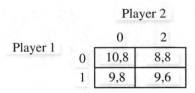

Fig. 1.19 Reduced normal-form game

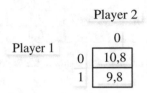

Fig. 1.20 Reduced normal-form game

While we could not identify any further strictly dominated strategies in part (a), we can now find weakly dominated strategies in this game. In particular, notice that, for player 2, an effort level of $x_2 = 0$ (in the left-hand column) weakly dominates $x_2 = 2$ (in the right-hand column). Indeed, a zero effort yields a payoff of 8 regardless of what player 1 does (i.e., in both rows), while $x_2 = 2$ yields a weakly lower payoff, i.e., 8 or 6. Thus, we can delete the column corresponding to an effort level of $x_2 = 2$ for player 2, allowing us to reduce the above payoff matrix to a single column matrix, as depicted in Fig. 1.20.

We can finally notice that player 1 finds that an effort of $x_1 = 0$ (top now) strictly dominates $x_1 = 1$ (in the bottom row), since player 1's payoff from a zero effort, 10, is strictly larger than that from an effort at $x_1 = 1$, 9. Hence, $x_1 = 0$ strictly dominates $x_1 = 1$, and thus $x_1 = 0$ also weakly dominates $x_1 = 1$. Intuitively, if player 2 exerts a zero effort (which is the only remaining strategy for player 2 after applying IDWDS), player 1, who starts with score advantage of 1.5, can anticipate that he will obtain the highest score in the class even if his effort level is also zero.

After deleting this weakly dominated strategy, we are left with a single cell that survives the application of IDWDS, corresponding to the strategy profile ($x_1 = 0$, $x_2 = 0$), in which both players exert the lowest amount of effort, as Fig. 1.21 depicts. After all, it seems that the incentive scheme the instructor proposed did not induce students to study harder for the exam, but instead to not study at all!![2]

---

[2]This is, however, a product of the score advantage of the most intelligent student. If the score advantage is null, or only 0.5, the equilibrium result after applying IDWDS changes, which is left as a practice.

Player 2

0

Player 1

0 | 10,8 |

Fig. 1.21 Single strategy profile surviving IDWDS

		Wife	
		Football	*Opera*
Husband	*Football*	3,1	0,0
	Opera	0,0	1,3

Fig. 1.22 Normal-form representation of the Battle of the Sexes game

Exercise 9—Applying IDSDS in the Battle of the Sexes Game[A]

A husband and a wife are leaving work, and do not remember which event they are attending to tonight. Both of them, however, remember that last night's argument was about attending either the football game (the most preferred event for the husband) or the opera (the most preferred event for the wife). To make matters worse, their cell phones broke, so they cannot call each other to confirm which event they are attending to. As a consequence, they must simultaneously and independently decide whether to attend to the football game or the opera.

The payoff matrix in Fig. 1.22, describes the preference of the husband (wife) for the football game (opera, respectively). Payoffs also indicate that both players prefer to be together rather than being alone (even if they are alone at their most preferred event). Apply IDSDS. What is the most precise prediction about how this game will be played when you apply IDSDS?

Answer

This game does not have strictly dominated strategies (note that there is not even any weakly dominated strategy), as we separately analyze for the husband and the wife below.

Husband: In particular, the husband prefers to go to the football game if his wife goes to the football game, but he would prefer to attend the opera if she goes to the opera. That is,

$$u_H(F,F) = 3 > 0 = u_H(O,F) \text{ and } u_H(O,O) = 1 > 0 = u_H(F,O)$$

Intuitively, the husband would like to coordinate with his wife by selecting the same event as her, i.e., a common feature in coordination games. Therefore, there is no event that yields an unambiguous larger payoff regardless of the event his wife attends to, i.e., there is no strictly dominant strategy for the husband.

Wife: A similar analysis is extensive to the wife, who would like to attend to the opera (her preferred event) obtaining a payoff of 3 only if her husband also attends the opera, i.e., $u_W(O, O) > u_W(O, F)$. Otherwise, she obtains a larger payoff attending to the football game, 1, rather than attending to the opera alone, 0, i.e., i.e., $u_W(F, F) > u_W(F, O)$. Hence, there is no event that provides her with unambiguously larger payoffs regardless of the event her husband selects, i.e., no event constitutes a strictly dominant strategy for the wife.

Hence, players have neither a strictly dominant nor a strictly dominated strategy. As a consequence, the application of IDSDS does not delete any strategy whatsoever, and our equilibrium outcome prescribes that any of the four strategy profiles in the above matrix could emerge, i.e., the husband could either attend the football game or the opera, and similarly, the wife could attend either the football game or the opera. (Similarly as in previous exercises, you can return to this exercise once you are done reading Chaps. 2 and 3, and find that the Nash equilibrium solution concept yields a more precise equilibrium prediction than IDSDS.)

Exercise 10—Applying IDSDS in Three-Player Games[B]

Consider the following anti-coordination game in Fig. 1.23, played by three potential entrants seeking to enter into a new industry, such as the development of software applications for smartphones. Every firm (labeled as A, B, or C) has the option of entering or staying out (i.e., remain in the industry they have been traditionally operating, such as, software for personal computers). The normal form game in Fig. 1.23 depicts the market share that each firm obtains, as a function of the entering decision of its rivals. Firms simultaneously and independently choose whether or not to enter. As usual in simultaneous-move games with three players, the triplet of payoffs describes the payoff for the row player (firm A) first, for the column player (firm B) second, and for the matrix player (firm C) third. Find the set

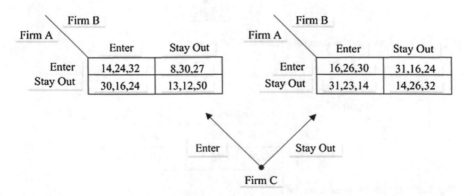

Fig. 1.23 Normal-form representation of a three-players game

of strategy profiles that survive the iterative deletion of strictly dominated strategies (IDSDS). Is the equilibrium you found using this solution concept unique?

Answer

We can start by looking at the payoffs for firm C (the matrix player). [Recall that the application of IDSDS is insensitive to the deletion order. Thus, we can start deleting strictly dominated strategies for the row, column or matrix player, and still reach the same equilibrium result.] In order to test for the existence of a dominated strategy for firm C, we compare the third payoff of every cell across both matrices. Figure 1.24 provides a visual illustration of this pairwise comparison across matrices.

We find that for firm C (matrix player), entering strictly dominates staying out, i.e., $u_C(s_A, s_B, E) > u_C(s_A, s_B, O)$ for any strategy of firm A, s_A, and firm B, s_B, $32 > 30$, $27 > 24$, $24 > 14$ and $50 > 32$ in the pairwise payoff comparisons depicted in Fig. 1.24 This allows us to delete the right-hand side matrix (corresponding to firm C choosing to stay out) since it is strictly dominated and thus would not be selected by firm C. We can, hence, focus on the left-hand matrix alone (where firm C chooses to enter), which we reproduce in Fig. 1.25.

We can now check that entering is strictly dominated for the row player (firm A), i.e., $u_A(E, s_B, E) < u_A(O, s_B, E)$ for any strategy of firm B, s_B, once we take into account that firm C selects its strictly dominant strategy of entering. Specifically, firm A prefers to stay out both when firm B enters (in the left-hand column, since $30 > 14$),

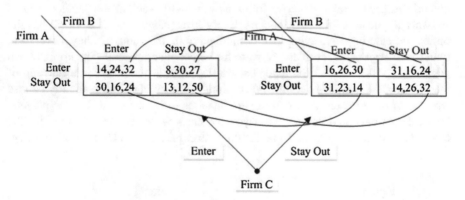

Fig. 1.24 Pairwise payoff comparison for firm C

Firm A \ Firm B	Enter	Stay Out
Enter	14,24,32	8,30,27
Stay Out	30,16,24	13,12,50

Fig. 1.25 Reduced normal-form game

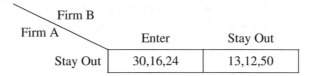

Fig. 1.26 Reduced normal-form game

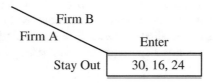

Fig. 1.27 An even more reduced normal-form game

and when firm B stays out (in the right-hand column, since 13 > 8). In other words, regardless of firm B's decision, firm A prefers to stay out. This allows us to delete the top row from the above matrix, since the strategy "Enter" would never be used by firm A, which leaves us with a single row and two columns, as illustrated in Fig. 1.26.

Once we deleted all but one strategy of firm C and one of firm A, the game becomes an individual-decision making problem, since only one player (firm B) must select whether to enter or stay out. Since entering yields a payoff of 16 to firm B, while staying out only entails 12, firm B chooses to enter. Firm B then regards staying out as a strictly dominated strategy, i.e., $u_B(O, E, E) > u_B(O, O, E)$ where we fix the strategies of the other two firms at their strictly dominant strategies: staying out for firm A and entering for firm C. We can, thus, delete the column corresponding to staying out in the above matrix, as depicted in Fig. 1.27.

As a result, the only cell (strategy profile) that survives the application of the iterative deletion of strictly dominated strategies (IDSDS) is that corresponding to (Stay Out, Enter, Enter), which predicts that firm A stays out, while both firms B and C choose to enter.

Exercise 11—Finding Dominant Strategies in games with $I \geq 2$ players and with Continuous Strategy Spaces[B]

There are I firms in an industry. Each can try to convince Congress to give the industry a subsidy. Let h_i denote the number of hours of effort put in by firm i, and let

$$C_i(h_i) = w_i(h_i)^2$$

be the cost of this effort to firm i, where w_i is a positive constant. When the effort levels of the I firms are given by the list $(h_1, \ldots, h_I)$, the value of the subsidy that gets approved is

$$S_i(h_i, h_j) = \alpha \sum_{i=1}^{I} h_i + \beta \prod_{j \neq i} h_j$$

where α and β are constants and $\alpha \geq 0$, and $\beta \geq 0$. Consider a game in which the firms decide simultaneously and independently how many hours they will each devote to this lobbying effort. Show that each firm has a strictly dominant strategy if and only if $\beta = 0$. What is firm i's strictly dominant strategy when condition $\beta = 0$ holds?

Answer
Firm i chooses the amount of lobbying h_i that maximizes its profit function π_i

$$\begin{aligned} \pi_i &= S_i(h_i, h_j) - C_i(h_i) \\ &= [\alpha \Sigma_i h_i + \beta(\Pi_i h_i)] - w_i(h_i)^2 \end{aligned}$$

Taking first-order conditions with respect to h_i yields

$$\alpha + \beta \left(\prod_{j \neq 1} h_j \right) - 2 w_i h_i = 0.$$

We can now rearrange, and solve for h_i to obtain firm i's best response function

$$h_i = \left[\alpha + \beta \left(\prod_{j \neq i} h_j \right) \right] \frac{1}{2 w_i}.$$

Notice that this best response function describes which is firm i's optimal effort level in lobbying activities as a function of other firms' effort levels, h_j, for every firm $j \neq i$. Interestingly, when $\beta = 0$ firm i's best response function becomes independent of his rival's effort, i.e., $h_i = \frac{\alpha}{2 w_i}$.

Dominant strategy: Note that, in order for firm i to have a dominant strategy, firm i must prefer to use a given strategy *regardless* of the particular actions selected by the other firms. In particular, when $\beta = 0$ firm i's best response function becomes $h_i = \frac{\alpha}{2 w_i}$, and thus it does not depend on the action of other firms, i.e., it is not a function of h_j. Therefore, firm i has a strictly dominant strategy, $h_i = \frac{\alpha}{2 w_i}$, when $\beta = 0$ since such action is independent on the other firms' actions.

Exercise 12—Equilibrium Predictions from IDSDS versus IDWDS^B

In previous exercises applying IDSDS we sometimes started finding strictly dominated strategies for the row player, while in other exercises we began identifying strictly dominated strategies for the column player (or the matrix player in games with three players). While the order of deletion does not affect the equilibrium outcome when applying IDSDS, it *can* affect the set of equilibrium outcomes when we delete weakly (rather than strictly) dominated strategies.

Use the game in Fig. 1.28 to show that the order in which weakly dominated strategies are eliminated can affect equilibrium outcomes.

Answer

First route: Taking the below payoff matrix, first note that, for player 1, strategy U weakly dominates D, since U yields a weakly larger payoff than D for any strategy (column) selected by player 2, i.e., $u_1(U, s_2) \geq u_1(D, s_2)$ for all $s_2 \in \{L, M, R\}$. In particular, U provides player 1 with the same payoff as D when player 2 selects L (a payoff of 2 for both U and D) and M (a payoff of 1 for both U and D), but a strictly higher payoff when player 2 chooses R (in the right-hand column) since $0 > -1$. Once we have deleted D because of being weakly dominated, we obtain the reduced-form matrix depicted in Fig. 1.29.

We can now turn to player 2, and detect that strategy L strictly dominates R, since it yields a strictly larger payoff than R, regardless of the strategy selected by player 1 (both when he chooses U in the top row, i.e., $1 > 0$, and when he chooses C in the bottom row, i.e., $2 > 1$), or more compactly $u_2(s_1, M) \geq u_2(s_2, R)$ for all s_1

2 1	L	M	R
U	2,1	1,1	0,0
C	1,2	3,1	2,1
D	2,-2	1,-1	-1,-1

Fig. 1.28 Normal-form game

2 1	L	M	R
U	2,1	1,1	0,0
C	1,2	3,1	2,1

Fig. 1.29 Reduced normal-form game after one round of IDWDS

$\in \{U, C\}$. Since R is strictly dominated for player 2, it is also weakly dominated. After deleting the column corresponding to the weakly dominated strategy R, we obtain the 2×2 matrix in Fig. 1.30.

At this point, notice that we are not done examining player 2, since you can easily detect that M is weakly dominated by strategy L. Indeed, when player 1 selects U (in the top row of the above matrix), player 2 obtains the same payoff from L and M, but when player 1 chooses C (in the bottom row), player 2 is better off selecting L, which yields a payoff of 2, rather than M, which only produces a payoff of 1, i.e., $u_2(s_1, M) \geq u_2(s_1, L)$ for all $s_1 \in \{U, C\}$. Hence, we can delete M because of being weakly dominated for player 2, leaving us with the (further reduced) payoff matrix in Fig. 1.31.

At this point, we can turn to player 1, and identify that U strictly dominates C (and, thus, it also weakly dominates C), since the payoff that player 1 obtains from U, 2, is strictly larger than that from C, 1. Therefore, after deleting C, we are left with a single strategy profile, (U, L), as depicted in the matrix of Fig. 1.32. Hence, using this particular order in our iterative deletion of weakly dominated strategies (IDWDS) we obtain the unique equilibrium prediction (U, L).

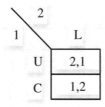

Fig. 1.30 Reduced normal-form game after two rounds of IDWDS

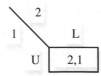

Fig. 1.31 Reduced normal-form game after three rounds of IDWDS

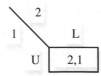

Fig. 1.32 Strategy surviving IDWDS (first route)

	L	M	R
U	2,1	1,1	0,0
C	1,2	3,1	2,1
D	2,-2	1,-1	-1,-1

(player 1 rows U, C, D; player 2 columns L, M, R)

Fig. 1.33 Normal-form game

	L	M
U	2,1	1,1
C	1,2	3,1
D	2,-2	1,-1

(player 1 rows U, C, D; player 2 columns L, M)

Fig. 1.34 Reduced normal-form game after one round of IDWDS

Second route: Let us now consider the same initial 3×3 matrix, which we reproduce in Fig. 1.33, and check if the application of IDWDS, but using a different deletion order (i.e., different "route"), can lead to a different equilibrium result than that found above. i.e., (U, L).

Unlike in our first route, let us now start identifying weakly dominated strategies for player 2. In particular, note that R is weakly dominated by M, since the former yields a weakly lower payoff than the latter (i.e., it provides a strictly higher payoff when player 1 chooses U in the top row, but the same payoff otherwise). That is, $u_2(s_1, M) \geq u_2(s_1, R)$ for all $s_1 \in \{U, C, D\}$. Once we delete R as being weakly dominated for player 2, the remaining matrix becomes that depicted in Fig. 1.34.

Turning to player 1, we cannot identify any other weakly dominating strategy. This equilibrium prediction using the second route of IDWDS is, hence, the six strategy profiles of Fig. 1.34, that is, $\{(U,L), (U,M), (C,L), (C,M), (D,L), (D,M)\}$. This equilibrium prediction is, of course, different (and significantly less precise) than what found when we started the application of IDWDS from player 1. Hence, equilibrium outcomes that arise from applying IDWDS are sensitive to the deletion order, while those emerging from IDSDS are not.

Pure Strategy Nash Equilibrium and Simultaneous-Move Games with Complete Information

2

Introduction

This chapter analyzes behavior in relatively simple strategic settings: simultaneous-move games of complete information. Let us define the two building blocks of this chapter: best responses and Nash equilibrium.

Best response. A strategy s_i^* is a best response of player i to a strategy profile s_{-i} selected by other players if it provides player i with a weakly larger payoff than any of his available strategies $s_i \in S_i$. Formally, strategy s_i^* is a best response to s_{-i} if and only if

$$u_i\left(s_i^*, s_{-i}\right) \geq u_i(s_i, s_{-i}) \text{ for all } s_i \in S_i.$$

We then say that strategy s_i^* is a best response to s_{-i}, and denote it as $s_i^* \in BR(s_{-i})$.

For instance, in a two-player game, s_1^* is a best response for player 1 to strategy s_2 selected by player 2 if and only if $u_1\left(s_1^*, s_2\right) \geq u_1(s_1, s_2)$ for all $s_1 \in S_1$ thus implying that $s_1^* \in BR_1(s_2)$.

We next define a Nash equilibrium by requiring that every player uses best responses to his opponents' strategies, i.e., players use mutual best responses.

Nash equilibrium. Strategy profile $s^* = (s_1^*, s_2^*, \ldots, s_N^*)$ is a Nash equilibrium if every player i's strategy is a best response to his opponents' strategies; that is, if for every player i his strategy s_i^* satisfies

$$u_i\left(s_i^*, s_{-i}^*\right) \geq u_i(s_i, s_{-i}^*) \text{ for all } s_i \in S_i$$

or, more compactly, strategy s_i^* is a best response to s_{-i}^*, i.e., $s_i^* \in BR_i(s_{-i}^*)$.

The original version of the chapter was revised: The erratum to the chapter is available at:
10.1007/978-3-319-32963-5_11

© Springer International Publishing Switzerland 2016
F. Munoz-Garcia and D. Toro-Gonzalez, *Strategy and Game Theory*,
Springer Texts in Business and Economics, DOI 10.1007/978-3-319-32963-5_2

In words, every player plays a best response to his opponents' strategies, and his conjectures about his opponents' behavior must be correct (otherwise, players could have incentives to modify their strategies and, thus, not be in equilibrium). As a consequence, players do not have incentives to deviate from their Nash equilibrium strategies; and we can understand such strategy profile as stable.

We initially focus on games where two players select between two possible strategies, such as the Prisoner's Dilemma game (where two prisoners decide to either cooperate or defect), and the Battle of the Sexes game (where a husband and a wife choose whether to attend the football game or the opera). Afterwards, we explain how to find best responses and equilibrium behavior in games where players choose among a continuum of strategies, such as in the Cournot game of quantity competition, and games where players' actions impose externalities on other players. Furthermore, we illustrate how to find best responses in games with more than two players, and how to identify Nash Equilibria in these contexts.

We finish this chapter with one application from law and economics about the incentives to commit crimes and to prosecute them through law enforcement, and a Cournot game in which the merging firms benefit from efficiency gains.[1]

Exercise 1—Prisoner's Dilemma[A]

Two individuals have been detained for a minor offense and confined in separate cells. The investigators suspect that these individuals are involved in a major crime, and separately offer each prisoner the following deal, as depicted in the matrix in Fig. 2.1.: If you confess while your partner doesn't, you will leave today without serving any time in jail; if you confess and your partner also confesses, you will serve 5 years in jail; if you don't confess and your partner does, you have to serve 15 years in jail (since you did not cooperate with the prosecutor but your partner provided us with evidence against you); finally, if none of you confess, you will serve one year in jail (since we only have limited evidence against you). If both players must simultaneously choose whether or not to confess, and they cannot coordinate their strategies, which is the Nash Equilibrium (NE) of the game?

Answer
Every player $i = \{1, 2\}$ has a strategy space of $S_i = \{C, NC\}$. In a NE, every player has complete information about all players' strategies and maximizes his own payoff, taking the strategy of his opponents as given. That is, every player selects his best response to his opponents' strategies. Let's start finding the best responses of player 1, for each of the possible strategies of player 2.

---

[1]While the Nash equilibrium solution concept allows for many applications in the area of industrial organization, we only explore some basic examples in this chapter, relegating many others to Chap. 5 (Applications to Industrial Organization).

Player 2

		confess	Not confess
Player 1	confess	-5, -5	0, -15
	Not confess	-15, 0	-1, -1

Fig. 2.1 Prisoner's dilemma game (Normal-form)

Player 1

If player 2 confesses (in the left-hand column), player 1's best response is to Confess, since his payoff from doing so, -5, is larger than that from not confessing, -15.[2] This is indicated in Fig. 2.2 by underlining the payoff that player 1 obtains from playing this best response, -5. If, instead, player 2 does Not confess (in the right-hand column), player 1's best response is to confess, given that his payoff from doing so, 0, is larger than that from not confessing, -1.[3] This is also indicated in Fig. 2.2 with the underlined best-response payoff 0. Hence, we can compactly represent player 1's best response ($BR_1(s_2)$) as $BR_1(C) = C$ to Confess, and $BR_1(NC) = C$ to not confess. Importantly, this implies that player 1 finds confess a strictly dominant strategy, as he chooses to confess regardless of what player 2 does.

Player 2

A similar argument applies for player 2. In particular, since the game is symmetric, we find that: (1) when player 1 confesses (in the top row), player 2's best response is to Confess, since $-5 > -15$; and (2) when player 1 does Not confess (in the bottom row), player 2's best response is to Confess, since $0 > -1$.[4] Hence, player 2's best response can be expressed as $BR_2(C) = C$ and $BR_2(NC) = C$, also indicating that Confess is a strictly dominant strategy for player 2, since he selects this

---

[2]A common trick many students use in order to be able to focus on the fact that we are examining the case in which player 2 confesses (in the left-hand column) is to cover with their hand (or a piece of paper) the columns in which player 2 selects strategies different from Confess (in this case, that means covering Not confess, but in larger matrices it would imply covering all columns except for the one we are analyzing at that point.) Once we focus on the column corresponding to Confess, player 1's best response becomes a straightforward comparison of his payoff from Confess, -5, and that from Not confess, -15, which helps us underline the largest of the two payoffs, i.e., -5.

[3]In this case, you can also focus on the column corresponding to Not confess by covering the column of Confess with your hand. This would allow you to easily compare player 1's payoff from Confess, 0, and Not confess, -1, underlining the largest of the two, i.e., 0.

[4]Similarly as for player 2, you can now focus on the row selected by player 1 by covering with your hand the row he did not select. For instance, when player 1 chooses Confess, you can cover the row corresponding to Not confess, which allows for an immediate comparison of the payoff when player 2 responds with Confess, -5, and when he does not, -15, and underline the largest of the two, i.e., -5. An analogous argument applies to the case in which player 1 selects Not confess, where you can cover the row corresponding to Confess with your hand.

Player 2

	confess	Not confess
confess	<u>-5</u>, -5	<u>0</u>, -15
Not confess	-15, 0	-1, -1

Player 1

Fig. 2.2 Prisoner's dilemma game (Normal-form)

strategy regardless of his opponent's strategy. Payoffs associated with player 2's best responses are underlined in Fig. 2.3. with red color.

We can now see that there is a single cell in which both players are playing a mutual best response, (C, C), as indicated by the fact that both players' payoffs are underlined (i.e., both players are playing best responses to each other's strategies). Intuitively, since we have been underling best response payoffs, a cell that has the payoffs of all players underlined entails that every player is selecting a best response to his opponent's strategies, as required by the definition of NE. Therefore, strategy profile (C, C) is the unique Nash Equilibrium (NE) of the game.

$$NE = (C, C)$$

Equilibrium vs. Efficiency. This outcome is, however, inefficient since it does not maximize social welfare (where social welfare is understood as the sum of both players' payoffs). In particular, if players could coordinate their actions, they would both select not to confess, giving rise to outcome (NC, NC), where both players' payoffs strictly improve relative to the payoff they obtain in the equilibrium outcome (C, C), i.e., they would only serve one year in jail rather than five years. This is a common feature in several games with intense competitive pressures, in which a conflict emerges between individual incentives (to confess in this example) and group/society incentives (not confess). Finally, notice that the NE is consistent with IDSDS. Indeed, since both players use strictly dominant strategies in the NE of the game, the equilibrium outcome according to NE coincides with that resulting from the application of IDSDS.

Player 2

	confess	Not confess
confess	<u>-5</u>, <u>-5</u>	<u>0</u>, -15
Not confess	-15, <u>0</u>	-1, -1

Player 1

Fig. 2.3 Prisoner's dilemma game (Normal-form)

Exercise 2—Battle of the Sexes[A]

A husband and a wife are leaving work, and do not remember which event they are attending to tonight. Both of them, however, remember that last night's argument was about either attending to the football game (the most preferred event for the husband) or the opera (the most preferred event for the wife). To make matters worse, their cell phones broke, so they cannot call each other to confirm which event they are attending to. As a consequence, they must simultaneously and independently decide whether to attend to the football game or the opera.

The payoff matrix in Fig. 2.4 describes the preference of the husband (wife) for the football game (opera, respectively), but also indicates that both players prefer to be together rather than being alone (even if they are alone at their most preferred event). Find the set of Nash Equilibria of this game.

Answer

Every player $i = \{H, W\}$ has strategy set $S_i = \{F, O\}$. In order to find the Nash equilibrium of this game, let us separately identify the best responses of each player to his opponent's strategies.

Husband

Let's first analyze the husband's best responses $BR_H(F)$ and $BR_H(O)$. If his wife goes to the football game (focusing our attention in the left-hand column), the husband prefers to also attend the football game since his payoff from doing so, 3, exceeds that from attending the opera by himself, 0. If, instead, his wife attends the opera (in the right-hand column), the husband prefers to attend the opera with her, given that his payoff from doing so, 1, while low (he dislikes opera!), is still larger than that from going to the football game alone, 0. Hence we can summarize the husband's best response as $BR_H(F) = F$ and $BR_H(O) = O$; as indicated in the underlined payoffs in the matrix of Fig. 2.5. Intuitively, the husband's best response is thus to attend the same event as his wife.

Wife

A similar argument applies to the wife, who also best responds by attending the same event as her husband, i.e., $BR_W(F) = F$ in the top row when her husband attends the football game, and $BR_W(O) = O$ in the bottom row when he goes to the opera; as illustrated in the payoffs underlined in red color in the matrix of Fig. 2.6.

		Wife	
		Football	Opera
Husband	Football	3, 1	0, 0
	Opera	0, 0	1, 3

Fig. 2.4 Battle of the sexes game (Normal-form representation)

Wife

	Football	Opera
Husband Football	$\underline{3}$, 1	0, 0
Opera	0, 0	$\underline{1}$, 3

Fig. 2.5 Battle of the sexes game—underlining best response payoffs for Husband

Wife

	Football	Opera
Husband Football	$\underline{3}$, $\underline{1}$	0, 0
Opera	0, 0	$\underline{1}$, $\underline{3}$

Fig. 2.6 Battle of the sexes game—underlining best response payoffs for Husband and Wife

Therefore, we found two strategy profiles in which both players are playing mutual best responses: (F, F) and (O, O), as indicated by the two cells in the matrix where both players' payoffs are underlined. Hence, this game has two pure-strategy Nash equilibria (*psNE*): (F, F), where both players attend the football game, and (O, O), where they both attend the opera. This can be represented formally as:

$$psNE = \{(F, F), (O, O)\}$$

Exercise 3—Pareto Coordination[A]

Consider the game in Fig. 2.7, played by two firms $i = \{1, 2\}$, each of them simultaneously and independently selecting to adopt either technology A or B. Technology A is regarded as superior by both firms, yielding a payoff of 2 to each firm if they both adopt it, while the adoption of technology B by both firms only entails a payoff of 1. Importantly, if firms do not adopt the same technology, both

Firm 2

	Technology A	Technology B
Firm 1 Technology A	2, 2	0, 0
Technology B	0, 0	1, 1

Fig. 2.7 Pareto coordination game (Normal-form)

	Firm 2	
	Technology A	Technology B
Technology A	2, 2	0, 0
Technology B	0, 0	1, 1

(Firm 1)

	Firm 2	
	Technology A	Technology B
Technology A	2, 2	0, 0
Technology B	0, 0	1, 1

(Firm 1)

Fig. 2.8 Pareto coordination game—underlining best response payoffs for Firm 1 (left matrix) and for both firms (right matrix)

obtain a payoff of zero. This can be explained because, even if firm i adopts technology A, such a technology is worthless if firm i cannot exchange files, new products and practices with the other firm $j \neq i$. Find the set of Nash Equilibria (NE) in this game.

Answer
Firm 1. Let's first examine firm 1's best response. Similarly as in the battle of the sexes game, firm 1's best response is to adopt the same technology as firm 2, i.e., $BR_1(A) = A$ when firm 2 chooses technology A, and $BR_1(B) = B$ when firm 2 selects technology B; as indicated in the payoffs underlined in blue color in the left-hand matrix of Fig. 2.8.
Firm 2. A similar argument applies to firm 2, since firms' payoffs are symmetric, i.e., $BR_2(A) = A$ and $BR_2(B) = B$; as depicted in the payoffs underlined in red color in the right-hand matrix.

Hence, we found two pure strategy Nash equilibria: (A, A) and (B, B), which are depicted in the matrix as the two cells were both players' payoffs are underlined.[5]

$$psNE = \{(A, A), (B, B)\}$$

Finally, note that, while either of the two technologies could be adopted in equilibrium, only one of them is efficient, (A, A), while the other equilibrium, (B, B), is inefficient, i.e., both firms would be better off if they could coordinate their simultaneous adoption of technology A.[6]

Exercise 4—Cournot game of Quantity Competition^A

Consider an industry with two firms competing in quantities, i.e., Cournot competition. For simplicity, assume that firms are symmetric in costs, $c > 0$, with no fixed costs and that they face a linear inverse demand $p(Q) = a - bQ$, where $a > c$,

---

[5]In both of these Nash equilibria, firms are playing mutual best responses, and thus no firm has incentives to unilaterally deviate.
[6]However, no firm has incentives to unilaterally move from technology B to A when its competitor is selecting technology B.

$b > 0$, and Q denotes aggregate output. Note that the assumption $a > c$ implies that the highest willingness to pay for the first unit is larger than the marginal cost that firms must incur in order to produce the first unit, thus indicating that a positive production level is profitable in this industry. If firms simultaneously and independently select their output level, q_1 and q_2, find the Nash Equilibrium (NE) of the Cournot game of quantity competition.

Answer

The profits of firm i are given by:

$$\pi_i = p(Q) \cdot q_i - cq_i$$

Given that $Q = q_1 + q_2$, every firm i chooses its production level q_i, taking the output level of its rival, q_j, as given. That is, every firm i solves

$$\max_{q_i} (a - bq_i - bq_j)q_i - cq_i$$

Taking first-order conditions with respect to q_i, we obtain

$$a - 2bq_i - bq_j - c = 0$$

and solving for q_i, we find firm i's best response function, $BR_i(q_j)$ that is,

$$q_i(q_j) = \frac{a - c}{2b} - \frac{1}{2}q_j$$

Figure 2.9 depicts the best response function of firm i, which originates at $\frac{a-c}{2b}$, indicating the production level that firm i sells when firm j is inactive, i.e., the monopoly output level; and decreases as its rival, firm j, produces a larger amount of output. Intuitively, firm i's and j's output are strategic substitutes, so that firm i is forced to sell fewer units when the market becomes flooded of firm j's products. When firm j's production is sufficiently large, i.e., firm j produces more than $\frac{a-c}{b}$ units, firm i is forced to remain inactive, i.e., $q_i^* = 0$.[7] This property is illustrated in the figure by the fact that firm i's best response function coincides with the horizontal axis (zero production) for all $q_j > \frac{a-c}{b}$.

A similar argument applies to firm j, obtaining best response function $q_j(q_i) = \frac{a-c}{2b} - \frac{1}{2}q_i$, as depicted in Fig. 2.10. (Note that we use the same axis, in order to be able to represent both best response functions in the same figure in our ensuing discussion.)

If we superimpose firm j's best response function on top of firm i's, we can visually see that the point where both functions cross each other represents the Nash Equilibrium of the Cournot game of quantity competition (Fig. 2.11).

---

[7]In order to obtain the output level of firm j that forces firm i to be inactive, set $q_i = 0$ on firm i's best response function, and solve for q_j. The output you obtain should coincide with the horizontal intercept of firm i's best response function in Fig. 2.9.

Fig. 2.9 Cournot game—
Best response function
of firm i

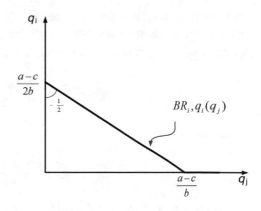

Fig. 2.10 Cournot game—
Best response function
of firm j

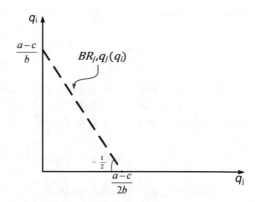

Fig. 2.11 Cournot game—
Best response functions and
Nash-equilibrium

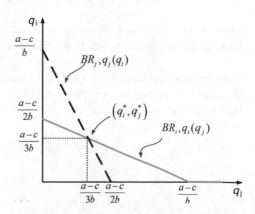

In order to precisely find the point in which both best response functions cross each other, let us simultaneously solve for q_i^* and q_j^* by, for instance, plugging $q_j(q_i) = \frac{a-c}{2b} = \frac{1}{2}q_i$ into $q_i(q_j)$, as we do next,

$$q_i^* = \frac{a-c}{2b} - \frac{1}{2}\left(\frac{a-c}{2b} - \frac{1}{2}q_i^*\right)$$

which simplifies to

$$q_i^* = \frac{a-c+bq_i^*}{4b}$$

and solving for q_i^*, we find the equilibrium output level for firm i,

$$q_i^* = \frac{a-c}{3b}$$

and that of firm j,

$$q_j^* = \frac{a-c}{2b} - \frac{1}{2}\frac{a-c}{3b} = \frac{a-c}{3b}.$$

As we can see from the results, both firms produce exactly the same quantities, since they both have the same technology (they both face the same production costs). Hence, the pure strategy Nash Equilibrium is:

$$psNE = \left\{q_i^*, q_j^*\right\} = \left\{\frac{a-c}{3b}, \frac{a-c}{3b}\right\}$$

(Notice that this exercise assumes, for simplicity, that firms are symmetric in their production costs. In subsequent chapters we investigate how firms' equilibrium production is affected when one of them exhibits a cost advantage ($c_i < c_j$). We also examine how firms' competition is affected when more than two firms interact in the same industry. see Chap. 5 for more details.)

Exercise 5—Games with Positive Externalities[B]

Two neighboring countries, $i = 1, 2$, simultaneously choose how many resources (in hours) to spend in recycling activities, r_i. The average benefit (π_i) for every dollar spent on recycling is:

$$\pi_i(r_i, r_j) = 10 - r_i + \frac{r_j}{2},$$

and the (opportunity) cost per hour for each country is 4. Country i's average benefit is increasing in the resources that neighboring country j spends on his recycling because a clean environment produces positive external effects on other countries.

Part (a) Find each country's best-response function, and compute the Nash Equilibrium (NE), (r_1^*, r_2^*)

Part (b) Graph the best-response functions and indicate the pure strategy Nash Equilibrium on the graph.

Part (c) On your previous figure, show how the equilibrium would change if the intercept of one of the countries' average benefit functions fell from 10 to some smaller number.

Answer

Part (a) Since the gains of recycling are given by $(\pi_i \cdot r_i)$, and the costs of the activity are $(4r_i)$, country 1's maximization problem consists of selecting the amount of hours devoted to recycling r_1 that solves:

$$\max_{r_1} \left(10 - r_1 + \frac{r_2}{2}\right) r_1 - 4r_1$$

Taking the first-order condition with respect to r_1

$$10 - 2r_1 + \frac{r_2}{2} - 4 = 0$$

Rearranging and solving for r_1 yields country 1's best-response function (BRF_1):

$$r_1(r_2) = 3 + \frac{r_2}{4}$$

Symmetrically, Country 2's best-response function is

$$r_2(r_1) = 3 + \frac{r_1}{4}$$

Inserting best-response function $r_2(r_1)$ into $r_1(r_2)$ yields

$$r_1 = 3 + \frac{3 + \frac{r_1}{4}}{4},$$

and, rearranging, we obtain an equilibrium level of recycling of $r_1^* = 4$ for country 1. Hence, country 2's equilibrium recycling level is

$$r_2^* = 3 + \frac{4}{4} = 4$$

Fig. 2.12 Positive
externalities—Best response
functions and
Nash-equilibrium

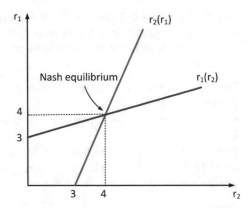

Note that, alternatively, countries' symmetry implies $r_1^* = r_2^*$. Hence, in BRF_1 we can eliminate the subscript (since both countries' recycling level coincides in equilibrium), and thus $r = 3 + \frac{r}{4}$, which, solving for r yields a symmetric equilibrium recycling of $r^* = 4$.

Hence, the psNE is given by:

$$psNE = \left\{ r_1^* = 4,\ r_2^* = 4 \right\}$$

Part (b) Both best response functions originate at 3 and increase with a positive slope of 1/4, as depicted in Fig. 2.12. Intuitively, countries' strategies are strategic complements, since an increase in r_2 induces Country 1 to strategically increase its own level of recycling, r_1, by 1/4.

Part (c) A reduction in the benefits from recycling produces a fall in the intercept of one of the countries' average benefit function, for example in Country 2. This change is indicated in Fig. 2.13 by the leftward shift (following the arrow) in

Fig. 2.13 Shift in best
response functions, change in
Nash-equilibrium

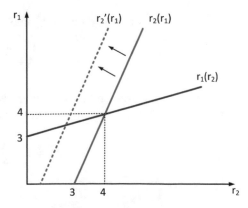

Country 2's best response function. In the new Nash Equilibrium, Country 2 recycles a lot less while Country 1 recycles a little less.

Exercise 6—Traveler's Dilemma[B]

Consider the following game, often referred to as the "traveler's dilemma." An airline loses two identical suitcases that belong to two different travelers. The airline is liable for up to $100 per suitcase. The airline manager, in order to obtain an honest estimate of each suitcase, separates each traveler i in a different room and proposes the following game: "Please write an integer $x_i \in [2, 100]$ in this piece of paper. As a manager, I will use the following reimbursement rule:

- If both of you write the same estimate, $x_1 = x_2 = x$, each traveler gets x.
- If one of you writes a larger estimate, i.e., $x_i > x_j$ where $i \neq j$, then:

  - The traveler who wrote the *lowest* estimate (traveler j) receives $x_j + k$, where $k > 1$; and
  - The traveler who wrote the *largest* estimate (traveler i) only receives $\max\{0, x_j - k\}$."

Part (a) Show that asymmetric strategy profiles, in which travelers submit different estimates, cannot be sustained as Nash equilibria.

Part (b) Show that symmetric strategy profiles, in which both travelers submit the same estimate, and such estimate is strictly larger than 2, cannot be sustained as Nash equilibria.

Part (c) Show that the symmetric strategy profile in which both travelers submit the same estimate $(x_1, x_2) = (2, 2)$ is the unique pure strategy Nash equilibrium.

Part (d) Does the above result still hold when the traveler writing the largest amount receives $x_j - k$ rather than $\max\{0, x_j - k\}$? Intuitively, since $k > 1$ by definition, a traveler can now receive a negative payoff if he submits the lowest estimate and $x_j < k$.

Answer

Part (a) We first show that asymmetric strategy profiles, (x_1, x_2) with $x_1 \neq x_2$, cannot be sustained as a Nash equilibrium. Consider, without loss of generality, that player 1 submits a higher estimate than player 2, $x_1 > x_2$. In this setting, it is easy to see that player 1 has incentives to deviate: he now obtains a payoff of $\max\{0, x_2 - k\}$, and he could increase his payoff by submitting an estimate that matches that of player 2, i.e., $x_1 = x_2$, which guarantees him a payoff of x_2 (as now the estimates from both travelers coincide), where

$$\max\{0, x_2 - k\} < x_2 \text{ for all } x_2 \text{ given that } k > 1.$$

Part (b) Using a similar argument, we can show symmetric strategy profiles in which both travelers submit the same estimate (but higher than two), i.e., (x_1, x_2), with $x_1 = x_2 > 2$, cannot be supported as Nash equilibria either. To see this, note that in such strategy profile every player i obtains a payoff $x_i = x_j$, but he can increase his payoff by deviating towards a lower estimate, i.e., $x_i = x_j - 1$ since estimates must be integer numbers. With such a deviation, player i's estimate becomes the lowest, and he thus obtains a payoff of

$$x_i + k = (x_j - 1) + k,$$

where $(x_j - 1) + k > x_j$ since $k > 1$ by definition.

Part (c) Hence, the only remaining strategy profile is that in which both travelers submit an estimate of 2, $x_1 = x_2 = 2$. Let us now check if it can be sustained as a Nash equilibrium, by showing that every player i has no profitable deviation. Every traveler i obtains a payoff of 2 under the proposed strategy profile. If player i deviates towards a higher price, traveler i would be now submitting the highest estimate, and thus would obtain a payoff of

$$\max\{0, x_j - k\} = \max\{0, 2 - k\},$$

where $\max\{0, 2 - k\} < 2$ since $k > 1$ by definition. That is, submitting a higher estimate reduces player i's payoff. Finally, note that submitting a lower estimate is not feasible since estimates must satisfy $x_i \in [2, 100]$ by definition.

Alternative approach: While the above analysis tests whether a specific strategy profile can/cannot be sustained as Nash Equilibrium of the game, a more direct approach would identify each player's best response function, and then find the point where player 1's and 2's best response functions cross each other, which constitutes the NE of the game. For a given estimate from player j, x_j, if player i writes an estimate lower than x_j, $x_i < x_j$, player i obtains a payoff of $x_i + k$, which is larger than the payoff he obtains from matching player j's estimate, i.e., $x_i = x_j = x$, as long as $x_i + k > x_j$, that is, if $k > x_j - x_i$. Intuitively, player i profitably undercuts player j's estimate if x_i is not extremely lower than x_j.[8] If, instead, player i writes a larger estimate than player j, $x_i > x_j$, his payoff becomes $\max\{0, x_j - k\}$, which is lower than his payoff from matching player j's estimate, i.e., $x_i = x_j = x$, since $\max\{0, x_j - k\} < x_j$.

In summary, player i does not have incentives to submit a higher estimate than player j's, but rather an estimate that is k-units lower than player j's estimate. Hence, player i's best response function can be written as

$$x_i(x_j) = \max\{2, x_j - k\}$$

---

[8]For instance, if $x_j = 5$ and $k = 2$, then player i has incentives to write an estimate of $x_i = 5 - 2 = 3$, but not lower than 3 since his payoff, $x_i + k$, is increasing in his own estimate x_i.

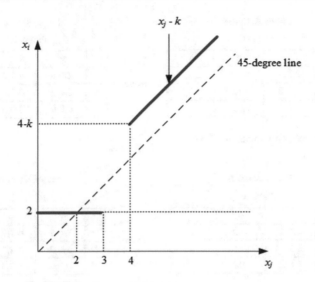

Fig. 2.14 Traveler's dilemma, best response functions

since the estimate that he writes must lie on the interval [2, 100]. This best response function is depicted in Fig. 2.14. Specifically, player i's best response function originates at $x_i = 2$ when player j submits $x_j = 2$; remains at $x_i = 2$ when $x_j = 3$ (since $k > 1$ entails that $3 - k < 2$); and becomes $x_i = \max\{2, 4 - k\}$ when $x_j = 4$, thus increasing in x_j. For instance, if $k = 2$, player i's best response function is $x_i = 2$ when player j's estimate is $x_j = 2, 3, 4$, but increases to $x_i = 3$ when player j's estimate is $x_j = 5$, and generally becomes $x_i(x_j) = x_j - 2$ for all $x_j > 4$. Graphically, this function has a flat segment for low values of x_j, but then increases in x_j in a straight line located k-units below the 45-degree line. A similar argument applies to player j's best response function. Hence, player 1's and 2's best response functions only cross at $x_i = x_j = 2$ (Fig. 2.14).

Part (d) Our above argument did not rely on the property of positive payoffs for the traveler submitting the highest estimate. Hence, all the previous proof applies to this reimbursement rule as well, implying that $x_1 = x_2 = 2$ is the unique pure strategy Nash equilibrium of the game.

Exercise 7—Nash Equilibria with Three Players[B]

Find all the Nash equilibria of the following three-player game (see Fig. 2.15), in which player 1 selects rows (a, b, or c), player 2 chooses columns (x, y, or z), and player 3 selects a matrix (either A in the left-hand matrix, or B in the right-hand matrix).

Player 3: matrix A   Player 3: matrix B
Player 2   Player 2

		x	y	z
Player 1	a	2,0,4	1,1,1	1,2,3
	b	3,2,3	0,1,0	2,1,0
	c	1,0,2	0,0,3	3,1,1

		x	y	z
Player 1	a	2,0,3	4,1,2	1,1,2
	b	1,3,2	2,2,2	0,4,3
	c	0,0,0	3,0,3	2,1,0

Fig. 2.15 Normal-form game with 3 players

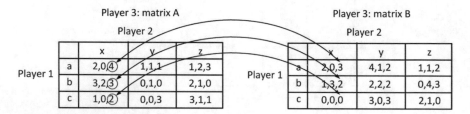

Fig. 2.16 Normal-form game with 3 players. BR_3 when player 2 chooses x

Answer

Player 3. Let's start by evaluating the payoffs for player 3 when Player 2 selects x (first column). The arrows in Fig. 2.16 help us keep track of player 3's pairwise comparison. For instance, when player 1 chooses a and player 2 selects x (in the top left-hand corner of either matrix), player 3 prefers to respond with matrix A, which gives him a payoff of 4, rather than with B, which only yields a payoff of 3. This comparison is illustrated by the top arrow in Fig. 2.16. A similar argument applies for the second arrow, which fixes the other players' strategy profile at (b, x), and for the third arrow, which fixes their strategy profile at (c, x). The highest payoff that player 3 obtains in each of these three pairwise comparisons is circled in Fig. 2.16.

Hence, we obtain that player 3's best responses are $BR_3(x, a) = A$, $BR_3(x, b) = A$, and $BR_3(x, c) = A$.

If player 2 selects y (in the second column of each matrix), player 3's pairwise comparisons are given by the three arrows in Fig. 2.17. In terms of best responses, this implies that $BR_3(y, a) = B$, $BR_3(y, b) = B$, and $BR_3(y, c) = \{A, B\}$. The highest payoff that player 3 obtains in each pairwise comparison are also circled in Fig. 2.17.

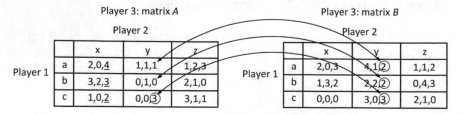

Fig. 2.17 Normal-form game with 3 players. BR_3 when player 2 chooses y

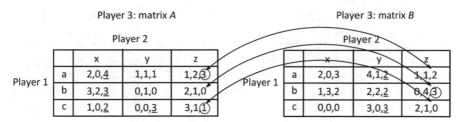

Fig. 2.18 Normal-form game with 3 players. BR_3 when player 2 chooses z

Player 3: matrix A

Player 2

		x	y	z
Player 1	a	2,0,<u>4</u>	1,1,1	1,2,<u>3</u>
	b	3,2,<u>3</u>	0,1,0	2,1,0
	c	1,0,<u>2</u>	0,0,<u>3</u>	3,1,<u>1</u>

Player 3: matrix B

Player 2

		x	y	z
Player 1	a	2,0,3	4,1,<u>2</u>	1,1,2
	b	1,3,2	2,2,<u>2</u>	0,4,<u>3</u>
	c	0,0,0	3,0,<u>3</u>	2,1,0

Fig. 2.19 Normal-form game with 3 players. Underlining player 3's response payoffs

If player 2 selects z (in the third column of each matrix), player 3's pairwise comparisons are depicted by the three arrows in Fig. 2.18 which in terms of best responses yields $BR_3(z, a) = A$, $BR_3(z, b) = B$, and $BR_3(z, c) = A$. Hence, the payoff matrix that arises after underlying (or circling) the payoff corresponding to the best responses of player 3 is the following (see Fig. 2.19)

Player 2. Let's now identify player 2's best responses as depicted in the circled payoffs of the matrices in Fig. 2.20. In particular, we take player 1's strategy as given (fixing the row) and player 3's as given (fixing the matrix) (where player 3 chooses matrix A). We obtain that player 2's best responses are $BR_2(a, A) = z$ when player 1 chooses a (in the top row), $BR_2(b, A) = x$ when player 1 selects b (in the middle row), and $BR_2(c, A) = z$ when player 1 chooses c (in the bottom row). Visually, notice that we are now fixing our attention on a matrix (strategy of player 3) and on a row (strategy of player 1), and horizontally comparing the payoff that player 2 obtains from selecting the left, middle or right-hand column. Similarly, when player 3 chooses matrix B, we obtain that player 2's best responses are $BR_2(a, B) = \{y, z\}$ when player 1 selects a (in the top row) since both y and z yield the same payoff, \$1, $BR_2(b, B) = x$ when player 1 chooses b (in the middle row), and $BR_2(c, B) = z$ when player 1 selects c (in the bottom row).[9]

Therefore, the matrices that arise after underlying the best response payoffs of player 2 are those in Fig. 2.21.

---

[9]Visually, this implies fixing your attention on the first row of the left-hand matrix, and horizontally search for which strategy of player 2 (column) provides this player with the highest payoff.

		Player 3: matrix A						Player 3: matrix B		
		Player 2						Player 2		
		x	y	z				x	y	z
Player 1	a	2,0,4	1,1,1	1,2,3	Player 1	a	2,0,3	4,1,2	1,1,2
	b	3,2,3	0,1,0	2,1,0		b	1,3,2	2,2,2	0,4,3
	c	1,0,2	0,0,3	3,1,1		c	0,0,0	3,0,3	2,1,0

Fig. 2.20 Normal-form game with 3 players. Circling best responses of player 2

		Player 3: matrix A						Player 3: matrix B		
		Player 2						Player 2		
		x	y	z				x	y	z
Player 1	a	2,0,4	1,1,1	1,2,3	Player 1	a	2,0,3	4,1,2	1,1,2
	b	3,2,3	0,1,0	2,1,0		b	1,3,2	2,2,2	0,4,3
	c	1,0,2	0,0,3	3,1,1		c	0,0,0	3,0,3	2,1,0

Fig. 2.21 Normal-form game with 3 players

Player 1. Let's finally identify player 1's best responses, given player's 2 strategy (fixing the column) and given player 3's strategy (fixing the matrix).

When player 3 chooses A (left matrix), player 1's best responses become $BR_1(x, A) = b$ when player 2 chooses x (left-hand column), $BR_1(y, A) = a$ when player 2 selects y (middle column), and $BR_1(z, A) = c$ when player 2 chooses z (right-hand column). Visually, notice that we are now fixing our attention on a matrix (strategy of player 3) and on a column (strategy of player 2), and vertically comparing the payoff that player 1 obtains from choosing the top, middle or bottom row[10]. Operating in an analogous fashion when player 3 chooses B (right-hand matrix), we obtain player 1's best responses: $BR_1(x, B) = a$ when player 2 chooses x (in the left-hand column), $BR_1(y, B) = a$ when player 2 selects y (middle column), and $BR_1(z, B) = c$ when player 2 chooses z (right-hand column).

We hence found three pure strategy Nash equilibria: (b, x, A), (c, z, A) and (a, y, B); as depicted in the cells where the payoffs of all players are underlined (Fig. 2.22), as these cells correspond to outcomes where players employ mutual best responses to each others' strategies (Fig. 2.23).

In summary, the Nash equilibria of this three player game are

$$psNE = \{(b, x, A), (c, z, A), (a, y, B)\}$$

---

[10]For instance, in finding $BR_1(x, A)$, we fix the matrix in which player 3 selects A (left matrix), and the column that player 2 selects x (left-hand column), and compare the payoffs that player 1 would obtain from responding with the first row (a), \$2, the second row (b), \$3, or with the third row (c), \$1. Hence, $BR_1(x, A) = b$. A similar argument applies to other best responses of player 1.

Player 3: matrix A   Player 3: matrix B

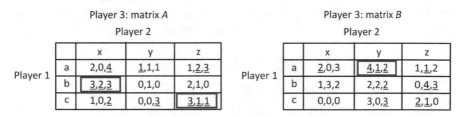

Fig. 2.22 Normal-form game with 3 players

Player 3: matrix A   Player 3: matrix B

Player 1	Player 2 x	y	z		Player 1	Player 2 x	y	z
a	2,0,4	1,1,1	1,2,3		a	2,0,3	4,1,2	1,1,2
b	3,2,3	0,1,0	2,1,0		b	1,3,2	2,2,2	0,4,3
c	1,0,2	0,0,3	3,1,1		c	0,0,0	3,0,3	2,1,0

Fig. 2.23 Normal-form game with 3 players. Nash equilibria

Exercise 8—Simultaneous-Move Games with $n \geq 2$ Players[B]

Consider a game with $n \geq 2$ players. Simultaneously and independently, the players choose between two options, X and Y. These options might represent, for instance, two available technologies for the n firms operating in an industry, e.g., selling smartphones with the Android operating system or, instead, opt for the newer Windows Phone operating system from Microsoft. That is, the strategy space for each player i is $S_i = \{X, Y\}$. The payoff of each player who selects X is:

$$2m_x - m_x^2 + 3$$

where m_x denotes the number of players who choose X. The payoff of each player who selects Y is

$$4 - m_y$$

where m_y is the number of players who choose Y. Note that $m_x + m_y = n$.

Part (a) For the case of only two players, $n = 2$, represent this game in its normal form, and find the pure-strategy Nash equilibria.

Part (b) Suppose now that $n = 3$. How many psNE does this game have?

Part (c) Consider now a game with $n > 3$ players. Identify an asymmetric psNE, i.e., an equilibrium in which a subset of players chooses X, while the remaining players choose Y.

Answer

Part (a) When both players choose X, $m_x = 2$ and $m_y = 0$, thus implying that every player's payoff is $2m_x - m_x^2 + 3$. Replacing $m_x = 2$, we obtain a payoff of

$$2(2) - (2)^2 + 3 = 3$$

for both players, as indicated in the cell corresponding to outcome (X, X) in the payoff matrix in Fig. 2.24. When, instead, both players choose Y, $m_y = 2$ and $m_x = 0$, and players' payoff becomes $4 - m_y = 4 - 2 = 2$; as depicted in outcome (Y, Y) of the payoff matrix. Finally, if only one player chooses X and another chooses Y, $m_x = 1$ and $m_y = 1$, this yield a payoff of

$$2m_x - m_x^2 + 3 = 2(1) - (1)^2 + 3 = 4$$

for the player who chose X, and $4 - m_y = 4 - 1 = 3$ for the player who chose Y; as represented in outcomes (X, Y) and (Y, X) in the payoff matrix (see cells away from the main diagonal in Fig. 2.24).

As usual, underlined payoffs represent a payoff corresponding to a player's best response, as we next separately describe for each player.

Player 1: In particular, player 1's best responses are $BR_1(X) = \{X, Y\}$ when player 2 chooses X (in the left-hand column) since player 1 is indifferent between responding with X (in the top row) or Y (in the bottom row) given that they both yield a payoff of $3. As a result, we underline both payoffs of $3 for player 1 in the column in which player 2 chooses X. If, instead, player 2 chooses Y (in the right-hand column), player 1's best response is $BR_1(Y) = \{X\}$, since player 1 obtains a higher payoff by selecting X ($4), than by choosing Y ($2). As a consequence, we underline the payoff of player 1 associated to his best response.

Player 2: Similarly, for player 2 we find that, when player 1 chooses X (in the top row), player 2 best responds with $BR_2(X) = \{X, Y\}$, since both X and Y yield a payoff of $3; while if player 1 selects Y (bottom row), player 2's best response is $BR_2(Y) = \{X\}$, since X yields a payoff of $4 while Y only entails a payoff of $2.

Therefore, since there are three cells where payoffs of all players have been underlined as the best responses, they represent strategy profiles where players' play mutual best responses, i.e. Nash equilibria of the game. There are, hence, three pure strategy Nash equilibrium in this game:

$$(X, X), (X, Y) \text{ and } (Y, X).$$

		Player 2	
		X	Y
Player 1	X	3,3	4,3
	Y	3,4	2,2

Fig. 2.24 Normal-form game with $n = 2$ players

Part (b) When introducing three players, the normal form representation of the game is depicted in the matrices of Figs. 2.25 and 2.26. (This is a standard three-player simultaneous-move game similar to that in the previous exercise).

In order to identify best responses, player 1 and 2 operate as in previous exercises, i.e., taking the action of player 3 (matrix player) as given, and comparing their payoffs across rows for player 1 and across columns for player 2. However, player 3 compares his payoffs across matrices, for a given strategy profile of player 1 and 2. In particular, this pairwise payoff comparison of player 3 is analogous to that depicted in Exercises 2.6 and 2.8 of this chapter. For instance, if player's 1 and 2 select (X, X), then player 3 obtains a payoff of only 0 if he were to select X as well (in the upper matrix), but a higher payoff of 3 if he, instead, selects Y (in the lower matrix). For this reason, we underline 3 in the third component of the cell corresponding to (X, X, Y), in the upper left-hand corner of the lower matrix. A similar argument applies for identifying other best responses of player 3, where we compare the third component of every cell across the two matrices. For instance, when player 1 and 2 select (X, Y), player 3 obtains a payoff of 3 if he chooses X (in the upper matrix), but only a payoff of 2 if he selects Y (in the lower matrix), which leads us to underline 3 in the third component of the payoff in the cell (X, Y) of the upper matrix.

Following a similar approach, we can see in Figs. 2.25 and 2.26 that there are three outcomes for which the payoffs of all players have been underlined (i.e., players are selecting mutual best responses). Specifically, the pure strategy Nash equilibria of the game with $n = 3$ players are:

$$psNE = \{(X, Y, X), (Y, X, X), (X, X, Y)\}$$

Part (c) When $n > 3$ players compete in this simultaneous-move game, the payoff from selecting strategy Y is

$$4 - m_y = 4 - (n - m_x)$$

		Player 2	
		X	Y
Player 1	X	0,0,0	3,3,3
	Y	3,3,3	2,2,4

Fig. 2.25 Normal-form game when player 3 chooses $s_3 = X$

		Player 2	
		X	Y
Player 1	X	3,3,3	4,2,2
	Y	2,4,2	1,1,1

Fig. 2.26 Normal-form game when player 3 chooses $s_3 = Y$

Fig. 2.27 Payoffs from
selecting strategy X and Y

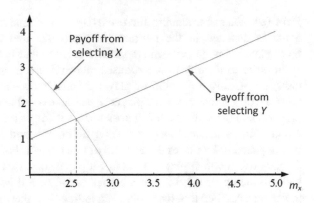

where the number of players selecting Y, m_y, is represented as those players who did not choose X, i.e., $m_y = n - m_x$. The payoff from selecting X is

$$2m_x - m_x^2 + 3$$

Hence, a player selects strategy X if and only if his payoff from selecting X is weakly higher than from choosing Y, that is

$$4 - (n - m_x) \geq 2m_x - m_x^2 + 3$$

Solving for m_x, yields that the number of players selecting strategy X is $m_x = \frac{1 \pm \sqrt{1 - 4(1-n)}}{2}$. For instance, in the case of $n = 5$ players, the above expression becomes $m_x = 2.56$ players, which implies that three players select X. (Note that the above result for m_x produces two roots, $m_x = 2.56$ and $m_x = -1.56$, but we only focus on the positive root.)

For illustration purposes, Fig. 2.27 depicts the payoff from selecting strategy Y when $n = 5$ interact, $4 - (5 - m_x) = m_x - 1$, and that from strategy X, $2m_x - m_x^2 + 3$. Intuitively, the payoff from strategy X is decreasing in the number of players choosing it, m_x (rightward movement in Fig. 2.27). Similarly, the payoff from selecting Y is also decreasing in the number of players choosing it, $n - m_x$; as depicted by leftward movements in Fig. 2.27. These incentives a negative network externality. For instance, settings in which a particular technology is very attractive when few other firms use it, but becomes less attractive as many other firms use it.

Exercise 9—Political Competition (Hoteling Model)[B]

Consider two candidates competing for office: Democrat (D) and Republican (R). While they can compete along several dimensions (such as their past policies, their endorsements from labor unions, their advertising, and even their looks!), we assume for simplicity that voters compare the two candidates according to only one dimension (e.g., the budget share that each candidate promises to spend on education). Voters' ideal policies are uniformly distributed along the interval [0, 1], and each votes for the candidate with a policy promise closest to the voter's ideal. Candidates simultaneously and independently announce their policy positions. A candidate's payoff from winning is 1, and from losing is −1. If both candidates receive the same number of votes, then a coin toss determines the winner of the election.

Part (a) Show that there exists a unique pure strategy Nash equilibrium, and that in involves both candidates proposals to promise a policy closest to the median voter.
Part (b) Show that with three candidates (democrat, republican, and independent), no pure strategy Nash equilibrium exists.

Answer

Part (a) Let $x_i \in [0, 1]$ denote candidate i's policy, where $i = \{D, R\}$. Hence, for a strategy profile (x_D, x_R) where, for instance, $x_D > x_R$, voters to the right-hand side of x_D vote democrat (since x_D is close to their ideal policy than x_R is), as well as half of the voters in the segment between x_D and x_R; as depicted in Fig. 2.28. In contrast, voters to the left-hand side of x_R and half of those in the segment between x_D and x_R vote republican. (The opposite argument applies for strategy profiles (x_D, x_R) satisfying $x_D < x_R$.)

We can now show that there exists a unique Nash in which both candidates announce $x_D = x_R = 0.5$. Our proof is similar to that in the Traveler's Dilemma game. First demonstrate that asymmetric strategy profiles where $x_D \neq x_R$ cannot be sustained as Nash equilibria of the game; second, to show that symmetric strategy profiles where $x_D = x_R = x$ but $x \neq 0.5$ cannot be supported as Nash equilibria either; and third, to demonstrate that symmetric strategy profile $x_D = x_R = 0.5$ can be sustained as Nash equilibrium of the game.

Let's first consider asymmetric strategy profiles where each candidate makes a different policy promise $x_i \neq x_j$, where $i = \{D, R\}$ and $j \neq i$:

Case 1 If $x_i < x_j < 0.5$, candidate i could increase his chances to win by positioning himself ε to the right of x_j (where $\varepsilon > 0$ is assumed to be small). Thus, any strategy profile where $x_i < x_j < 0.5$ cannot be supported as a Nash equilibrium.

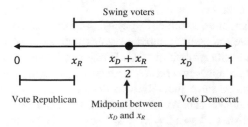

Fig. 2.28 Allocation of voters

Case 2 If $0.5 < x_i < x_j$, candidate j could increase his chances to win by positioning himself ε to the left of x_i. Thus, any strategy profile where $0.5 < x_i < x_j$ cannot be supported as a Nash equilibrium either.

Case 3 If $x_i < 0.5 < x_j$, candidate i could increase his chances to win by positioning him self ε to the left of x_j. Thus, any case where $x_i < 0.5 < x_j$ cannot be supported as a Nash equilibrium. (Note that the candidate j would also want to deviate to ε right to the candidate i). Thus, there cannot be *asymmetric* Nash equilibria.

Let us now consider *symmetric* strategy profiles where both candidates make the same policy promise, but their common policy differs from 0.5.

Case 1 If $x_D = x_R < 0.5$, a tie occurs, and each candidate wins the election with probability 1/2. However, candidate D could win the election with certainty by positioning himself ε to the right of x_R. (A similar argument applies to candidate R, who would also have incentives to position himself ε to the right of x_D.) Thus, any strategy profile where $x_D = x_R < 0.5$ cannot be supported as a Nash equilibrium.

Case 2 A similar argument applies if $0.5 < x_D = x_R$, where a tie occurs and every candidate wins the election with probability 1/2. However, candidate R could increase his chances of winning by positioning himself ε to the left of x_D. Thus, any strategy profile where $0.5 < x_D = x_R$ cannot be supported as a Nash equilibrium either.

Finally, if both candidates choose the same policy, $x_D = x_R = x$, and such common policy is $x = 1/2$, each candidate receives half of the votes, and wins the election with probability 0.5. In this setting, however, neither candidate has incentives to deviate; otherwise his votes would fall from half of the electorate, guaranteeing him to lose the election. Therefore, there exists only one Nash equilibrium, in which $x_D = x_R = 0.5$.

Part (b) Suppose that a Nash equilibrium exists with a triplet of policy proposals (x_D^*, x_R^*, x_I^*), where D denotes democrat, R republican, and I independent. We will next show that: (1) symmetric strategy profiles in which all candidates make the same that proposal, $x_D^* = x_R^* = x_I^*$, cannot be sustained as Nash equilibria; (2) asymmetric strategy profiles where two candidates choose the same proposal, but a third candidate differs, cannot be supported as equilibria either; and (3) asymmetric strategy profiles in which all three candidates choose different proposals cannot be sustained as equilibria; ultimately entailing that no pure strategy equilibrium exists.

First case. Consider, first, symmetric policy proposals $x_D^* = x_R^* = x_I^*$. All candidates, hence, receive the same number of votes (one third of the electorate), and each candidate wins with probability 1/3. While candidates didn't have incentives to alter their policy promises in a setting with two candidates, with three of them we can see that candidates have incentives to deviate from such strategy profile. In particular, any candidate can win the election by moving to the right (if their common policy satisfies $x_D^* = x_R^* = x_I^* < 2/3$) or moving to the left (if their common policy satisfies $x_D^* = x_R^* = x_I^* > 1/3$); as depicted in Fig. 2.29. Similarly,

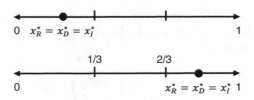

Fig. 2.29 Allocation of voters

when the location of the three candidates satisfies $x_D^* = x_R^* = x_I^* > 1/2$ each candidate has incentives to deviate towards the left, while if $x_D^* = x_R^* = x_I^* < 1/2$ each candidate has incentives to deviate to the right.

Second case. Consider now strategy profiles in which two candidates choose the same policy, $x_i^* = x_j^* = x^*$, but the third candidate differs, $x^* \neq x_k^*$, where $i \neq j \neq k$. If their policies satisfy $x^* < x_k^*$, then candidate k has incentives to approach x^*, i.e., $x^* + \varepsilon$, as such position increases his votes. Similarly, if $x^* > x_k^*$, candidate k has incentives to approach x^*, i.e., $x^* - \varepsilon$, which increases his votes. Since we found that at least one player has a profitable deviation, the above strategy profile cannot be sustained as a Nash equilibrium.

Third case. Finally, consider strategy profiles where all three candidates make different policy promises, $x_i^* \neq x_j^* \neq x_k^*$. The candidate that is located the farthest on the right will be able to win by moving ε to the right of its closest competitor; implying that the original strategy profile cannot be equilibrium. (A similar argument applies to the other candidates, such as that located the farthest to the left, who could win by moving to the left of its closest competitor.) Therefore, there exists no pure strategy Nash Equilibrium in this game.

Exercise 10—Tournaments^B

Several strategic settings can be modeled as a tournament, whereby the probability of winning a certain prize not only depends on how much effort you exert, but also on how much effort other participants in the tournament exert. For instance, wars between countries, or R&D competitions between different firms in order to develop a new product, not only depend on a participant's own effort, but on the effort put by its competitors. Let's analyze equilibrium behavior in these settings. Consider that the benefit that firm 1 obtains from being the first company to launch a new drug is \$36 million. However, the probability of winning this R&D competition against its rival (i.e., being the first to launch the drug) is $\frac{x_1}{x_1 + x_2}$, which increases with this firm's own expenditure on R&D, x_1, relative to total expenditure by both firms, $x_1 + x_2$. Intuitively, this suggests that, while spending more than its rival, i.e., $x_1 > x_2$, increases firm 1's chances of being the winner, the fact that $x_1 > x_2$ does not guarantee that firm 1 will be the winner. That is, there is still some randomness as to which firm will be the first to develop the new drug, e.g., a firm can spend more resources than its rival but be "unlucky" because its laboratory exploits a few weeks before being able to develop the drug. For simplicity, assume that firms' expenditure cannot exceed 25, i.e., $x_i \in [0, 25]$. The cost is simply x_i, so firm 1's profit function is

$$\pi_1(x_1, x_2) = 36\left(\frac{x_1}{x_1 + x_2}\right) - x_1$$

and there is an analogous profit function for firm 2:

$$\pi_2(x_1, x_2) = 36\left(\frac{x_2}{x_1 + x_2}\right) - x_2$$

You can easily check that these profit functions are increasing and concave in a firm's own expenditure. Intuitively, this indicates that, while profits increase in the firm's R&D, the first million dollar is more profitable than the 10th million dollar, e.g., the innovation process is more exhausted.

Part (a) Find each firm's best-response function.

Part (b) Find a symmetric Nash equilibrium, i.e., $x_1^* = x_2^* = x^*$.

Answer

Part (a) Firm 1's optimal expenditure is the value of x_1 for which the first derivative of its profit function equals zero. That is,

$$\frac{\partial \pi_1(x_1, x_2)}{\partial x_1} = 36\left[\frac{x_1 + x_2 - x_1}{(x_1 + x_2)^2}\right] - 1 = 0$$

Rearranging, we find

$$36\left[\frac{x_2}{(x_1 + x_2)^2}\right] - 1 = 0$$

which simplifies to

$$36x_2 = (x_1 + x_2)^2$$

and further rearranging

$$6\sqrt{x_2} = x_1 + x_2$$

Solving for x_1, we obtain firm 1's best response function

$$x_1(x_2) = 6\sqrt{x_2} - x_2$$

Figure 2.30 depicts firm 1's best response function, $x_1(x_2) = 6\sqrt{x_2} - x_2$ as a function of its rival's expenditure, x_2 in the horizontal axis for the admissible set $x_2 \in [0, 25]$.

It is straightforward to show that, for all values of $x_2 \in [0, 25]$, firm 1's best response also lies in the admissible set $x_1 \in [0, 25]$. In particular, the maximum of BR_1 occurs at $x_2 = 9$ since

$$\frac{\partial BR_1(x_2)}{\partial x_2} = \frac{\partial\left[6\sqrt{x_2} - x_2\right]}{\partial x_2} = 3(x_2)^{-\frac{1}{2}} - 1$$

Hence, the point at which this best response function reaches its maximum is that in which its derivative is zero, i.e., $3(x_2)^{-1/2} - 1 = 0$, which yields a value of $x_2 = 9$. At this point, firm 1's best response function informs us that firm 1 optimally spends $6\sqrt{9} - 9 = 9$. Finally, note that the best response function is concave in its rival expenditure, x_2, since

$$\frac{\partial^2 BR_1(x_2)}{\partial x_2^2} = -\frac{3}{2}(x_2)^{-\frac{3}{2}} < 0.$$

By symmetry, firm 2's best response function is $x_2(x_1) = 6\sqrt{x_1} - x_1$.
Part (b) In a symmetric Nash equilibrium $x_1^* = x_2^* = x^*$. Hence, using this property in the best-response functions found in part (c), yields

$$x^* = 6\sqrt{x^*} - x^*$$

Rearranging, we obtain $2x^* = 6\sqrt{x^*}$, and solving for x^*, we find $x^* = 9$. Hence, the unique symmetric Nash equilibrium has each firm spending 9. As Fig. 2.31 depicts, the points at which the best response function of player 1 and 2 cross each other occur at the 45-degree line (so the equilibrium is symmetric). In particular, those points are the origin, i.e., $(0, 0)$, but this case is uninteresting since it implies that no firm spends money on R&D, and $(9, 9)$.

Fig. 2.30 Tournament—
Firm 1's best response
function

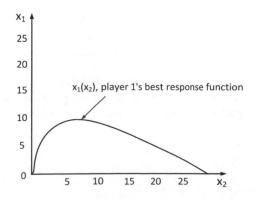

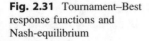

Fig. 2.31 Tournament–Best response functions and Nash-equilibrium

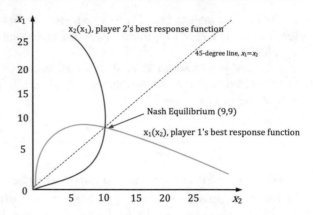

Exercise 11—Lobbying[A]

Consider two interest groups, A and B, seeking to influence a government policy, each with opposed interest (group A's most preferred policy is 0, while that of group B is 1). Each group simultaneously and independently chooses a monetary contribution to government officials, $s_i \in [0, 1]$, where $i = \{A, B\}$. The policy (x) that the government implements is a function of the contributions from both interest groups, as follows:

$$x(s_A, s_B) = \frac{1}{2} - s_A + s_B$$

Hence, if interest groups contribute zero (or if their contributions coincide, thus canceling each other), the government implements its ideal policy, $\frac{1}{2}$. [In this simplified setting, the government is not a strategic player acting in the second stage of the game, since its response to contributions is exogenously described by policy function $x(s_A, s_B)$.] Finally, assume that the interest groups have the following utility functions:

$$u_A(s_A, s_B) = -[x(s_A, s_B)]^2 - s_A$$

$$u_B(s_A, s_B) = -[1 - x(s_A, s_B)]^2 - s_B$$

which decrease in the contribution to the government, and in the squared distance between their ideal policy (0 for group A, and 1 for group B) and the implemented policy $x(s_A, s_B)$. Find the Nash equilibrium of this simultaneous-move game.

Answer
Substituting the policy function into the utility function of every group, we obtain

$$u_A(s_A,\ s_B) = -\left[\frac{1}{2} - s_A + s_B\right]^2 - s_A$$

$$u_B(s_A,\ s_B) = -\left[1 - \left(\frac{1}{2} - s_A + s_B\right)\right]^2 - s_B$$

Taking first order conditions with respect to s_A in the utility function of group A yields

$$2\left[\frac{1}{2} - s_A + s_B\right] - 1 = 0$$

Rearranging, we obtain $1 - 2s_A + 2s_B - 1 = 0$, which, solving for s_A, yields the best-response function for interest group A, $s_A(s_B) = s_B$.

Similarly, taking first order conditions with respect to s_B in the utility function of group B, we find

$$2\left[1 - \left(\frac{1}{2} - s_A + s_B\right)\right] - 1 = 0$$

which simplifies to $-s_A + s_B = 0$, thus yielding the best-response function for interest group B, $s_B(s_A) = s_A$. Graphically, both best response functions coincide with the 45-degree line, and completely overlap to one another. As a consequence, the set of pure strategy NEs is given by all the points in the 45-degree line, i.e., all points satisfying $s_A = s_B$, or, more formally, the set

$$\{(s_A,\ s_B) \in [0,\ 1]^2 : s_A = s_B\}.$$

Furthermore, since both interest groups are contributing the same amount to the government, their contributions cancel out, and the government implements its preferred policy, $\frac{1}{2}$. Finally, note that the strategic incentives in this game are similar to those in other Pareto Coordination games with symmetric NEs. While the game has multiple NEs in which both groups choose the same contribution level, the NE in which both groups choose a zero contribution Pareto dominates all other NEs with positive contributions.

Exercise 12—Incentives and Punishment[B]

Consider the following "law and economics" game, between a criminal and the government. The criminal selects a level of crime, $y \geq 0$, and the government chooses a level of law enforcement $x \geq 0$. Both choices are simultaneous and independent, and utility functions of the government (G) and the criminal (C) are, respectively,

$$u_G = -\frac{y^2}{x} - xc^4 \qquad \text{and} \qquad u_C = \frac{1}{1+xy}\sqrt{y}$$

Intuitively, the government takes into account that crime, y, is harmful for society (i.e., y enters negatively into the government's utility function u_G), and that each unit of law enforcement, x, is costly to implement, at a cost of c^4 per unit. In contrast, the criminal enjoys $\sqrt{y}$ if he is not caught, which definitely occurs when $x = 0$ (i.e., his utility becomes $u_c = \sqrt{y}$ when $x = 0$), while the probability of not being caught is $\frac{1}{1+xy}$.

Part (a) Find each player's best-response function. Depict these best-response functions, with x on the horizontal axis and y on the vertical axis.
Part (b) Compute the Nash equilibrium of this game.
Part (c) Explain how the equilibrium levels of law enforcement, x and crime, y, found in part (b) change as the cost of law enforcement, c, increases.

Answer

Part (a) First, note that the government, G, selects the level of law enforcement, x, that solves

$$\max_{x} \quad -\frac{y^2}{x} - xc^4$$

Taking first-order conditions with respect to x yields

$$\frac{y^2}{x^2} - c^4 = 0$$

Rearranging and solving for x, we find the government's (G) best response function, BR_G, to be

$$x(y) = \frac{y}{c^2}.$$

Intuitively, the government's level of law enforcement, x, increases in the amount of criminal activity, y, and decreases in the cost of every unit of law enforcement, c.

Second, the criminal, C, selects the level of crime, y, that solves

$$\max_{y} \frac{1}{1+xy} \sqrt{y}$$

Taking first-order conditions with respect to y yields

$$\frac{1}{2y^{1/2}(1+xy)} - \frac{y^{1/2}x}{(1+xy)^2} = 0$$

Rearranging and solving for y, we find the criminal's best response function, BR_C, to be

$$y(x) = \frac{1}{x}.$$

which decreases in the level of law enforcement chosen by the government, x.

The government's best response function, BR_G, and the criminal's best response function, BR_C, are represented in Fig. 2.32. Note that BR_C (i.e., $y(x) = \frac{1}{x}$) is clearly decreasing in x but becomes flatter as x increases, i.e., $\frac{dy(x)}{dx} = -\frac{1}{x^2} < 0$ and $\frac{d^2y(x)}{dx^2} = \frac{2}{x^3} > 0$. In order to depict the government's best response function BR_G, $x(y) = \frac{y}{c^2}$, it is convenient to solve for y which yields $y = c^2x$. As depicted in Fig. 2.32, BR_G originates at $(0, 0)$ and has a slope of c.

Part (b) We find the values of x and y that simultaneously solve both players' best response functions $x = \frac{y}{c^2}$ and $y = \frac{1}{x}$. For instance, you can plug the second expression into the first expression. This yields $x^* = \frac{1/x^*}{c^2}$, which, solving for x^*, entails an equilibrium level of law enforcement of $x^* = \frac{1}{c}$. Therefore, the equilibrium level of crime is $y(\frac{1}{c}) = \frac{1}{1/c} = c$. The Nash equilibrium is, hence, $x^* = \frac{1}{c}$ and $y^* = c$; as illustrated in the point where best response function BR_G crosses BR_C in Fig. 2.32.

Fig. 2.32 Incentives and Punishment

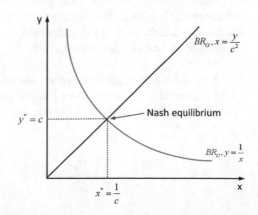

Fig. 2.33 Incentives and
Punishment-Comparative
statics

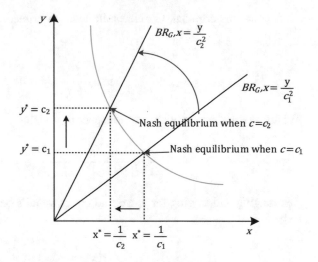

Part (c) As Fig. 2.33 illustrates, an increase in the cost of enforcement c pivots the government's best response function, BR_G leftward, with center at the origin (recall that c is the slope of BR_G). In contrast, the criminal's best response function is unaffected (since it is independent on c). This pivoting effect produces a new crossing point that lies to the northwest of the original Nash equilibrium, entailing a higher level of criminal activity (y) and a lower level of enforcement (x).

Exercise 13—Cournot mergers with Efficiency Gains[B]

Consider an industry with three identical firms each selling a homogenous good and producing at a constant cost per unit c with $1 > c > 0$. Industry demand is given by $p(Q) = 1-Q$, where $Q = q_1 + q_2 + q_3$. Competition in the marketplace is in quantities.

Part (a) Find the equilibrium quantities, price and profits.
Part (b) Consider now a merger between two of the three firms, resulting in duopolistic structure of the market (since only two firms are left: the merged firm and the remaining firm). The merger might give rise to efficiency gains, in the sense that the firm resulting from the merger produces at a cost $e \cdot c$, with $e \leq 1$ (whereas the remaining firm still has a cost c):

  i. Find the post-merger equilibrium quantities, price and profits.
 ii. Under which conditions does the merger reduce prices?
iii. Under which conditions is the merger beneficial to the merging firms?

Answer

Part (a)

Each firm $i = \{1, 2, 3\}$ has a profit of

$$\pi_i = (1 - Q - c)q_i.$$

Hence, since $Q \equiv q_1 + q_2 + q_3$, profits can be rewritten as:

$$\pi_i = \big(1 - (q_i + q_j + q_k) - c\big)q_i,$$

The first-order conditions are given by

$$1 - 2q_i - q_j - q_k - c = 0$$

since firms are symmetric $q_i = q_j = q_k = q$ in equilibrium, that is

$$1 - 2q - q - q - c = o$$

or

$$1 - 4q - c = 0.$$

Solving for q at the symmetric equilibrium yields a Cournot output of,

$$q_c = \frac{1 - c}{4}$$

Hence, equilibrium prices are $p_C = 1 - \frac{1-c}{4} - \frac{1-c}{4} - \frac{1-c}{4} = 1 - 3\left(\frac{1-c}{4}\right) = \frac{1+3c}{4}$ and every firm i's equilibrium profits are $\pi_C = \left(\frac{1+3c}{4} - c\right)\frac{1-c}{4} = \frac{(1-c)^2}{16}$.

Part (b)

i. After the merger, two firms are left: firm 1, with cost $e \cdot c$, and firm 3, with cost c. Hence, the two profit functions are now given by:

$$\pi_1 = (1 - Q - ec)q_1.$$

$$\pi_3 = (1 - Q - c)q_3$$

Taking first order conditions of π_1 with respect to q_1 yields

$$1 - 2q_1 - q_3 - ec = 0$$

and, solving for q_1, we obtain firm 1's best response function

$$q_1(q_3) = \frac{1 - ec}{2} - \frac{1}{2}q_3$$

Similarly taking first-order conditions of firm 3's profits, π_3, with respect to q_3 yields

$$1 - 2q_3 - q_1 - c = 0$$

which, solving for q_3, provides us with firm 3's best response function

$$q_3(q_1) = \frac{1 - c}{2} - \frac{1}{2}q_1$$

Plugging $q_3(q_1)$ into $q_1(q_3)$, yields

$$q_1^* = \frac{1 - ec}{2} - \frac{1}{2}\left(\frac{1 - c}{2} - \frac{1}{2}q_1^*\right)$$

Rearranging and solving for q_1^*, we obtain firm 1's equilibrium output

$$q_1^* = \frac{1 - c(2e - 1)}{3}$$

Plugging this output level into firm 3's best response function yields an equilibrium output of

$$q_3^* = \frac{1 - c(2 - e)}{3}$$

Note that the outsider firm can sell a positive output at equilibrium only if the merger does not give rise to strong cost savings: that is $q_3 \geq 0$ if $e \geq \frac{2c-1}{c}$ (if $c < 1/2$, then the previous payoff becomes $\frac{2c-1}{c} < 0$, implying that $e \geq \frac{2c-1}{c}$ holds for all $e \geq 0$, ultimately entailing that the outsider firm will always sell at the equilibrium. We hence concentrate on values of c that satisfy $c > 1/2$.) Figure 2.34 illustrates cutoff $e > \frac{2c-1}{c}$, where $c > 1/2$. The region of (e, c)-combinations above this cutoff indicate parameters for which the merger is not sufficiently cost saving to induce the outside firm to produce positive output levels. The opposite occurs when the cost-saving parameter, c, is lower than $(2c - 1)/c$, thus indicating that the merger is so cost saving that the nonmerged firm cannot profitably compete against the merged firm.

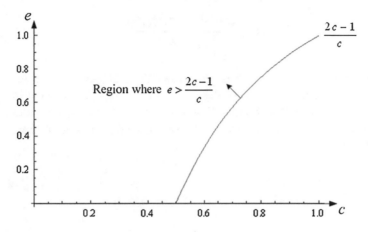

Fig. 2.34 Positive production after the merger

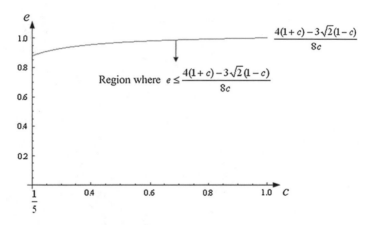

Fig. 2.35 Profitable mergers if e is low enough

- The equilibrium price is $p_m = 1 - \frac{1-c(2e-1)}{3} - \frac{1-c(2-e)}{3} = \frac{1+c(1+e)}{3}$, and equilibrium profits are given by $\pi_1 = \frac{(1-c(2e-1))^2}{9}$ and $\pi_3 = \frac{(1-c(2-e))^2}{9}$.

ii. Prices decrease after the merger only if there are sufficient efficiency gains: that is, $p_m \leq p_c$ can be rewritten as $e \leq \frac{5c-1}{4c}$. Note that if $c < 1/5$, then $\frac{5c-1}{4c} < 0$, implying that $e \leq \frac{5c-1}{4c}$ cannot hold for any $e \geq 0$. As a consequence, $p_m > p_c$, and prices will never fall no matter how strong efficiency gains, e, are

iii. To see if the merger is profitable, we have to study the inequality $\pi_1 \geq 2\pi_c$, which, solving for e, yields

$$e \leq \frac{4(1+c) - 3\sqrt{2}(1-c)}{8c}$$

In other words, the merger is profitable only if it gives rise to enough cost savings. Figure 2.35 depicts the cutoff of e where costs are restricted to $c \in \left[\frac{1}{5}, 1\right]$. Notice that if cost savings are sufficiently strong, i.e., parameter e is sufficiently small as depicted in the region below the cutoff, the merger is profitable.

Mixed Strategies, Strictly Competitive Games, and Correlated Equilibria

Introduction

This chapter analyzes how to find equilibrium behavior when players are allowed to randomize, helping us to identify mixed strategy Nash equilibria (msNE). Finding this type of equilibrium completes our analysis in Chap. 2 where we focused on Nash equilibria involving pure strategies (not allowing for randomizations). However, as shown in Chap. 2, some games do not have a pure strategy Nash equilibrium. As this chapter explores, allowing players to randomize (mix) will help us identify a msNE in this type of games. For other games, which already have pure strategy Nash equilibria (such as the Battle of the Sexes game), we will find that allowing for randomizations gives rise to one more equilibrium outcome. Let us next define mixed strategies and msNE.

Mixed strategy. Consider that every player i has a finite strategy space $S_i = \{s_1, s_2, \ldots, s_m\}$ with its associated probability simplex ΔS_i, which we can understand as the set of all probability distributions over S_i. (For instance, if the strategy space has only two elements ($m = 2$), then the simplex is graphically represented by a [0,1] segment in the real line; while if the strategy space has three elements ($m = 3$) the simplex can be graphically depicted by a Machina's triangle in $\mathbb{R}^2$.) Hence, a mixed strategy is an element of the simplex (a point in the previous examples for $m = 2$ and $m = 3$) that we denote as $\sigma_i \in \Delta S_i$,

$$\sigma_i = \{\sigma_i(s_1), \sigma_i(s_2), \ldots, \sigma_i(s_m)\}$$

where $\sigma_i(s_k)$ denotes the probability that player i plays pure strategy s_k, and satisfies $\sigma_i(s_k) \geq 0$ for all k (positive or zero probabilities for each pure strategy) and $\sum_{k=1}^{m} \sigma_i(s_k) = 1$ (the sum of all probabilities must be equal to one).

We can now use this definition of a mixed strategy to define a msNE, as follows.

The original version of the chapter was revised: The erratum to the chapter is available at: 10.1007/978-3-319-32963-5_11

msNE. Consider a strategy profile $\sigma = (\sigma_1, \sigma_2, \ldots, \sigma_N)$ where σ_i denotes a mixed strategy for player i. We then say that strategy profile σ is a msNE if and only if

$$u_i(\sigma_i, \sigma_{-i}) \geq u_i(s_i', \sigma_{-i}) \quad \text{for all } s_i' \in S_i \text{ and for every player } i.$$

Intuitively, mixed strategy σ_i is a best response of player i to the strategy profile σ_{-i} selected by other players.[1]

The chapter starts with games of two players who choose among two available strategies. A natural examples is the Battle of the Sexes game, where husband and wife simultaneously and independently choose whether to attend to a football game or the opera. We then extend our analysis to settings with more than two players, and to contexts in which players are allowed to choose among more than two strategies. We illustrate how to find the convex hull of Nash equilibrium payoffs (where players use pure or mixed strategies), and afterwards explain how to find correlated equilibria when players use a publicly observable random variable. For completeness, we also examine an exercise which illustrates the difference between the equilibrium payoffs that players can reach when playing the mixed strategy Nash equilibrium of a game and when playing the correlated equilibrium, showing that they do not necessarily coincide. At the end of the chapter, we focus on strictly competitive games: first, describing how to systematically check whether a game is strictly competitive; and second, explaining how to find "maxmin strategies" (also referred to as "security strategies") using a two-step graphical procedure.

Exercise 1—Game of Chicken[A]

Consider the game of Chicken (depicted in Fig. 3.1), in which two players driving their cars against each other must decide whether or not to swerve.

Part (a) Is there any strictly dominated strategy for Player 1? And for Player 2?
Part (b) What are the best responses for Player 1? And for Player 2?
Part (c) Can you find any pure strategy Nash Equilibrium (psNE) in this game?
Part (d) Find the mixed strategy Nash Equilibrium (msNE) of the game. *Hint:* denote by p the probability that Player 1 chooses Straight and by $(1 - p)$ the probability that he chooses to Swerve. Similarly, let q denote the probability that Player 2 chooses Straight and $(1 - q)$ the probability that she chooses to Swerve.

---

[1]Note that we compare player i's expected payoff from mixed strategy σ_i, $u_i(\sigma_i, \sigma_{-i})$, against his expected payoff from selecting pure strategy s_i', $u_i(s_i', \sigma_{-i})$. We could, instead, compare $u_i(\sigma_i, \sigma_{-i})$ against $u_i(\sigma_i', \sigma_{-i})$, where $\sigma_i' \neq \sigma_i$. However, for player i to play mixed strategy σ_i', he must be indifferent between at least two pure strategies, e.g., s_i' and s_i''. Otherwise, player i would not be mixing, but choosing a pure strategy. Hence, his indifference between pure strategies s_i' and s_i'' entails that $u_i(s_i', \sigma_{-i}) = u_i(s_i'', \sigma_{-i})$, implying that it suffices to check if mixed strategy σ_i yields a higher expected payoff than all pure strategies that player i could use in his randomization. (This is a convenient result, as we will not need to compare the expected payoff of mixed strategy σ_i against all possible mixed strategies $\sigma_i' \neq \sigma_i$; which would entail studying the expected payoff from all feasible randomizations.)

		Player 2	
		Straight	Swerve
Player 1	Straight	0,0	3,1
	Swerve	1,3	2,2

Fig. 3.1 Game of Chicken

Part (e) Show your result of part (d) by graphically representing every player i's best response function $BRF_i(s_j)$, where $s_j = \{Swerve, Straight\}$ is the strategy selected by player $j \neq i$.

Answer

Part (a) No, there are no strictly dominated strategies for any player. In particular, Player 1 prefers to play Straight when Player 2 plays Swerve getting a payoff of 3 instead of 2. However, he prefers to Swerve when Player 2 plays Straight, getting a payoff of 1 by swerving instead of a payoff of 0 if he chooses to go straight. Hence, there is no strictly dominated strategy for player 1, i.e., a strategy that Player 1 would never use regardless of what strategy his opponent selects. Since payoffs are symmetric, a similar argument applies to Player 2.

Part (b) *Player 1.* If player 2 chooses straight (fixing our attention in the left-hand column), player 1's best response is to Swerve, since his payoff from doing so, 1, is larger than that from choosing Straight. Hence,

$$BR_1(Straight) = Swerve, \text{obtaining a payoff of } 1$$

If player 2 instead chooses to Swerve (in the right-hand column), player 1's best response is Straight, since his payoff from Straight, 3, exceeds that from Swerve, 2. That is,

$$BR_1(Swerve) = Straight, \text{obtaining a payoff of } 3$$

Player 2. A similar argument applies to player 2's best responses; since players are symmetric in their payoffs. Hence,

$$BR_2(Straight) = Swerve, \text{obtaining a payoff of } 1$$
$$BR_2(Swerve) = Straight, \text{obtaining a payoff of } 3$$

These best responses illustrate that the Game of Chicken is Anti-Coordination game, where each player responds by choosing exactly the opposite strategy of his/her competitor.

Part (c) Note that in strategy profiles (Swerve, Straight) and (Straight, Swerve), there is a mutual best response by both players. To see this more graphically, the matrix in Fig. 3.2 underlines the payoffs corresponding to each player's selection of his/her best response. For instance, $BR_1(Straight) = Swerve$ implies that, when

		Player 2	
		Straight	Swerve
Player 1	Straight	0,0	<u>3,1</u>
	Swerve	<u>1,3</u>	2,2

Fig. 3.2 Nash equilibria in the Game of Chicken

Player 2 chooses Straight (in the left-hand column), Player 1 responds Swerving (in the bottom left-hand cell), obtaining a payoff of 1; while when Player 2 chooses Swerve (in the right-hand column) $BR_1(Swerve) = Straight$, and thus Player 1 obtains a payoff of 3 (top right-hand side cell). A similar argument applies to Player 2. There are, hence, two cells where the payoffs of all players have been underlined, i.e., mutual best responses (Swerve, Straight) and (Straight, Swerve). These are the two pure strategy Nash equilibria of the Chicken game.

Part (d) *Player 1.* Let q denote the probability that Player 2 chooses Straight, and $(1 - q)$ the probability that she chooses Swerve as depicted in Fig. 3.3.

In this context, the expected value that Player 1 obtains from playing Straight (in the top row) is:

$$EU_1(Straight) = 0q + 3(1 - q)$$

since, fixing our attention in the top row, Player 1 gets a payoff of zero when Player 2 chooses Straight (which occurs with probability q) or a payoff of 3 when Player 2 selects Swerve (which happens with the remaining probability $1 - q$). If, instead, Player 1 chooses to Swerve (in the bottom row), his expected utility becomes

$$EU_1(Swerve) = 1q + 2(1 - q)$$

since Player 1 obtains a payoff of 1 when Player 2 selects Straight (which occurs with the probability q) or a payoff of 2 when Player 2 chooses Swerve (which happens with probability $1 - q$). Player 1 must be indifferent between choosing Straight or Swerve. Otherwise, he would not be randomizing, since one pure strategy would generate a larger expected payoff than the other, leading Player 1 to select such a pure strategy. We therefore need that

			Player 2	
			Straight	Swerve
			q	1 - q
Player 1	Straight	p	0,0	<u>3,1</u>
	Swerve	1 - p	<u>1,3</u>	2,2

Fig. 3.3 Searching for a mixed strategy equilibrium in the Game of Chicken

$$EU_1(Straight) = EU_1(Swerve)$$

$$0q + 3(1 - q) = 1q + 2(1 - q)$$

rearranging and solving for probability q yields

$$3 - 3q = q + 2 - 2q \Rightarrow q = \frac{1}{2}$$

That is, Player 2 must be choosing Straight with 50 % probability, i.e., $q = \frac{1}{2}$, since otherwise Player 1 would not be indifferent between Straight and Swerve.

Player 2. When he chooses Straight (in the left-hand column), he obtains an expected utility of

$$EU_2(Straight) = 0p + 3(1 - p)$$

since he can get a payoff of zero when player 1 chooses Straight (which occurs with probability p, as depicted in Fig. 3.3), or a payoff of 3 when Player 1 selects to Swerve (which happens with probability $1 - p$). When, instead, Player 2 chooses Swerve (directing our attention to the right-hand column), he obtains an expected utility of

$$EU_2(Swerve) = 1p + 2(1 - p)$$

since his payoff is 1 if player 1 chooses Straight (which occurs with probability p) or a payoff of 2 if Player 1 selects to Swerve (which happens with probability $1 - p$). Therefore, Player 2 is indifferent between choosing Straight or Swerve when

$$EU_2(Straight) = EU_2(Swerve)$$

$$p0 + 3(1 - p) = 1p + 2(1 - p)$$

Solving for probability p in the above indifference condition, we obtain

$$1 = 2p \Rightarrow p = \frac{1}{2}$$

Hence, the mixed strategy Nash equilibrium of the Chicken game prescribes that every player randomizes between driving Straight and Swerve half of the time. More formally, the mixed strategy Nash equilibrium (msNE) is given by

$$msNE = \left\{ \left(\frac{1}{2} Straight, \frac{1}{2} Swerve \right), \left(\frac{1}{2} Straight, \frac{1}{2} Swerve \right) \right\}$$

where the first parenthesis indicates the probability distribution over Straight and Swerve for Player 1, and the second parenthesis represents the analogous probability distribution for Player 2.

Part (e) In Fig. 3.4, we first draw the thresholds which specify the mixed strategy Nash equilibrium (msNE): $p = 1/2$ and $q = 1/2$. Then, we draw the two pure strategy Nash equilibria found in question (c):

- One psNE in which Player 1 plays Straight and Player 2 plays Swerve (which in probability terms means $p = 1$ and $q = 0$), graphically depicted at the lower right-hand corner of Fig. 3.4 (in the southeast); and
- Another psNE in which Player 1 plays Swerve and Player 2 plays Straight (which in probability terms means $p = 0$ and $q = 1$), graphically depicted at the upper left-hand corner of Fig. 3.4 (in the northwest).

Player 1's best response function. Once we have drawn these pure strategy equilibria, notice that from Player 1's best response function, $BR_1(q)$, we know that, if $q < \frac{1}{2}$, then $EU_1(Straight) > EU_1(Swerve)$, thus implying that Player 1 plays Straight using pure strategies, i.e., $p = 1$. Intuitively, when Player 1 knows that Player 2 is rarely selecting Straight, i.e., $q < \frac{1}{2}$, his best response is to play Straight, i.e., $p = 1$ for all $q < \frac{1}{2}$, as illustrated in the vertical segment of $BR_1(q)$ in the right-hand side of Fig. 3.4. In contrast, $q > \frac{1}{2}$ entails $EU_1(Straight) < EU_1(Swerve)$, and therefore $p = 0$. In this case, Player 2 is likely playing Straight, leading Player 1 to respond with Swerve, i.e., $p = 0$, as depicted in the vertical segment of $BR_1(q)$ that overlaps the vertical axis in the left-hand side of Fig. 3.4.

Player 2's best response function. A similar analysis applies to Player 2's best response function, $BR_2(p)$, depicted in Fig. 3.5: (1) when $p < \frac{1}{2}$, Player 2's expected utility comparison satisfies $EU_2(Straight) > EU_2(Swerve)$, thus implying $q = 1$,

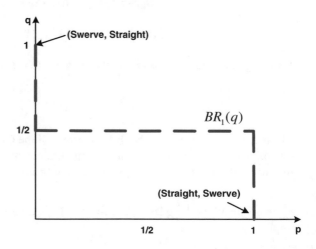

Fig. 3.4 Game of Chicken— Best response function of Player 1

Fig. 3.5 Game of Chicken—
Best response function of
Player 2

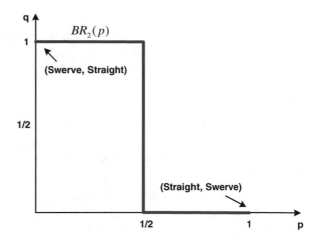

Fig. 3.6 Game of Chicken—
Best response functions,
psNE and msNE

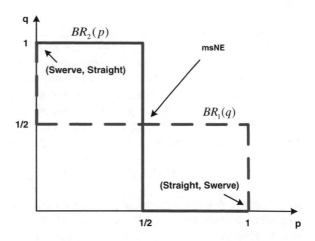

as depicted in the horizontal segment of $BR_2(p)$ at the top left-hand side of Fig. 3.5; and (2) when $p > \frac{1}{2}$, $EU_2(Straight) < EU_2(Swerve)$, ultimately leading to $q = 0$, as illustrated by the horizontal segment of $BR_2(p)$ that overlaps the horizontal axis in the right-hand side of the figure.

Superimposing both players' best response functions, we obtain Fig. 3.6, which depicts the two psNE of this game (southeast and northwest corners), as well as the msNE (strictly interior point, where players are randomizing their strategies).

Exercise 2—Lobbying Game[A]

Let us consider the following lobbying game in Fig. 3.7 where two firms simultaneously and independently decide whether to lobby Congress in favor a particular bill. When both firms (none of them) lobby, congress' decisions are unaffected,

Firm 2

		Lobby	Not Lobby
Firm 1	Lobby	-5,-5	15,0
	Not Lobby	0,15	10,10

Fig. 3.7 Lobbying game

implying that each firm earns a profit of 10 if none of them lobbies (-5 if both choose to lobby, respectively). If, instead, only one firm lobbies its payoff is 15 (since it is the only beneficiary of the policy), while that of the firm that did not lobby collapses to zero.

Part (a) Find the pure strategy Nash equilibria of the lobbying game.
Part (b) Find the mixed strategy Nash equilibrium of the lobbying game.
Part (c) Graphically represent each player's best response function.

Answer

Part (a) Let us first identify each firm's best response function. Firm 1's best response is $BR_1(Lobby) = Not\ lobby$ and $BR_1(Not\ lobby) = Lobby$, and similarly for Firm 2, thus implying that the lobbying game represents an anti-coordination game, such as the Game of Chicken, with two psNE: (Lobby, Not lobby) and (Not lobby, Lobby). Let us now analyze the mixed strategy equilibria of this game.
Part (b) *Firm 1.* Let q be the probability that Firm 2 chooses Lobby, and $(1-q)$ the probability that she chooses Not Lobby. The expected profit of Firm 1 playing Lobby (fixing our attention on the top row of the matrix) is

$$EU_1(Lobby) = -5q + 15(1-q) = 15 - 20q$$

since Firm 1 obtains a payoff of -5 when Firm 2 also chooses to lobby (which happens with probability q) or a payoff of 15 when Firm 2 does not lobby (which occurs with probability $1 - q$). When, instead, firm 1 chooses Not lobby (directing our attention to the bottom row of Fig. 3.7), its expected profit is:

$$EU_1(Not\ Lobby) = 0q + 10(1-q) = 10 - 10q$$

given that Firm 1 obtains a payoff of zero when Firm 2 lobbies and a profit of 10 when Firm 2 does not lobby. If Firm 1 is mixing between Lobbying and Not lobbying, it must be indifferent between Lobbying and Not lobbying. Otherwise, it would select the pure strategy that yields the highest expected payoff. Hence, we must have that

$$EU_1(Lobby) = EU_1(Not\ Lobby)$$

rearranging and solving for probability q yields

$$15 - 20q = 10 - 10q \Rightarrow q = 1/2$$

Therefore, for Firm 1 to be indifferent between Lobbying and Not lobbying, it must be that Firm 2 chooses to Lobby with a probability of $q = 1/2$.

Firm 2. A similar argument applies to Firm 2. If Firm 2 chooses to lobby (fixing the right-hand column), it obtains an expected profit of

$$EU_2(Lobby) = -5p + 15(1 - p)$$

since Firm 2 obtains a profit of -5 when Firm 1 also lobbies (which occurs with probability p), but a profit of 15 when Firm 1 does not lobby (which happens with probability $1 - p$). When Firm 2, instead selects to Not lobby (in the right-hand column of Fig. 3.7), it obtains an expected profit of

$$EU_2(Not\, Lobby) = 0p + 10(1 - p)$$

Hence, firm 2, must be indifferent between Lobbying and Not lobbying, that is,

$$EU_2(Lobby) = EU_2(Not\, Lobby)$$

solving for probability p yields

$$-5p + 15(1 - p) = 0p + 10(1 - p) \Rightarrow p = 1/2$$

Hence, Firm 2 is indifferent between Lobbying and Not lobbying as long as Firm 1 selects to Lobby with a probability $p = 1/2$. Then, in the lobbying game the mixed strategy Nash equilibrium (msNE) prescribes that:

$$msNE = \left\{ \left(\frac{1}{2}Lobby, \frac{1}{2}No\, Lobby \right), \left(\frac{1}{2}Lobby, \frac{1}{2}No\, Lobby \right) \right\}$$

Part (c) Figure 3.8 below depicts every player's best response function, and the crossing points of both best response functions identify the two psNE of the game $(p, q) = (0, 1)$, illustrating (Not lobby, Lobby), $(p, q) = (1, 0)$ which corresponds to (Lobby, Not lobby), and the msNE of the game $(p, q) = \left(\frac{1}{2}, \frac{1}{2} \right)$ found above.

In order to understand the construction of this figure, let us next analyze each player's best response function separately.

Player 1: For any $q > 1/2$, Firm 1 decides to play No Lobbying (in pure strategies), thus implying that $p = 0$ for any $q > 1/2$, as depicted by the vertical segment of $BR_1(q)$ (dashed line) that overlaps the vertical axis (left-hand side of Fig. 3.9). Intuitively, if Firm 2 is likely lobbying, i.e., $q > 1/2$, Firm 1 prefers not to lobby. Similarly, for any $q < 1/2$, Firm 2 is rarely lobbying, which induces firm 1 to lobby, and therefore $p = 1$, as illustrated by the vertical segment of $BR_1(q)$ in the right-hand side of Fig. 3.9.

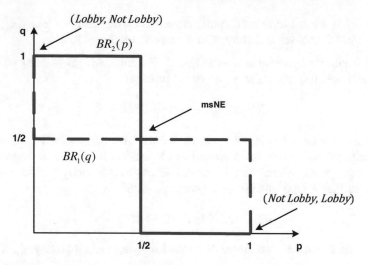

Fig. 3.8 Lobbying game—Best response functions, psNE and msNE

Fig. 3.9 Lobbying game—
Best response function, firm 1

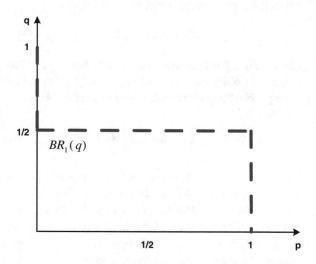

Player 2: For any $p > 1/2$, Firm 2 is better off Not lobbying than Lobbying, and hence $q = 0$ for any $p > 1/2$, as $BR_2(p)$ represents in the horizontal segment that overlaps the horizontal axis (see right-hand side of Fig. 3.10). Intuitively, if Firm 1 is likely lobbying, $p > \frac{1}{2}$, Firm 2 prefers Not to lobby. In contrast, for any $p < \frac{1}{2}$, Firm 2 is better off Lobbying than Lobbying, what implies that $q = 1$ for any p below $1/2$. Intuitively, Firm 1 is likely not lobbying, making lobbying more attractive for Firm 2. This result is graphically depicted by the horizontal segment of $BR_2(p)$ in the top left-hand side of Fig. 3.10.

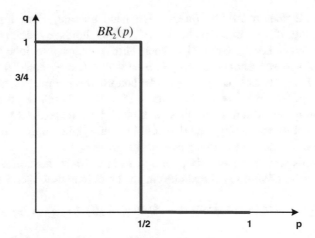

Fig. 3.10 Lobbying game—Best response function, firm 2

Superimposing $BR_1(q)$, as depicted in Fig. 3.9, and $BR_2(p)$, as illustrated in Fig. 3.10, we obtain the crossing points of both best response functions, as indicated in Fig. 3.8, which represent the two psNEs and one msNE in the Lobbying game.

Exercise 3—A Variation of the Lobbying Game[B]

Consider a variation of the above lobbying game. As depicted in the payoff matrix of Fig. 3.11, if Firm 1 lobbies but Firm 2 does not, Congress decision yields a profit of $x - 15$ for Firm 1, where $x > 25$, and does not yield any profits for Firm 2.

Part (a) Find the psNE of the game
Part (b) Find the msNE of the game
Part (c) Given the msNE you found in part (b), what is the probability that the outcome (Lobby, Not lobby) occurs?
Part (d) How does your result in part (c) varies as x increases? Interpret.

Answer
Part (a) Let us first examine Firm 1's best response function. In particular, $BR_1(Lobby) = No\ Lobby$, since when Firm 2 lobbies (in the left-hand column), Firm 1 obtains a larger payoff not lobbying, i.e., zero, than lobbying, -5. When,

		Firm 2	
		Lobby	No Lobby
Firm 1	Lobby	-5,-5	x-15,0
	No Lobby	0,15	10,10

Fig. 3.11 A variation of the Lobbying game

instead, Firm 2 does not lobby (in the right-hand column) Firm 1 gets a larger payoff lobbying if $x - 15 > 10$, or $x > 25$, which holds by definition. Hence, $BR_1(Not\ Lobby) = Lobby$. Similarly, when Firm 1 chooses to lobby (in the top row), Firm 2's best response is $BR_2(Lobby) = Not\ Lobby$, since $0 > -5$; and when Firm 1 selects not to lobby (in the bottom row) Firm 2's best response becomes $BR_2(Not\ Lobby) = Lobby$, since $15 > 10$. Therefore, there are two strategy profiles in which firms play a mutual best response to each other's strategies: (Lobby, Not Lobby) and (Not Lobby, Lobby), i.e., only one firm lobbies in equilibrium. These are the two psNE of the game.

Part (b) (in this part of the exercise, we operate in a similar fashion as in Exercise 2.) If Firm 1 is indifferent between Lobbying and Not lobbying, then it must be that

$$EU_1(Lobby) = EU_1(Not\ lobby), \text{ or} -5q + (x - 15)(1 - q) = 0q + 10(1 - q)$$

Rearranging terms and solving for probability q, yields $q = \frac{25-x}{20-x}$

Similarly, Firm 2 must be indifferent between Lobbying and Not lobbying (otherwise, it would select a pure strategy). Hence,

$$EU_2(Lobby) = EU_2(Not\ lobby), \text{ or} -5p + 15(1 - p) = 0p + 10(1 - p)$$

And solving for probability p, we obtain $p = \frac{1}{2}$. (This comes at no surprise since the payoffs of Firm 2 coincide with those in Exercise 3.2, thus implying that the probability p that makes Firm 2 indifferent between Lobbying and Not lobbying also coincides with that found in Exercise 3.2, i.e., $p = \frac{1}{2}$. However, that of Firm 1 changed, since the payoffs of Firm 1 have been modified.[2]

Hence, the msNE is

$$\left\{ \left(\frac{1}{2}Lobby, \frac{1}{2}Not\ Lobby \right), \left(\frac{25 - x}{20 - x}Lobby, 1 - \frac{25 - x}{20 - x}Not\ Lobby \right) \right\}$$

Part (c) The probability that outcome (Lobby, No Lobby) occurs is given by the probability that Firm 1 lobbies, p, times the probability that Firm 2 does not lobby, $1 - q$, i.e., $p(1 - q)$. Since we know that $p = \frac{1}{2}$, and $q = \frac{25-x}{20-x}$, then $p(1 - q) = \frac{1}{2}\left(1 - \frac{25-x}{20-x}\right)$, or simplifying,

---

[2]Nonetheless, in the specific case in which $x = 30$, the payoff matrix in Fig. 3.11 coincides with that in Exercise 3.2, and the probability q that makes Firm 1 indifferent reduces to $q = \frac{25-30}{20-30} = \frac{-5}{-10} = \frac{1}{2}$, as that in Exercise 3.2.)

Fig. 3.12 Probability of outcome (*Lobby, Not Lobby*) as a function of x

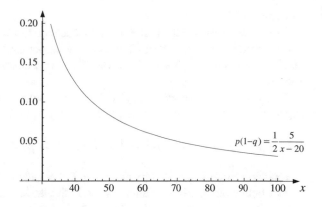

$$p(1-q) = \frac{1}{2}\left(\frac{5}{x-20}\right)$$

Part (d) As x increases, the probability of outcome (Lobby, Not Lobby) decreases, i.e.,

$$\frac{\partial[p(1-p)]}{\partial x} = -\frac{5}{2(x-20)^2} < 0.$$

which is negative given that $x > 25$ by definition. Figure 3.12 depicts $p(1-q)$ for different values of x (starting with $x = 25$, since $x > 25$ by definition), and illustrates that $p(1-q)$ decreases in x.

In addition, as x becomes larger, the aggregate payoff in outcome (Lobby, Not Lobby), i.e., $(x-15)+0$, exceeds the aggregate payoff from any other strategy profile, thus becoming socially optimal. In particular, when $(x-15)+0 > 10+10$, i.e., $x > 35$, this outcome becomes socially efficient. Intuitively, the profit that Firm 1 obtains when it is the only firm lobbying, $x - 15$, is so high that aggregate profits become larger than in the case in which no firm lobbies, $10+10$. Otherwise, it is socially optimal that no firm lobbies.

Exercise 4—Mixed Strategy Equilibrium with $n > 2$ Players[B]

Consider exercise 2.8 (psNE with n Players) from Chap. 2, where every player had to select between X and Y. Recall that the payoff of each player who selects X is $2m_x - m_x^2 + 3$ where m_x is the number of players who choose X, while the payoff of each player who selects Y is $4 - m_y$, where m_y is the number of players who choose Y, and $m_x + m_y = n$.

In Chap. 2, we found the three psNE of the game when $n = 3$. Still assuming that $n = 3$, determine whether this game has a symmetric msNE in which each player selects X with probability p and Y with the remaining probability $1 - p$.

Answer
The matrices in Figs. 3.13 and 3.14 reproduce the normal form game when $n = 3$, as described in exercise 2.8 of Chapter 2. (In particular, Fig. 3.13 depicts the payoff matrix when player 3 chooses X, while Fig. 3.14 illustrates the matrix when player 3 chooses Y.)

If every player chooses X with probability p and Y with probability $1 - p$, the expected utility that player 1 obtains from playing X is:

$$EU_1(X) = p^2 0 + p(1-p)3 + (1-p)p3 + (1-p)^2 4 = p(1-p)6 + 4(1-p)^2$$

First, note that if player 1 is choosing X we must only pay attention to the top row of both matrices. In particular, when both player 2 and 3 play X, which occurs with probability $p \times p = p^2$, then Player 1 obtains a payoff of 0 from playing X (see top left-hand cell in the upper matrix). If, instead, only one of his opponents plays X, which happens with probability $p(1 - p)$, his payoff becomes 3. (This holds both if player 2 plays X while player 3 chooses Y, and vice versa.) Finally, when both of his opponents play Y, which occurs with probability $(1 - p)^2$, his payoff from selecting X is 4 (see the top right-hand cell in the lower matrix).

And player 1's expected utility from playing Y is:

$$EU_1(Y) = p^2 3 + p(1-p)2 + (1-p)p2 + (1-p)^2 1$$
$$= 3p^2 + 4(1-p)p + (1-p)^2$$

In this case, we fix our attention on the bottom row of both matrices: when both players 2 and 3 choose X, player 1's payoff is 3; when one of his opponents selects X, player 1's payoff is 2; and when both players 2 and 3 choose Y, then player 1's payoff is only 1.

Player 2

		X	Y
		X	Y
Player 1	X	0,0,0	3,3,3
	Y	3,3,3	2,2,4

Fig. 3.13 Player 3 chooses X

Player 2

		X	Y
		X	Y
Player 1	X	3,3,3	4,2,2
	Y	2,4,2	1,1,1

Fig. 3.14 Player 3 chooses Y

Player 2

		L	C	R
	T	3, 2	4, 3	1, 4
Player 1	C	1, 3	7, 0	2, 1
	B	2, 2	8,- 5	2, 0

Fig. 3.15 A 3×3 payoff matrix

Hence, Player 1 is indifferent between choosing X and Y for values of p that satisfy $EU_1(X) = EU_1(Y)$. That is,

$$p(1-p)6 + 4(1-p)^2 = 3p^2 + 4(1-p)p + (1-p)^2$$

and simplifying,

$$2p^2 + 4p - 3 = 0$$

Solving for p, we find two roots of p: either $p = -1 - \sqrt{2.5} < 0$, which cannot be a solution to our problem, since we need $p \in [0, 1]$, or $p = -1 + \sqrt{2.5} = 0.58$, which is the solution to our problem.

A similar argument applies to the randomization of players 2 and 3, since their payoffs are symmetric. Hence, every player in this game randomizes between X and Y (using mixed strategies) assigning probability $p = 0.58$ to strategy X, and $1 - p = 0.42$ to strategy Y.

Exercise 5—Randomizing Over Three Available Actions[B]

Consider the following game where Player 1 has three available strategies (Top, Center, Bottom) and Player 2 has also three options (Left, Center, Right) as shown in the normal form game of Fig. 3.15.

Find all Nash equilibria (NEs) of this game.

Answer
First, observe that the pure strategy C of player 2 is strictly dominated by strategy R. Indeed, regardless of the pure strategy that player 1 chooses (i.e., independently on the row he selects), player 2's payoffs is strictly higher with R than with C, that is $3 < 4$, $0 < 1$ and $-5 < 0$. Hence, strategy C is never going to be part of a NE in pure or mixed strategies. We can then eliminate strategy C (middle column) for player 2, as depicted in the reduced game in Fig. 3.16.

Underlying best response payoffs, we find that there is no psNE, i.e., there is no cell where the payoff of all players are underlined as best responses. Let's next analyze msNE.

Player 2

		L	R
T		$\underline{3}, 2$	$1, \underline{4}$
Player 1 C		$1, \underline{3}$	$\underline{2}, 1$
B		$2, \underline{2}$	$\underline{2}, 0$

Fig. 3.16 Reduced payoff matrix (after deleting column C)

Let us assign probability q to player 2 choosing L and $(1 - q)$ to him selecting R; as depicted in the matrix of Fig. 3.17. Similarly, let p_1 represent the probability that player 1 chooses T, p_2 the probability he selects C, and $1 - p_1 - p_2$ the probability of him playing B.

Let us next find the expected payoffs that each player obtains from choosing each of his pure strategies, anticipating that his rival uses the above randomization profile. First, for player 1, we obtain:

$$EU_1(T) = 3q + 1(1 - q) = 1 + 2q$$
$$EU_1(C) = 1q + 2(1 - q) = 2 - q$$
$$EU_1(B) = 2q + 2(1 - q) = 2$$

And for player 2 we have that:

$$EU_2(L) = 2p_1 + 3p_2 + 2(1 - p_1 - p_2)$$
$$= 2p_1 + 3p_2 + 2 - 2p_1 - 2p_2$$
$$= 2 + p_2$$

and

$$EU_2(R) = 4p_1 + 1p_2 + 0(1 - p_1 - p_2)$$
$$= 4p_1 + p_2$$

Player 2

| | | | q | $1 - q$ |
			L	R
p_1		**T**	3, 2	1, 4
Player 1 p_2		**C**	1, 3	2, 1
$1 - p_1 - p_2$		**B**	2, 2	2, 0

Fig. 3.17 Assigning probabilities to each pure strategy

Hence if player 2 mixes, he must be indifferent between L and R, entailing:

$$2 + p_2 = 4p_1 + p_2$$

$$p_1 = 1/2$$

implying that player 1 mixes between C and B with a 50 % probability on each. We have, thus, found the mixing strategy of player 1, and we can then turn to player 2 (which entails analyzing the expected payoffs of player 1). We next consider different mixed strategies for player 1, showing that only some of them can be sustained in equilibrium.

Mixing between T and C alone. If player 1 mixes between T and C alone (assigning no probability weight to B), then he must be indifferent between T and C, as follows:

$$EU_1(T) = EU_1(C)$$
$$1 + 2q = 2 - q, \text{ which yields } q = \tfrac{1}{3}$$

Hence, player 1 randomizes between T and C, assigning a probability of $q = \tfrac{1}{3}$ to T and $1 - q = \tfrac{2}{3}$ to C.

Mixing between T and B alone. If, instead, player 1 mixes between T and B (assigning no probability to C), then his indifference condition must be:

$$EU_1(T) = EU_1(B)$$
$$1 + 2q = 2, \text{ which yields } q = 1/2$$

implying that player 2 mixes between L and R with 50 % on each. We have then identified a msNE:

$$\left\{ \left(\tfrac{1}{2}T, 0C, \tfrac{1}{2}B \right), \left(\tfrac{1}{2}L, \tfrac{1}{2}R \right) \right\}$$

In words, player 1 evenly mixes between T and B, i.e. $p_1 = 1/2$ and $p_2 = 0$, while player 2 evenly mixes between L and R.

Mixing between C and B alone. Finally, if player 1 mixes between C and B, it must be that he is indifferent between their expected payoffs

$$EU_1(C) = EU_1(B)$$
$$2 - q = 2, \text{ which yields } q = 0$$

Therefore, player 1 cannot randomize between C and B alone; otherwise, player 2 would play R with certainty ($q = 0$), driving player 1 to best respond with B or C (both of them yield the same payoff, 2); a contradiction.

Mixing among all three actions. Note that a msNE in which player 1 randomizes between all three strategies cannot be sustained since for that mixing strategy he would need:

$$EU_1(T) = EU_1(C) = EU_1(B)$$
$$1 + 2q = 2 - q = 2$$

Providing us with two equations $1 + 2q = 2$ and $2 - q = 2$, which cannot simultaneously hold, i.e., $1 + 2q = 2$ entails $q = 1/2$ while $2 - q = 2$ yields $q = 0$.

Exercise 6—Pareto Coordination Game[B]

Consider the Pareto coordination game depicted in Fig. 3.18. Determine the set of psNE and msNE, and depict both players' best response correspondences.

Answer

Let us first analyze the set of psNE of this game. If Player 2 chooses L (in the left-hand column), then Player 1's best response is $BR_1(L) = U$; while if he chooses R (in the right-hand column), Player 1 responds with $BR_1(R) = D$. Let us now analyze Player 2's best responses. If Player 1 chooses U (in the top row), then Player 2's best response is $BR_2(U) = L$; while if Player 1 chooses D (in the bottom row), then Player 2 responds with $BR_2(D) = R$. The payoff matrix in Fig. 3.19 underlines the payoff associated to every player's best responses. Since there are two cells in which all players payoffs are underlined (indicating that they are playing mutual best responses), there exist 2 pure strategy Nash equilibria: $psNE = \{(U, L), (D, R)\}$.

Let us now examine the msNE of the game. Let p denote the probability that Player 1 chooses U and q the probability that Player 2 selects L. For Player 1 to be indifferent between U and D, we need

Player 2

		L	R
Player 1	U	9,9	0,8
	D	8,0	7,7

Fig. 3.18 Pareto coordination game

Player 2

		L	R
Player 1	U	9,9	0,8
	D	8,0	7,7

Fig. 3.19 Underlining best response payoffs in the Pareto coordination game

$$EU_1(U) = EU_1(D)$$

that is,

$$9q + 0(1 - q) = 8q + 7(1 - q)$$

And solving for q yields $q^* = \frac{7}{8}$. Similarly, for Player 2 to be indifferent between L and R, we need

$$EU_2(L) = EU_2(R)$$

or

$$9p + 0(1 - p) = 8p + 7(1 - p)$$

And solving for p, we obtain $p^* = \frac{7}{8}$. Hence, the msNE of this game is

$$msNE = \left\{ \left(\frac{7}{8}U, \frac{1}{8}D \right), \left(\frac{7}{8}L, \frac{1}{8}R \right) \right\}$$

We can now construct the best response correspondences for each player:

- For Player 1 we have that if $q < \frac{7}{8}$, then Player 1's best response is $BR_1(q) = D$. Intuitively, in this setting, Player 2 is not likely to select L, but rather R. Hence, player 1's expected payoff from D is larger than that of U, thus implying $p = 0$. If, instead, $q > \frac{7}{8}$, Player 1's best response becomes $BR_1(q) = U$, thus entailing $p = 1$. Finally, if q is exactly $q = \frac{7}{8}$, then Player 1's best response is to randomize between U and D, given that he is indifferent between both of them. This is illustrated in Fig. 3.20, where $BR_1(q)$ implies $p = 0$ along the vertical axis of all $q < \frac{7}{8}$ (in the left-hand side of the figure), becomes flat at exactly $q = \frac{7}{8}$, and turns vertical again (i.e., $p = 1$ in the right-hand side of the figure) for all $q > \frac{7}{8}$.
- Similarly, for Player 2, we have that: (1) if $p < \frac{7}{8}$, his best response becomes $BR_2(p) = R$, thus implying $q = 0$; (2) if $p > \frac{7}{8}$, his best response is $BR_2(p) = L$, thus entailing $q = 1$; and (3) if $p = \frac{7}{8}$, Player 2 randomizes between U and D. Graphically, $BR_2(p)$ lies at $q = 0$ along the horizontal axis for all $p < \frac{7}{8}$ (as illustrated in the left-hand side of Fig. 3.20), becomes vertical at exactly $p = \frac{7}{8}$, and turns horizontal again for all $p > \frac{7}{8}$ (i.e., $q = 1$ at the upper right-hand part of the figure).

Therefore, $BR_1(q)$ and $BR_2(p)$ intersect at three points; illustrating two psNE

$$(p, q) = (0, 0) \Rightarrow psNE(D, R)$$

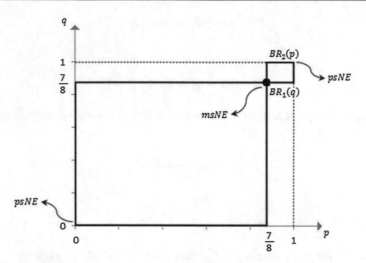

Fig. 3.20 Pareto coordination game—Best response functions

$$(p, q) = (1, 1) \Rightarrow psNE(U, L)$$

and one msNE:

$$(p, q) = \left(\frac{7}{8}, \frac{7}{8}\right) \Rightarrow msNE\left\{\left(\frac{7}{8}U, \frac{1}{8}D\right), \left(\frac{7}{8}L, \frac{1}{8}R\right)\right\}$$

Exercise 7—Mixing Strategies in a Bargaining Game[C]

Consider the sequential-move game in Fig. 3.21. The game tree describes a bargaining game between Player 1 (proposer) and Player 2 (responder). Player 1 makes a take-it-or-leave-it offer to Player 2, specifying an amount $s = \{0, \frac{v}{2}, v\}$ out of an initial surplus v, i.e., no share of the pie, half of the pie, or all of the pie, respectively. If Player 2 accepts such a distribution, Player 2 receives the offer s, while Player 1 keeps the remaining surplus $v - s$. If Player 2 rejects, both players get a zero payoff.

Part (a) Describe the strategy space for every player.
Part (b) Provide the normal form representation of this bargaining game.
Part (c) Does any player have strictly dominated pure strategies?
Part (d) Does any player have strictly dominated mixed strategies?

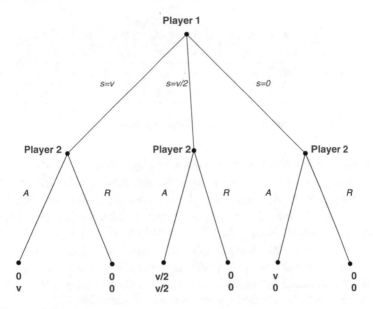

Fig. 3.21 Bargaining game

Answer

Part (a) Strategy set for player 1

$$S_1 = \left\{0, \frac{v}{2}, v\right\}$$

Strategy set for player 2

$$S_2 = \{AAA, AAR, ARR, RRR, RRA, RAA, ARA, RAR\}$$

For every triplet, the first component specifies Player 2's response upon observing that Player 1 makes offer $s = v$ (in the left-hand side of Fig. 3.21), the second component is instead his response to an offer $s = \frac{v}{2}$, and the third component describes Player 2's response to an offer $s = 0$ (in the right-hand side of Fig. 3.21).

Part (b) Normal form representation: Using the three strategies for Player 1 and the eight available strategies for Player 2 found in part (a), the 3×8 matrix of Fig. 3.22 represents the normal form representation of this game.

Part (c) No player has any strictly dominated pure strategy:

Player 1. For Player 1, we find that $s = \frac{v}{2}$ yields a weakly (not strictly) higher payoff than $s = v$, that is $u_1\left(s = \frac{v}{2}, s_2\right) \geq u_1(s = v, s_2)$ for all strategies of Player 2, $s_2 \in S_2$ (i.e., some columns in the above matrix), which is satisfied with strict equality for some strategies of Player 2, such as *ARR*, *RRR* or *RRA*.

		Player 2							
		AAA	AAR	ARR	RRR	RRA	RAA	ARA	RAR
Player 1	$s = v$	$0,v$	$0,v$	$0,v$	$0,0$	$0,0$	$0,0$	$0,v$	$0,0$
	$s = v/2$	$v/2,v/2$	$v/2,v/2$	$0,0$	$0,0$	$0,0$	$v/2,v/2$	$0,0$	$v/2,v/2$
	$s = 0$	$v,0$	$0,0$	$0,0$	$0,0$	$v,0$	$v,0$	$v,0$	$0,0$

Fig. 3.22 Normal-form representation of the Bargaining game

Similarly, $s = 0$ yields a weakly (but not strictly) higher payoff than $s = \frac{v}{2}$. That is, $u_1(s = v, s_2) \geq u_1\left(s = \frac{v}{2}, s_2\right)$ for all $s_2 \in S_2$, with strict equality for some $s_2 \in S_2$, such as ARR and RRR. Similarly, $s = \frac{v}{2}$ yields a weakly higher payoff than $s = 0$ for some strategies of Player 2, such as RRR, but a strictly larger payoff for other strategies, such as RAR. Hence, there is no weakly dominated strategy for Player 1.

Player 2. Similarly, for Player 2, $u_2(s_2, s_1) \geq u_2\left(s_2', s_1\right)$ for any two strategies of Player 2 $s_2 \neq s_2'$ and for all $s_1 \in S_1$ with strict equality for some $s_1 \in S_1$.

Part (d) Once we have shown that there is no strictly dominated pure strategy, we focus on the existence of strictly dominated *mixed* strategies.

We know that Player 1 is never going to mix assigning a strictly positive probability to his pure strategy $s = v$ (i.e., offering the whole pie to Player 2) given that it will reduce for sure his expected payoff, for any strategy with which Player 2 responds. Indeed, such strategy yields a strictly lower (or equal) payoff than other of his available strategies, such as $s = 0$ or $s = \frac{v}{2}$.

If Player 1 mixes between $s = 0$ and $s = \frac{v}{2}$, we can see that he is going to obtain a mixed strategy σ_1 that yields a expected utility, $u_1(\sigma_1, s_1)$, which exceeds his utility from selecting the pure strategy $s = v$. That is,

$$u_1(\sigma_1, s_1) \geq u_1(s = v, s_2) \text{ for all } s_2 \in S_2$$

with strict equality for $s_2 = ARR$ and $s_2 = RRR$, but strict inequality (yielding a strictly higher expected payoff) for all other strategies of player 2. We can visually check this result in Fig. 3.22 by noticing that $s = v$, in the top row, yields a zero payoff for any strategy of player 2. However, a linear combination of strategies $s = \frac{v}{2}$ and $s = 0$, in the middle and bottom rows, yields a positive expected payoff for columns AAA, AAR, RRA, RAA, ARA and RAR; since all of them contain at least one positive payoff for Player 1 in the middle or bottom row. However, in the remaining columns (ARR and RRR), Player 1's payoff is zero both in the middle and bottom row; thus implying that his expected payoff, zero, coincides with his payoff from playing the pure strategy $s = v$ in the top row. A similar argument applies to Player 2. Therefore, there doesn't exist any strictly dominated mixed strategy.

Player 2

		L	R
Player 1	U	6,6	2,7
	D	7,2	0,0

Fig. 3.23 Normal-form game

Exercise 8—Depicting the Convex Hull of Nash Equilibrium Payoffs[C]

Consider the two-player normal form game depicted in Fig. 3.23.

Part (a) Compute the set of Nash equilibrium payoffs (allowing for both pure and mixed strategies).
Part (b) Draw the convex hull of such payoffs.
Part (c) Is there any correlated equilibrium that yields payoffs outside this convex hull?

Answer

Part (a) Set of Nash equilibrium payoffs.
psNE. Let us first analyze the set of psNE. Player 1's best responses are $BR_1(L) = D$ when player 2 chooses L (in the left-hand column), and $BR_1(R) = U$ when player 2 selects R (in the right-hand column). Player 2's best responses are $BR_2(U) = R$ when player 1 chooses U (in the top row) and $BR_2(D) = L$ when player 1 selects D (in the bottom row). Hence, there are two cells in which all players are playing mutual best responses. That is, there are two psNEs, $(U, R), (D, L)$, with associated equilibrium payoffs of:

$$(2, 7) \text{ for psNE } (U, R), \text{ and}$$
$$(7, 2) \text{ for psNE } (D, L).$$

msNE. Let us now examine the msNE of this game. Player 1 randomizes with probability p, as depicted in Fig. 3.24.
Hence, Player 2 is indifferent between L and R, if

$$EU_2(L) = EU_2(R)$$
$$6p + 2(1 - p) = 7p + 0(1 - p)$$

And solving for probability p, we obtain $p = \frac{2}{3}$
Similarly, Player 2 randomizes with probability q. Therefore, Player 1 is indifferent between his two strategies, U and D, if

$$EU_1(U) = EU_1(D)$$
$$6q + 2(1 - q) = 7q + 0(1 - q)$$

Player 2

			q	1-q
			L	R
Player 1	p	U	6,6	2,7
	1-p	D	7,2	0,0

Fig. 3.24 Normal-form game

And solving for probability q, we find $q = \frac{2}{3}$. Therefore, the msNE is given by:

$$msNE = \left\{ \left(\frac{2}{3}U, \frac{1}{3}D \right), \left(\frac{2}{3}L, \frac{1}{3}R \right) \right\}$$

The expected payoff that Player 1 obtains from this symmetric msNE is:

$$EU_1(p^*, q^*) = \frac{2}{3} \underbrace{\left[\frac{2}{3} \cdot 6 + \frac{1}{3} \cdot 7 \right]}_{Player\,2\,plays\,L} + \frac{1}{3} \underbrace{\left[\frac{2}{3} \cdot 2 + \frac{1}{3} \cdot 0 \right]}_{Player\,2\,plays\,R}$$

$$= \frac{38}{9} + \frac{4}{9} = \frac{14}{3}$$

And similarly, the expected payoff of Player 2 is

$$EU_2(p^*, q^*) = \frac{2}{3} \underbrace{\left[\frac{2}{3} \cdot 6 + \frac{1}{3} \cdot 7 \right]}_{Player\,1\,plays\,U} + \frac{1}{3} \underbrace{\left[\frac{2}{3} \cdot 2 + \frac{1}{3} \cdot 0 \right]}_{Player\,1\,plays\,D}$$

$$= \frac{38}{9} + \frac{4}{9} = \frac{14}{3}$$

Hence the expected payoffs for playing this msNE are given by $EU_i(p^*, q^*) = 14/3$ for both players $i = 1, 2$.

Best Responses. Figure 3.25 depicts the best response for each player. For Player 1, note that: (1) when $q < \frac{2}{3}$, his best response becomes $BR_1(q) = U$, thus implying $p = 1$, as indicated in the vertical segment of $BR_1(q)$ on the right-hand side of the figure; (2) when $q = \frac{2}{3}$, his best response is to randomize between U and D, as illustrated in the horizontal line at exactly $q = \frac{2}{3}$; finally (3) when $q > \frac{2}{3}$, his best response is $BR_1(q) = D$, thus entailing $p = 0$, as depicted in the vertical segment of $BR_1(q)$ in the left-hand side of the figure which overlaps with the vertical axis for all $q > \frac{2}{3}$. A similar analysis applies to the best response function of player 2, $BR_2(p)$, where $BR_2(p) = L$ when $p < \frac{2}{3}$, this implying $q = 1$ (as depicted in the horizontal segment at the top of Fig. 3.25), while his best response becomes

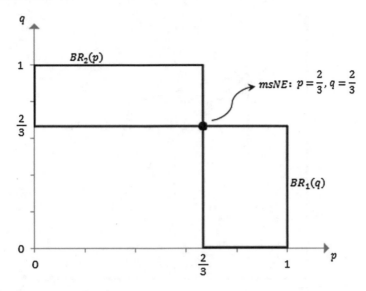

Fig. 3.25 Best response functions

$BR_2(p) = R$ when $p > \frac{2}{3}$, which entails $q = 0$, as indicated in the horizontal line that coincides with the p-axis for all $p > \frac{2}{3}$.

Part (b) *Convex hull of equilibrium payoffs*. From our previous analysis, equilibrium payoffs are $(2,7)$ for the psNE (U, D), and $(7, 2)$ for the psNE (D, L). Figure 3.26 depicts these two payoffs pairs (see northwest and southeast corners, respectively). In addition, it also illustrates the expected payoff from the msNE, $\frac{14}{3} = 4.66$ for each player, and the convex hull of Nash equilibrium payoffs (shaded area).

Part (c) Let us now analyze the correlated equilibrium. Assume that a trusted party tells each player what to do based on the outcome of the following experiment (using a publicly observable random variable, such as a three-sided dice whose outcome all players can observe; as illustrated in Fig. 3.27).

Hence, player 1's expected payoff at the correlated equilibrium is

$$\frac{1}{3} \cdot 2 + \frac{1}{3} \cdot 7 + \frac{1}{3} \cdot 6 = \frac{15}{3} = 5$$

since player 1 obtains a payoff of 2 when players choose outcome (U, R), a payoff of 7 when they select (D, L), and a payoff of 6 in outcome (U, L). And similarly for player 2's expected payoff

$$\frac{1}{3} \cdot 7 + \frac{1}{3} \cdot 2 + \frac{1}{3} \cdot 6 = \frac{15}{3} = 5$$

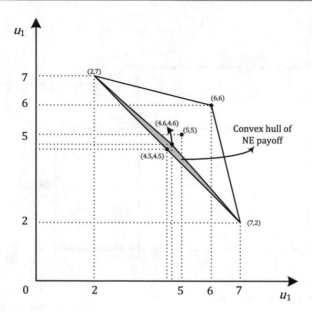

Fig. 3.26 Convex hull of equilibrium payoffs

Outcome	Probability
U,R	1/3
D,L	1/3
U,L	1/3

Fig. 3.27 Experiment

Therefore, as Fig. 3.26 shows, the correlated equilibrium yields a payoff pair $(5, 5)$ outside this convex hull of Nash equilibrium payoffs. However, notice that there exists a Nash equilibrium (potentially involving mixed strategies) in which each player obtains a expected payoff of $EU_1 = EU_2 = 4.5$, which lies at the southwest frontier of the convex hull.

The equation of the line connecting points $(2, 7)$ and $(7, 2)$ is $u_2 = 9 - u_1$. To see this, recall that the slope of a line can be found in this context with $m = \frac{2-7}{7-2} = -1$, while the vertical intercept is found by inserting either of the two points on the equation. For instance, using $(2, 7)$ we find that $7 = b - 2$ which, solving for b, yields the vertical intercept $b = 9$. It is then easy to check that point $(4.5, 4.5)$ lies on this line since $4.5 = 9 - 4.5$ holds with equality.

		Game 1 Player 2				*Game 2* Player 2	
		L	*R*			*L*	*R*
Player 1	*U*	2, 1	0, 0		*U*	4, 4	1, 5
	D	0, 0	1, 2		*D*	5, 1	0, 0

Fig. 3.28 Two normal-form games

Exercise 9—Correlated Equilibrium^C

Consider the two games depicted in Fig. 3.28, in which player 1 chooses rows and player 2 chooses columns:

Part (a) In each case, find the set of Nash equilibria (pure and mixed) and depict their equilibrium payoff pairs.

Part (b) Characterize one correlated equilibrium where the correlating device is a publicly observable random variable, and illustrate the payoff pairs on the figure you developed in part (a).

Part (c) Characterize the set of all correlated equilibria, and depict their payoff pairs.

Answer

Part (a) *Game 1*. Let's first characterize the best response of each player in Game 1. If Player 1 plays U (in the top row), Player 2's best response is $BR_2(U) = L$; while if player 1 plays D (in the bottom row), Player 2's best response is $BR_2(D) = R$. For Player 1's best responses, we find that if Player 2 plays L (in the left-hand column), then $BR_1(L) = U$; whereas if Player 2 plays R (in the right-hand column), $BR_1(R) = D$. These best responses yield payoffs depicted with the bold underlined numbers in the matrix of Fig. 3.29.

As a consequence, we have two PSNE: $\{(U, L), (D, R)\}$, with corresponding equilibrium payoffs pairs (2, 1) and (1, 2).

In terms of mixed strategies, Player 2 randomizes by choosing a probability q that makes Player 1 indifferent between choosing Up or Down, as depicted in Fig. 3.30. Hence, if Player 1 is indifferent between U and D, we must have that

$$EU_1(U) = EU_1(D)$$

	L	*R*
U	**2, 1**	0, 0
D	0, 0	**1, 2**

Fig. 3.29 Underlining best response payoffs of Game 1

		q L	$1\text{-}q$ R
p	U	2, 1	0, 0
$1\text{-}p$	D	0, 0	1, 2

Fig. 3.30 Searching for msNE in Game 1

which implies,

$$2q + 0 \cdot (1 - q) = 0q + 1 \cdot (1 - q)$$

rearranging, and solving for probability q, yields

$$2q = 1 - q \leftrightarrow q = \frac{1}{3}$$

Similarly, Player 1 randomizes by choosing a probability p such that makes Player 2 indifferent between left or right,

$$EU_2(L) = EU_2(R)$$

or

$$1 \cdot p + 0(1 - p) = 0p + 2(1 - p)$$

rearranging, and solving for probability p, we obtain

$$p = 2 - 2p \leftrightarrow p = \frac{2}{3}$$

Hence, the msNE of Game 1 is

$$msNE = \left\{ \left(\frac{2}{3} U, \frac{1}{3} D \right), \left(\frac{1}{3} L, \frac{2}{3} R \right) \right\}$$

And the associated expected payoffs in this MSNE are given by,

$$EU_1\left(\sigma_1^*, \sigma_2^*\right) = \frac{2}{3} \left[\frac{1}{3} \cdot 2 + \frac{2}{3} \cdot 0 \right] + \frac{1}{3} \left[\frac{1}{3} \cdot 0 + \frac{2}{3} \cdot 1 \right] = \frac{2}{3} \cdot \frac{2}{3} + \frac{1}{3} \cdot \frac{2}{3} = \frac{4}{9} + \frac{2}{9} = \frac{6}{9}$$
$$= \frac{2}{3}$$

$$EU_2\left(\sigma_1^*, \sigma_2^*\right) = \frac{1}{3} \left[\frac{2}{3} \cdot 1 + \frac{1}{3} \cdot 0 \right] + \frac{2}{3} \left[\frac{2}{3} \cdot 0 + \frac{1}{3} \cdot 2 \right] = \frac{2}{9} + \frac{2}{3} \cdot \frac{2}{3} = \frac{2}{9} + \frac{4}{9} = \frac{6}{9} = \frac{2}{3}$$

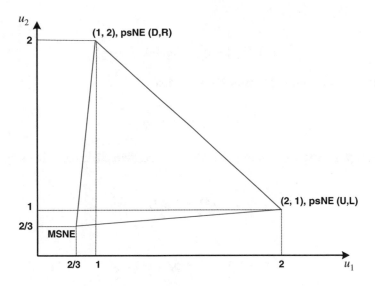

Fig. 3.31 Convex hull of equilibrium payoffs (Game 1)

	L	R
U	4, 4	**1**, **5**
D	**5**, **1**	0, 0

Fig. 3.32 Underlining best response payoffs in Game 2

Figure 3.31 depicts the convex hull representing all Nash equilibrium payoffs of game 1.

Game 2. Let's now characterize the best responses of each player in Game 2: if Player 1 plays U (in the top row), then Player 2's best response becomes $BR_2(U) = R$; while if player 1 plays D (in the bottom row), Player 2 responds with $BR_2(D) = L$. For Player 1's best responses, we find that if Player 2 plays L (in the left-hand column), then $BR_1(L) = D$; whereas if Player 2 plays R (in the right-hand column), $BR_1(R) = U$. These best responses yield the payoffs depicted with the bold underlined numbers in the matrix of Fig. 3.32.

Hence we have two cells in which all players' payoffs were underlined, thus indicating mutual best responses in the Nash Equilibrium of the game. This game, therefore, has two psNE: $\{(U, R), (D, L)\}$, with associated equilibrium payoff pairs of (1, 5) and (5, 1).

Let us now analyze the set of msNE in this game. Similarly as in Game 1, Player 2 mixes between L and R with associated probabilities q and $1 - q$ to make Player 1 indifferent between his two pure strategies (U and D), as follows:

$$EU_1(U) = EU_1(D)$$

or

$$4q + 1 \cdot (1 - q) = 5q + 0 \cdot (1 - q)$$

rearranging, and solving for probability q, yields

$$4q + 1 - q = 5q \leftrightarrow q = \frac{1}{2}$$

Similarly, Player 1 mixes to make Player 2 indifferent between choosing left or right,

$$EU_2(L) = EU_2(R)$$

or

$$4p + 1(1 - p) = 5p + 0(1 - p)$$

rearranging, and solving for probability p, we obtain

$$4p + 1 - p = 5p \leftrightarrow p = \frac{1}{2}$$

Therefore, the msNE of game 2 is:

$$msNE = \left\{ \left(\frac{1}{2}U, \frac{1}{2}D \right), \left(\frac{1}{2}L, \frac{1}{2}R \right) \right\}$$

And the associated payoffs of this msNE are:

$$EU_1\left(\sigma_1^*, \sigma_2^*\right) = \frac{1}{2}\left[\frac{1}{2} \cdot 4 + \frac{1}{2} \cdot 1\right] + \frac{1}{2}\left[\frac{1}{2} \cdot 5 + \frac{1}{2} \cdot 0\right] = \frac{1}{2} \cdot \frac{5}{2} + \frac{1}{2} \cdot \frac{5}{2} = \frac{10}{4} = \frac{5}{2}$$

$$EU_2\left(\sigma_1^*, \sigma_2^*\right) = \frac{1}{2}\left[\frac{1}{2} \cdot 4 + \frac{1}{2} \cdot 1\right] + \frac{1}{2}\left[\frac{1}{2} \cdot 0 + \frac{1}{2} \cdot 5\right] = \frac{1}{2} \cdot \frac{5}{2} + \frac{1}{2} \cdot \frac{5}{2} = \frac{10}{4} = \frac{5}{2}$$

Figure 3.33 illustrates the convex hull representing all Nash equilibrium payoffs of Game 2.

Part (b) *Game 1*. First, note that intuitively, by using a correlating device (a publicly observable random variable) we construct convex combinations of the two PSNE of the game. For example, we can determine that the publicly observable random variable is a coin flip, where Heads makes Player 1 play U and Player 2 play L; while Tails make Player 1 play D and Player 2 play R. Hence, the probabilities assigned to every strategy profile for Game 1 are summarized in Fig. 3.34.

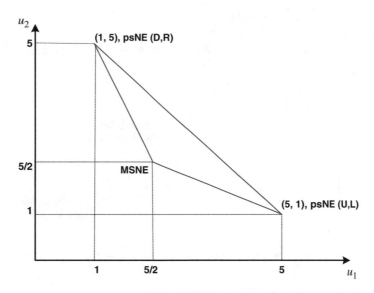

Fig. 3.33 Convex hull of equilibrium payoffs (Game 2)

	L	R
U	$\frac{1}{2}$	0
D	0	$\frac{1}{2}$

Fig. 3.34 Probabilities assigned to each strategy profile in a correlated equilibrium (Game 1)

Given the above probabilities, the expected payoffs for each player in Game 1 are:

$$EU_1 = \frac{1}{2} \cdot U_1(U,L) + \frac{1}{2} \cdot U_1(D,R) = \frac{1}{2} \cdot 2 + \frac{1}{2} \cdot 1 = \frac{3}{2}$$

$$EU_2 = \frac{1}{2} \cdot U_2(U,L) + \frac{1}{2} \cdot U_2(D,R) = \frac{1}{2} \cdot 1 + \frac{1}{2} \cdot 2 = \frac{3}{2}$$

Graphically, we can see that the correlated equilibrium payoffs are located in the midpoint of the line connecting the two psNE payoffs, i.e., $\alpha u_2(U,L) + (1 - \alpha)u_2(D,R)$, as depicted in Fig. 3.35, where $\alpha = 1/2$ since the correlating device assigns the same probability to Nash equilibria (U, L) and to (D, R).

Game 2. The probabilities that the publicly observable random variable assigns to each strategy profile (in pure strategies) in Game 2 are summarized in Fig. 3.36.

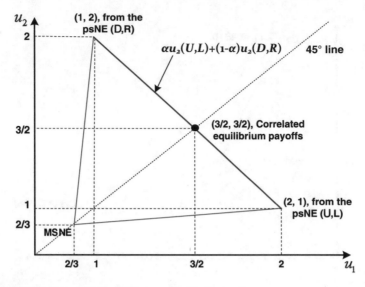

Fig. 3.35 Nash and correlated equilibrium payoffs (Game 1)

	L	R
U	$\dfrac{1}{2}$	0
D	0	$\dfrac{1}{2}$

Fig. 3.36 Probabilities assigned to each strategy profile in a correlated equilibrium (Game 2)

Yielding expected payoffs for each player of

$$EU_1 = \frac{1}{2} \cdot U_1(U,R) + \frac{1}{2} \cdot U_1(D,L) = \frac{1}{2} \cdot 1 + \frac{1}{2} \cdot 5 = \frac{6}{2} = 3$$

$$EU_2 = \frac{1}{2} \cdot U_2(U,R) + \frac{1}{2} \cdot U_2(D,L) = \frac{1}{2} \cdot 5 + \frac{1}{2} \cdot 1 = \frac{6}{2} = 3$$

And the graphical representation on the frontier of the convex hull is (Fig. 3.37). Similarly as for Game 1, correlated equilibrium payoffs lie at the midpoint of the NE payoffs, since the correlating device assigns the same probability to Nash equilibria (U, R) and to (D, L).

Part (c) Let us now allow for the correlated equilibrium to assign a more general probability distribution given by p, q, r to strategy profiles $(U, L), (U, R)$ and (D, L), respectively; as depicted in the table of Fig. 3.38.
Player 1 Consider that Player 1 receives the order of playing U. Then, his conditional probability of being at cell (U, L) is $\frac{p}{p+q}$, while his conditional probability

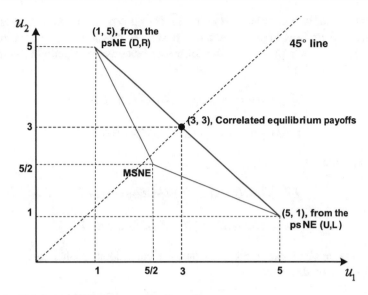

Fig. 3.37 Correlated equilibrium payoffs (Game 2)

	L	R
U	p	q
D	r	$1\text{-}p\text{-}q\text{-}r$

Fig. 3.38 Probabilities assigned to each strategy profile

of being at (U, R) is $\frac{q}{p+q}$. (Note that, in order to find both conditional probabilities, we use the standard formula of Bayes' rule.)

If instead, Player 1 receives the order of playing D, then his conditional probability of being at cell (D, L) is $\frac{r}{1-p-q}$, while his conditional probability of being at (D, R) is $\frac{1-p-q-r}{1-p-q}$.[3]

Player 2 If Player 2 receives the order of playing L we have that his conditional probability of being at cell (U, L) is $\frac{p}{p+r}$, while his conditional probability of being at (D, L) is $\frac{r}{p+r}$.

If instead, Player 2 receives the order of playing R, his conditional probability of being at cell (U, R) is $\frac{q}{1-p-r}$, and his conditional probability of being at (D, R) is $\frac{1-p-q-r}{1-p-r}$.[4]

---

[3]The denominator is given by the probability of outcomes (D, L) and (D, R), i.e., $r+(1-p-q-r)=1-p-q$; while the numerator reflects, respectively, the probability of outcome (D, L), r, or outcome (D, R), $1-p-q-r$; as depicted in the matrix of Fig. 3.38.

[4]Similarly as for Player 1, the denominator represents the probability of outcomes (U, R) and (D, R), i.e., $q+(1-p-q-r)=1-p-r$; while the numerator indicates, respectively, the probability of outcome (U, R), q, or (D, R), $1-q-p-r$; as illustrated in the matrix of Fig. 3.38.

We know that in a correlated equilibrium the players must prefer to stick to the rule that tells them how to play (i.e., there must be no profitable deviations from this rule), which we can represent by the following incentive compatibility conditions for Player 1

$$EU_1(U|he\ is\ told\ U) \geq EU_1(D|he\ is\ told\ U)$$

$$EU_1(D|he\ is\ told\ D) \geq EU_1(U|he\ is\ told\ D)$$

and for Player 2,

$$EU_2(L|he\ is\ told\ L) \geq EU_2(R|he\ is\ told\ L)$$

$$EU_2(R|he\ is\ told\ R) \geq EU_2(L|he\ is\ told\ R)$$

Let's now apply these conditions, and the conditional probabilities we found above, to each of the games:

Game 1

1. $EU_1(U|he\ is\ told\ U) \geq EU_1(D|he\ is\ told\ U) \Leftrightarrow \dfrac{1}{p+q} \cdot 2p \geq \dfrac{1}{p+q} \cdot q,$ which simplifies to $2p \geq q$.

2. $EU_1(D|he\ is\ told\ D) \geq EU_1(U|he\ is\ told\ D) \Leftrightarrow \dfrac{1}{1-p-q} \cdot (1-p-q-r) \geq \dfrac{1}{1-p-q} \cdot 2r$, which reduces to $1-p-q \geq 3r$

3. $EU_2(L|he\ is\ told\ L) \geq EU_2(R|he\ is\ told\ L) \Leftrightarrow \dfrac{1}{p+r} \cdot p \geq \dfrac{1}{p+r} \cdot 2r$, which can be rearranged as $p \geq 2r$.

4. $EU_2(R|he\ is\ told\ R) \geq EU_2(L|he\ is\ told\ R) \Leftrightarrow \dfrac{2}{1-p-r} \cdot (1-p-q-r) \geq \dfrac{1}{1-p-r} \cdot q$, which simplifies to $1-p-r \geq \frac{3}{2}q$.

Summarizing, the set of correlated equilibria for Game 1 can be sustained for any probability distribution $(p,\ q,\ r)$ that satisfies:

$$2p \geq q \quad and \quad 1-p-q \geq 3r \ \ for\ Player\ 1, \quad and$$

$$p \geq 2r \quad and \quad 1-p-r \geq \frac{3}{2}q \ \ for\ Player\ 2.$$

As a remark, note that the numerical example we analyzed in the previous part of the exercise (where $q = 0$, $r = 0$, $p = 1/2$ and $1 - p - q - r = 1/2$) satisfies these conditions. In particular, for Player 1, condition $2p \geq q$ reduces to $1 \geq 0$, while condition $1 - p - q \geq 3r$ simplifies to $\frac{1}{2} \geq 0$ in this case. Similarly, for Player 2, condition $p \geq 2r$ reduces to $1 \geq 0$, whereas condition $1 - p - r \geq \frac{3}{2}q$ simplifies to $\frac{1}{2} \geq 0$ in this numerical example.

Game 2. Applying the same methodology to Game 2 yields:

1. $EU_1(U|he\ is\ told\ U) \geq EU_1(D|he\ is\ told\ U) \Leftrightarrow \dfrac{1}{p+q} \cdot (4p+q) \geq \dfrac{1}{p+q} \cdot 5q$,

   which simplifies to $q \geq p$

2. $EU_1(D|he\ is\ told\ D) \geq EU_1(U|he\ is\ told\ U) \Leftrightarrow \dfrac{1}{1-p-q} \cdot 5r \geq \dfrac{1}{1-p-q} \cdot$

   $(4r+1-p-q-r)$, which can be rearranged to $2r \geq 1-p-q$

3. $EU_2(L|he\ is\ told\ L) \geq EU_2(R|he\ is\ told\ L) \Leftrightarrow \dfrac{1}{p+r} \cdot (4p+r) \geq \dfrac{1}{p+r} \cdot 5p$, which

   reduces to $r \geq p$

4. $EU_2(R|he\ is\ told\ R) \geq EU_2(L|he\ is\ told\ R) \Leftrightarrow \dfrac{1}{1-p-r} \cdot 5q \geq \dfrac{1}{1-p-r} \cdot$

   $(4q+1-p-q-r)$, which simplifies to $2q \geq 1-p-r$

Summarizing, the set of all correlated equilibrium in Game 2 are given by probability distributions (p, q, r) that satisfy the following four constraints:

$$q \geq p \quad \text{and} \quad 2r \geq 1-p-q \ \text{for Player 1, and}$$
$$r \geq p \ \text{and} \ 2q \geq 1-p-r \ \text{for Player 2}$$

By definition, these probability distribution over the four outcomes of these 2×2 games are defining a polytope in Δ_3 (the equivalent of a simplex in Δ_2 but depicted in 3 dimensions) that will have associated a set of EU_i which will be a superset of the convex hull of NE payoffs.

Exercise 10—Relationship Between Nash and Correlated Equilibrium Payoffs[C]

Consider the relationship between Nash and correlated equilibrium payoffs. Can there be a correlated equilibrium where both players receive a lower payoff than their lowest symmetric Nash equilibrium payoff? Explain or give an example.

Answer
Yes, there is such a correlated equilibrium. In particular, it is easy to find a correlated equilibrium in Exercise 3.8 (an anti-coordination game) in which both players obtain a lower payoff than in the lowest Nash equilibrium payoff of the

		Player 2	
		L	R
Player 1	U	6,6	2,7
	D	7,2	0,0

Fig. 3.39 Payoff matrix

Player 2

		L	R
Player 1	U	0	2/5
	D	2/5	1/5

Fig. 3.40 Probabilities assigned to each strategy profile

convex hull (where each player obtains $EU_1 = EU_2 = 4.5$ in the MSNE of the game). For completeness, Fig. 3.39 reproduces the payoff matrix of exercise 3.8.

Consider a correlated equilibrium with the probability distribution summarized in Fig. 3.40.

Let us now check that no player has incentives to unilaterally deviate from this correlated equilibrium. If Player 1 is told to play U, it must be that the outcome arising from the above probability distribution in the correlated equilibrium is (U, R), since (U, L) does not receive a positive probability. In this setting, Player 1's expected utility from selecting U is 2, while that from unilaterally deviating towards D is only zero. (See the payoff matrix in Fig. 3.39 where, fixing Player 2's strategy at the left-hand column, player 1's payoff is higher when he sticks to the order of playing U, which yields a payoff of 2, than deviating to D, which only entails a payoff of zero.) Hence, Player 1 does not have incentives to deviate. Similarly, if Player 1 is told to play D, then he does not know whether the realization of the above probability distribution is outcome (D, L) or (D, R). His expected payoff from agreeing to select D (in the bottom row of the payoff matrix) is

$$EU_1(D) = \frac{\frac{2}{5}}{\frac{2}{5} + \frac{1}{5}} \cdot 7 + \frac{\frac{1}{5}}{\frac{2}{5} + \frac{1}{5}} \cdot 0 = \frac{2}{3} \cdot 7 = 4.66$$

(Note that the first ratio identifies the probability of outcome (D, L), conditional on D occurring, i.e., a straightforward application of Bayes' rule. Similarly, the second term identifies the conditional probability of outcome (D, R), given that D occurs.).

If, instead, Player 1 deviates to U, his expected utility becomes

$$EU_1(U) = \frac{\frac{2}{5}}{\frac{2}{5} + \frac{1}{5}} \cdot 6 + \frac{\frac{1}{5}}{\frac{2}{5} + \frac{1}{5}} \cdot 2 = \frac{2}{3} \cdot 6 + \frac{1}{3} \cdot 2 = 4.66$$

Therefore, Player 1 does not have strict incentives to deviate in this setting either. By symmetry, we can conclude that player 2 does not have incentives to deviate from the correlated equilibrium.

Let us now find the expected payoff in this correlated equilibrium

Fig. 3.41 Nash and
Correlated equilibrium
payoffs

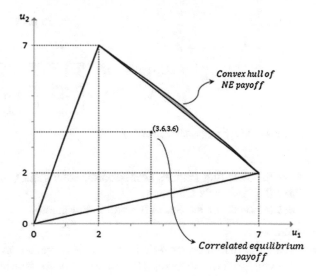

$$EU_1 = \overbrace{\frac{2}{5} \cdot 2}^{P_1 playsU} + \frac{3}{5} \overbrace{\left[\underbrace{\frac{2}{3} \cdot 7}_{P_2 playsL} + \underbrace{\frac{1}{3} \cdot 0}_{P_2 playsR} \right]}^{P_1 playsD} = 3.6$$

And similarly for player 2,

$$EU_2 = \overbrace{\frac{2}{5} \cdot 2}^{P_2 playsL} + \frac{3}{5} \overbrace{\left[\underbrace{\frac{2}{3} \cdot 7}_{P_1 playsU} + \underbrace{\frac{1}{3} \cdot 0}_{P_1 playsD} \right]}^{P_2 playsR} = 3.6$$

Figure 3.41 depicts the payoff pair (3.6, 3.6) arising in this correlated equilibrium, and compares it with the NE payoff pairs (as illustrated in the shaded convex hull which we obtained in Exercise 3.8). As required, our example illustrates that players obtain a lower payoff in the correlated equilibrium, 3.6, than in any of the symmetric Nash equilibria of the game (where the lowest payoff both players can obtain is 4.5.)

Exercise 11—Identifying Strictly Competitive Games[A]

For the following games, determine which of them satisfy the definition of strictly competitive games:

Player 2

H T

Player 1	H	1,-1	-1,1
	T	-1,1	1,-1

Fig. 3.42 Matching pennies game

Part (a) Matching Pennies (Anti-coordination game),
Part (b) Prisoner's Dilemma
Part (c) Battle of the Sexes (Coordination game).

Answer
Recall the definition of strictly competitive games as those strategic situations involving players with completely opposite interests/incentives. We provide a more formal definition below.

Strictly Competitive game (Definition): A two player, strictly competitive game requires that, for every two strategy profiles s and s', if player i prefers s then his opponent, player j, must prefer s'. That is,

$$\text{If} \quad u_i(s) > u_i(s') \quad \text{then} \quad u_j(s) < u_j(s') \quad \text{for all} \quad i = \{1,2\} \text{ and } j \neq i$$

Intuitively, if player i is better off when the outcome of the game changes from s' to s, $u_i(s) > u_i(s')$, player j must be worse off from such a change; a result that holds for any two strategy profiles s and s' we compare. Alternatively, a game is not strictly competitive if we can identify at least two strategy profiles s and s' for which players' incentives are alined, i.e., both players prefer s to s', or both prefer s' to s.

Part (a) Matching Pennies Game. Consider the matching pennies game depicted in Fig. 3.42, where H stands for *Heads* and T denotes *Tails*.

To establish if this is a strictly competitive game, we must compare the payoff for player 1 under every pair of strategy profiles, and that of player 2 under the same two strategy profiles.

When comparing strategy profiles (H, H) and (H, T), i.e., the cells in the top row of the matrix, we find that player 1's payoffs satisfy

$$u_1(H,H) = 1 > -1 = u_1(H,T)$$

while that of player 2 exhibits the opposite ranking,

$$u_2(H,H) = -1 < 1 = u_2(H,T).$$

When comparing strategy profiles (H, H) and (T, T) on the main diagonal of the matrix, we find that player 1 is indifferent between both strategy profiles, i.e.,

$$u_1(H,H) = 1 = 1 = u_1(T,T)$$

and so is player 2 since

$$u_2(H,H) = -1 = -1 = u_2(T,T)$$

An analogous result applies for the comparison of strategy profiles (H, T) and (T, H), where player 1 is indifferent since

$$u_1(H,T) = -1 = -1 = u_1(T,H)$$

and so is player 2 given that

$$u_2(H,T) = 1 = 1 = u_2(T,H).$$

Finally, when comparing (H, T) and (T, T), in the right-hand column of the matrix, we obtain that player 1 prefers (T, T) since

$$u_1(H,T) = -1 < 1 = u_1(T,T)$$

while player 2 prefers (H, T) given that

$$u_2(H,T) = 1 > -1 = u_2(T,T)$$

Hence, for any two strategy profiles $s = (s_1, s_2)$ and $s' = (s'_1, s'_2)$, we have shown that if $u_1(s) > u_1(s')$ then $u_2(s') > u_2(s)$, or if $u_1(s) = u_1(s')$ then $u_2(s') = u_2(s)$. Therefore, the Matching pennies game is strictly competitive.

Part (b) Prisoner's Dilemma. Consider the Prisoner's Dilemma game in Fig. 3.43, where C stands for *Confess* and NC denotes *Not confess*. For this game we also start comparing the payoffs that player 1 and player 2 obtain at every pair of strategy profiles.

When comparing (C, NC) and (NC, C), we find that player 1 prefers the latter, i.e.,

$$u_1(C,NC) = -10 < 0 = u_1(NC,C).$$

Player 2

		C	NC
Player 1	C	-5 , -5	-10 , 0
	NC	0 , -10	-1 , -1

Fig. 3.43 Prisoner's dilemma game

while player 2 prefer the former, i.e.,

$$u_2(C, NC) = 0 > -10 = u_2(NC, C)$$

Similarly, when comparing (C, C) and (C, NC), in the left-hand column of the matrix, we find that player 1 prefers the latter, i.e.,

$$u_1(C, C) = -5 < 0 = u_1(C, NC)$$

while player 2 prefer the former, i.e.,

$$u_2(C, C) = -5 > -10 = u_2(C, NC).$$

In all our comparisons of strategy profiles thus far, we have obtained that strategy profiles that are favored by one player are opposed by the other player, as required in strictly competitive games. However, when comparing (C, C) and (NC, NC), on the main diagonal of the matrix, we find that player's preferences become aligned, since both players prefer (NC, NC) to (C, C). Indeed, player 1 finds that

$$u_1(NC, NC) = -1 > -5 = u_1(C, C).$$

Similarly, player 2 also prefers (NC, NC) over (C,C) since

$$u_2(NC, NC) = -1 > -5 = u_2(C, C).$$

Hence, this game is *not* strictly competitive because we could find a pair of strategy profiles, namely (C, C) and (NC, NC), for which *both* players are better off at (NC, NC) than at (C, C). That is, if $s = (NC, NC)$ and $s' = (C, C)$, then we obtain that $u_1(s) < u_1(s')$ and $u_2(s) < u_2(s')$, which violates the definition of strictly competitive games.

Note: As long as we can show that there are two strategy profiles for which the definition of strictly competitive games does not hold, i.e., for which players' preferences are aligned, then we can claim that the game is not strictly competitive, without having to continue checking whether player's preferences are aligned or misaligned over all remaining pairs of strategy profiles.

		Wife	
		F	O
Husband	F	3 , 1	0 , 0
	O	0 , 0	1 , 3

Fig. 3.44 Battle of the Sexes game

Part (c) Battle of the sexes. The payoff matrix in Fig. 3.44 describes the Battle of the Sexes game, where F denotes going to the *Football* game, while O represents going to the *Opera*.

In this game, we can easily find a pair of strategy profiles for which the definition of strictly competitive games does not hold. In particular, for strategy profiles (O, F) and (F, F), both of them in the left-hand column of the matrix, we have that both the husband (row player) and the wife (column player) prefer strategy profile (F, F) over (O, F). In particular, the husband finds that

$$u_H(O, F) = 0 < 3 = u_H(F, F)$$

and similarly for the wife, since

$$u_W(O, F) = 0 < 1 = u_W(F, F).$$

Intuitively, the husband prefers that both players attend his preferred choice (football game), than being alone at the opera. Similarly, the wife prefers to attend to an event she dislikes (football game) with her husband, than being alone at the football game, as described in strategy profile (O, F). Hence, we found a pair of strategy profiles for which players' preferences are aligned, implying that the battle of the sexes game is *not* a strictly competitive game.

Exercise 12—Maxmin Strategies[C]

Consider the game in Fig. 3.45, where player 1 chooses Top or Bottom and player 2 selects Left or Right.

Part (a) Find every player's maxmin strategy. Draw every player's expected utility, for a given strategy of his opponent.
Part (b) What is every player's expected payoff from playing her maxmin strategy?
Part (c) Find every player's Nash equilibrium strategy, both using pure strategies (psNE) and using mixed strategies (msNE).
Part (d) What is every player's expected payoff from playing her Nash equilibrium strategy?
Part (e) Compare player's payoff when they play maxmin and Nash equilibrium strategies (from parts (b) and (d), respectively). Which is higher?

Player 2

		Left	*Right*
Player 1	*Top*	6 , 0	0 , 6
	Bottom	3 , 2	6 , 0

Fig. 3.45 Normal-form game

Player 2

q $1-q$

                Left   *Right*

Player 1

		Left	Right
p	*Top*	6 , 0	0 , 6
$1-p$	*Bottom*	3 , 2	6 , 0

Fig. 3.46 Normal-form game with assigned probabilities

Answer

Part (a) Let p denote the probability with which player 1 plays Top, and thus $1-p$ represent the probability that he plays Bottom. Similarly, q and $(1-q)$ denote the probability that player 2 plays Left (Right, respectively); as depicted in Fig. 3.46. *Player 1.* Let us first analyze player 1. When Player 2 chooses Left (fixing our attention on the left-hand column) Player 1's expected utility from randomizing between Top and Bottom with probabilities p and $1-p$, respectively, is:

$$EU_1(p|Left) = 6p + 3(1-p) = 6p + 3 - 3p = 3p + 3$$

Figure 3.47 depicts this expected utility, with a vertical intercept at 3, and a positive slope of 3, which implies that, at $p = 1$, EU_1 becomes 6.

And when Player 2 chooses Right (in the right-hand column) Player 1's expected utility becomes

$$EU_1(p|Right) = 0p + 6(1-p) = 6 - 6p$$

which, as Fig. 3.47 illustrates, originates at 6 and decreases in p, crossing the horizontal axis when $p = 1$. In particular, $EU_1(p|Right)$ and $EU_1(p|Left)$ cross at

Fig. 3.47 Expected utilities for player 1

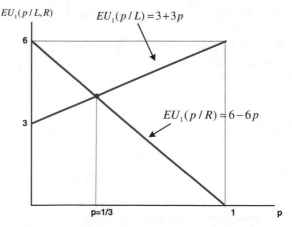

$$3q + 3 = 6 - 6p \Leftrightarrow p = 1/3$$

Recall the intuition behind maxmin strategies: player 1 first anticipates that his opponent, player 2, seeks to minimize player 1's payoff (since by doing so, player 2 increases his own payoff in a strictly competitive game). Graphically, this implies that player 2 chooses Left for all $p \leq \frac{1}{3}$ and Right for all $p > \frac{1}{3}$, yielding a lower envelope of expected payoffs for player 1 represented by equation $3 + 3p$ from $p = 0$ to $p = \frac{1}{3}$, and by equation $6 - 6p$, for all $p > \frac{1}{3}$; as depicted in Fig. 3.48. Taking such lower envelope of expected payoffs as given (i.e., assuming that Player 2 will try to minimize Player 1's payoffs), Player 1 chooses the probability p that maximizes his expected payoff, which occurs at the point where $EU_1(p|Left)$ crosses $EU_1(p|Right)$. This point graphically represents the highest point of the lower envelope, as Fig. 3.48 depicts, i.e., his maxmin strategy, as it maximizes the minimum of his payoffs.

Player 2. Let us now examine Player 2 using a similar procedure. When Player 1 chooses Top (in the top row), Player 2's expected utility becomes

$$EU_2(q|Top) = 0q + 6(1 - q) = 6 - 6q$$

which originates at 6 and decreases in q, crossing the horizontal axis when $q = 1$; as Fig. 3.49 illustrates. And when Player 1 chooses Bottom (in the bottom row), Player 2's expected utility is

$$EU_2(q|Bottom) = 2q + 0(1 - q) = 2q$$

which originates at (0, 0) and increases with a positive slope of 2, thus reaching a height of 2 when $q = 1$; also depicted in Fig. 3.49. Hence, the expected utilities $EU_2(q|Top)$ and $EU_2(q|Bottom)$ cross at:

$$6 - 6q = 2q \Leftrightarrow p = 3/4$$

Fig. 3.48 Lower envelope of expected payoffs (player 1)

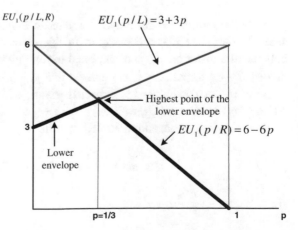

Fig. 3.49 Expected utilities
of player 2

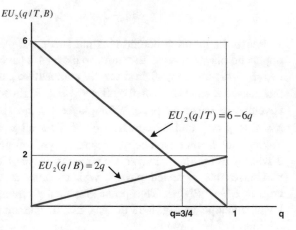

Fig. 3.50 Lower envelope of
expected payoffs (player 2)

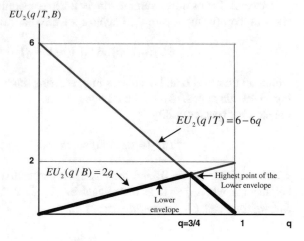

   A similar intuition as for player 1 applies to Player 2: considering that Player 1
seeks to minimize Player 2's payoff in this strictly competitive game, Player 2
anticipates a lower envelope of expected payoffs given by $2q$ from $q = 0$ to $q = \frac{3}{4}$,
and by $6 - 6q$ for all $q > \frac{3}{4}$; as depicted in Fig. 3.50. Hence, Player 2 chooses the
value of q that maximizes his expected payoff, which occurs at the point where
$EU_2(q|Top)$ crosses $EU_2(q|Bottom)$, i.e., $q = \frac{3}{4}$ at his maxmin strategy.
   Summarizing, the maxmin strategy profile is:

$$Player\ 1 : \left\{ \frac{1}{3}Top, \frac{2}{3}Bottom \right\},\ and$$

$$Player\ 2 : \left\{ \frac{3}{4} Left, \frac{1}{4} Right \right\}$$

Part (b) In order to find player 1's expected utility in the maxmin strategy profile, we can use either $EU_1(p|Left)$ or $EU_1(p|Right)$, since they both reach the some height at $p = \frac{1}{3}$, given that they cross at exactly this point.

$$EU_1 \left(p = \frac{1}{3} | Left \right) = 3 + 3p = 3 + 3 \left(\frac{1}{3} \right) = 3 + 1 = 4$$

And similarly for the expected utility of Player 2 in this maximin strategy profile

$$EU_2 \left(q = \frac{3}{4} | Top \right) = 6 - 6q = 6 - 6 \left(\frac{3}{4} \right) = 1.5$$

Part (c) *psNE.* Let us first check for psNEs in this game. For player 1, we can immediately identify that his best responses are $BR_1(L) = T$ and $BR_1(R) = B$, where he does not have a dominant strategy. The payoffs that player 1 obtains when selecting the above best responses are underlined in the payoff matrix of Fig. 3.51 in the (Top, Left) and (Bottom, Right) cells. Similarly, for player 2, we find that his best responses are $BR_2(T) = R$ and $BR_2(B) = L$, indicated by the underlined payoff in cells (Top, Right) and (Bottom, Left). There does not exists a mutual best response by both players, i.e., in the payoff matrix there is no cell with the payoffs of all players being underlined. Hence, there is no psNE (Fig. 3.51).

msNE. Let us now search for msNE (recall that such equilibrium must exist given that we could not find any psNE). Player 1 must be indifferent between Top and Bottom; otherwise, he would be selecting one of these strategies in pure strategies. Hence, player 1 must face $EU_1(Top) = EU_1(Bottom)$, and the value of probability q that makes player 1 indifferent between T and B is

$$EU_1(Top) = EU_1(Bottom)$$

$$6q = 3q + 6(1 - q) \Leftrightarrow q = 2/3$$

Player 2

			q	1-q
			Left	*Right*
Player 1				
P		*Top*	6 , 0	0 , 6
1-p		*Bottom*	3 , 2	6 , 0

Fig. 3.51 Normal-form game

Similarly, player 2 must be indifferent between Left and Right, since otherwise he would choose one of them in pure strategies. Therefore, the value of probability p that makes player 2 indifferent between L and R is

$$EU_2(L) = EU_2(R)$$

$$2(1 - p) = 6p \Leftrightarrow p = 1/4$$

Hence, the msNE of this game is

$$\left\{ \left(\frac{1}{4} Top, \frac{3}{4} Bottom \right), \left(\frac{2}{3} Left, \frac{1}{3} Right \right) \right\}$$

Part (d) Let us now find the expected utility that player 1 obtains in the previous msNE:

$$EU_1 = 6pq + 0p(1 - q) + 3q(1 - p) + 6(1 - q)(1 - p)$$

$$= 6\left(\frac{1}{4}\right)\left(\frac{2}{3}\right) + 3\left(\frac{2}{3}\right)\left(\frac{3}{4}\right) + 6\left(\frac{1}{3}\right)\left(\frac{3}{4}\right) = 1 + \frac{3}{2} + \frac{1}{2} = 4$$

And similarly for player 2, his expected utility in the msNE becomes

$$EU_2 = 0pq + 6p(1 - q) + 2q(1 - p) + 0(1 - p)(1 - q)$$

$$= 6\left(\frac{1}{4}\right)\left(\frac{1}{3}\right) + 2\left(\frac{2}{3}\right)\left(\frac{3}{4}\right) = 1.5$$

Part (e) Player 1's expected utility from playing the msNE of the game, 4, coincides with that from playing his maxmin strategy, 4. A similar argument applies to Player 2, who obtains an expected utility of 1.5 under both strategies.

Sequential-Move Games with Complete Information

4

Introduction

In this chapter we move from simultaneous-move to sequential-move games, and describe how to solve these games by using backward induction, which yields the set of Subgame Perfect Nash equilibria (SPNE). Intuitively, every player anticipates the optimal actions that players acting in subsequent stages will select, and chooses his actions accordingly. Formally, we say that every player's action is "sequentially rational," meaning that his strategy represents an optimal action for player i at any stage of the game at which he is called on to move, and given the information he has at that point. From a practical standpoint, "sequential rationality" requires every player to put himself in the shoes of players acting in the last stage of the game and predict their optimal behavior on that late stage; then, anticipating the behavior of players acting in the last stage, study the optimal choices of players in the previous to last stage; and similarly for all previous stages until the first move.

We first analyze games with two players, and then extend our analysis to settings with more than two players, such as a Stackelberg game with three firms sequentially competing in quantities. We consider several applications from industrial organization as an economic motivation of settings in which players act sequentially. For completeness, we also explore contexts in which players are allowed to choose among more than two strategies, such as bargaining games (in which proposals are a share of the pie, i.e., a percentage from 0 % to 100 %), or the Stackelberg game of sequential quantity competition (in which each firm chooses an output level from a continuum). In bargaining settings, we analyze equilibrium behavior not only when players are selfish, but also when they exhibit inequity aversion and thus avoid unequal payoff distributions. Finally, we study a moral hazard game in which a manager offers a menu of contracts in order to induce a high effort level from his employee.

The original version of the chapter was revised: The erratum to the chapter is available at: 10.1007/978-3-319-32963-5_11

F. Munoz-Garcia and D. Toro-Gonzalez, *Strategy and Game Theory*,
Springer Texts in Business and Economics, DOI 10.1007/978-3-319-32963-5_4

Exercise 1—Ultimatum Bargaining Game[B]

In the ultimatum bargaining game, a proposer is given a pie, normalized to a size $1, and he is asked to make a monetary offer, x, to the responder who, upon receiving the offer, only has the option to accept or reject it (as if he received an "ultimatum" from the proposer). If the offer x is accepted, then the responder receives it while the proposer keeps the remainder of the pie $1-x$. However, if he rejects it, both players receive a zero payoff. Operating by backward induction, the responder should accept any offer x from the proposer (even if it is low) since the alternative (reject the offer) yields an even lower payoff (zero). Anticipating such as a response, the proposer should then offer one cent (or the smallest monetary amount) to the responder, since by doing so the proposer guarantees acceptance and maximizes his own payoff. Therefore, according to the subgame perfect equilibrium prediction in the ultimatum bargaining game, the proposer should make a tiny offer (one cent or, if possible, an amount approaching zero), and the responder should accept it, since his alternative (reject the offer) would give him a zero payoff. However, in experimental tests of the ultimatum bargaining game, subjects who are assigned the role of proposer rarely make offers close to zero to the subject who plays as a responder. Furthermore, sometimes subjects in the role of the responder reject positive offers, which seems to contradict our equilibrium predictions. In order to explain this dissonance between theory and experiments, many scholars have suggested that players' utility function is not as selfish as that specified in standard models (where players only care about the monetary payoff they receive). Instead, the utility function should also include social preferences, measured by the difference between the payoff a player obtains and that of his opponent, which gives rise to envy (when the monetary amount he receives is lower than that of his opponent) or guilt (when his monetary payoff is higher than his opponent's). In particular, suppose that the responder's utility is given by

$$u_R(x, y) = x + \alpha(x - y),$$

where x is the responder's monetary payoff, y is the proposer's monetary payoff, and α is a positive constant. That is, the responder not only cares about how much money he receives, x, but also about the payoff inequality that emerges at the end of the game, $\alpha(x - y)$, which gives rise to either envy, if $x < y$, or guilt, if $x > y$. For simplicity, assume that the proposer is selfish, i.e., his utility function only considers his own monetary payoffs $u_P(x, y) = y$, as in the basic model.

Part (a) Use a game tree to represent this game in its extensive form, writing the payoffs in terms of m, the monetary offer of the proposer, and parameter α.
Part (b) Find the subgame perfect equilibrium. Describe how equilibrium payoffs are affected by changes in parameter α.
Part (c) Depict the equilibrium monetary amount that the proposer keeps, and the payoff that the responder receives, as a function of parameter α.

Answer

Part (a) First, player 1 offers a division of the pie, m, to player 2, who either accepts or rejects it. However, payoffs are not the same as in the standard ultimatum bargaining game. While the utility of player 1 (proposer) is just the remaining share of the pie that he does not offer to player 2, i.e., $1 - m$, the utility of player 2 (responder) is

$$m + \alpha[m - (1 - m)] = m + \alpha(2m - 1)$$

where m is the payoff of the responder, and $1 - m$ is the payoff of the proposer. We depict this modified ultimatum bargaining game in Fig. 4.1.

Part (b) Operating by backward induction, we first focus on the last mover (player 2, the responder). In particular, player 2 accepts any offer m from player 1 such that:

$$m + \alpha(2m - 1) \geq 0,$$

since the payoff he obtains from rejecting the offer is zero. Solving for m, this implies that player 2 accepts any offer m that satisfies $m \geq \frac{\alpha}{1+2\alpha}$. Anticipating such a response from player 2, player 1 offers the minimal m that generates acceptance, i.e., $m^* = \frac{\alpha}{1+2\alpha}$, since by doing so player 1 can maximize the share of the pie he keeps. This implies that equilibrium payoffs are:

$$(1 - m^*, m^*) = \left(\frac{1+\alpha}{1+2\alpha}, \frac{\alpha}{1+2\alpha} \right)$$

The proposer's equilibrium payoff is decreasing in α since its derivative with respect to α is

$$\frac{\partial(1 - m^*)}{\partial \alpha} = -\frac{1}{(1+2\alpha)^2}$$

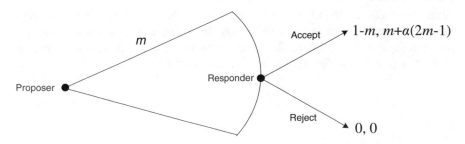

Fig. 4.1 Ultimatum bargaining game (extensive-form)

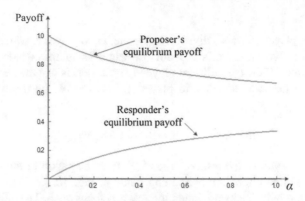

Fig. 4.2 Equilibrium payoff in ultimatum bargaining game with social preferences

which is negative for all $\alpha > 0$. In contrast, the responder's equilibrium payoff is increasing in α given that its derivative with respect to α is

$$\frac{\partial m^*}{\partial \alpha} = \frac{1}{(1 + 2\alpha)^2}$$

which is positive for all $\alpha > 0$.

Part (c) In Fig. 4.2 we represent how the proposer's equilibrium payoff (in red color) decreases in α, and how that of the responder (in green color) increases in α. Intuitively, when the responder does not care about the proposer's payoff, $\alpha = 0$, the proposer makes an equilibrium offer of $m^* = 0$, which is accepted by the (selfish) responder, as in the SPNE of the standard ultimatum bargaining game without social preferences. However, when α increases, the minimum offer that the proposer must make to guarantee acceptance, m^*, increases. Intuitively, the responder will not accept very unequal offers since his concerns for envy are relatively strong.

When $\alpha = 0.5$ the equilibrium split becomes $(1 - m^*, m^*) = \left(\frac{1+0.5}{1+2\times0.5}, \frac{0.5}{1+2\times0.5}\right) = \left(\frac{1.5}{2}, \frac{0.5}{2}\right) = (0.75, 0.25)$, thus indicating that the proposer offers more to the responder as he cares more about the payoff difference, and when the proposer cares the most about the payoff difference, $\alpha = 1$, the equilibrium split becomes $(2/3, 1/3)$. However, for all the values of α between zero and one, the proposer's payoff is higher than that of the responder.

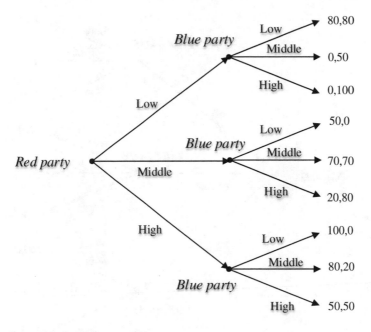

Fig. 4.3 Sequential electoral competition

Exercise 2—Electoral competition^A

Consider the game tree in Fig. 4.3, describing a sequential-move game played between two political parties. The Red party acts first choosing an advertising level: Low, Middle or High. The Blue party observes the Red party's advertising and responds with its own advertising level (Low, Middle or High as well). Find the SPNE of the game.

Answer

We use backwards induction to find the SPNE of the game. First we examine the optimal response of the Blue party in the last stage of the game, which is to choose High regardless of the Red party's level of advertising in the first stage of the game. Indeed, after observing Low (see top of the tree) $100 > 80 > 50$; after observing Middle (center of the game) $80 > 70 > 0$; and after observing High (see bottom of the tree) $50 > 20 > 0$. As a consequence, a High advertising level is a strictly dominant strategy for the Blue party. Figure 4.4 highlights the branches that the Blue party chooses as its optimal responses in each contingency.

Anticipating the Blue candidate's response of High for all advertising levels of the Red party, the Red party compares its payoffs from each advertising level, as depicted in Fig. 4.5. Specifically, the Red party's payoffs from choosing High becomes 50, from Middle is 20, and from Low is 0. Hence, the Red party chooses High, and the unique SPNE is {High, (High, High, High)} where the first component reflects the Red party's choice (first mover), while the triplet (High, High, High) indicates that the Blue party (second mover) chooses a High advertising level

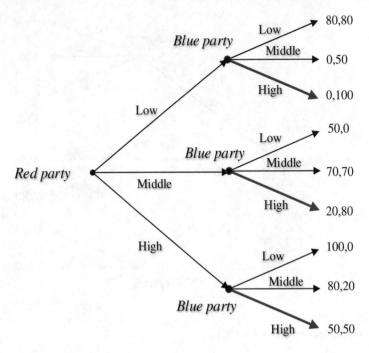

Fig. 4.4 Sequential electoral competition—optimal responses of the last mover

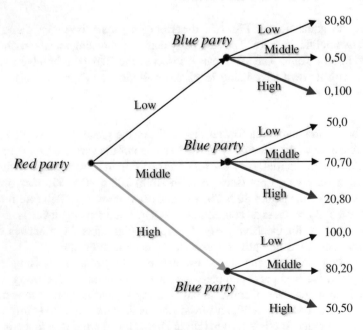

Fig. 4.5 Sequential electoral competition—optimal actions

regardless of the Red party's advertising level in the first stage of the game. As a remark, note that (High, High) is the equilibrium outcome of this sequential-move game, but not the SPNE of the game. Instead, the SPNE needs to specify optimal actions both on- and off-the-equilibrium path. In this case, this means describing optimal action both after the Red party chooses Low or Middle (off-the-equilibrium).

Exercise 3—Electoral Competition with a Twist[A]

Consider the game tree in Fig. 4.6, describing a sequential-move game played between two political parties. The game is similar to that in Exercise 4.2, but with different payoffs. The Red party acts first choosing an advertising level. The Blue party acts after observing the Red party's advertising level (Low, Middle or High) and responds with its own advertising level. Find the SPNE of the game.

Answer
Second mover First, we examine the payoffs received by the Blue party in the lower part of the game tree. When analyzing each sub-game independently by comparing the payoffs for the Blue party (the number on the right side of each payoff pair). Starting from the subgame on the left-hand side of the tree, when the Red party chooses Low, we can see that for the Blue party is better to choose High because 100 > 80 > 50. In the case of the subgame in the center of the figure, when the Red party chooses Middle, the Blue party have incentives to choose also High, because 80 > 70 > 30. In the case the Red party chooses High, the Blue party responds with

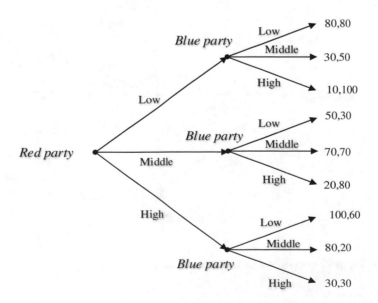

Fig. 4.6 Sequential electoral competition with a twist

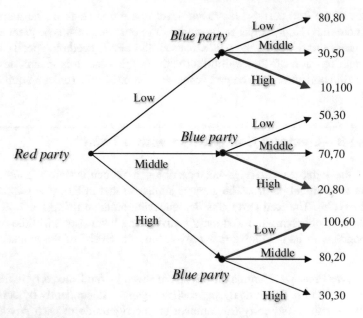

Fig. 4.7 Sequential electoral competition with a twist—optimal responses

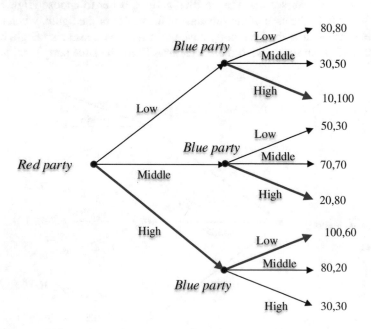

Fig. 4.8 Sequential electoral competition with a twist—optimal actions

Low advertisement expenditure since $60 > 30 > 20$. We shade the branches corresponding to these optimal responses (High, High, Low) in Fig. 4.7.

First mover Once the best responses at the subgames for the Blue party are identified, we proceed to identify the best strategy for the Red party by comparing its payoffs in each case. In case the Red party chooses Low advertisement level, the response for the Blue party is to choose High, implying that the Red party gets 10. If the Red party chooses Middle, the payoff given the High response of Blue party is 20. Finally, if Red party's strategy is to choose High, then its payoff is 100, since the best response by Blue is to choose Low. Hence, since $100 > 20 > 10$, the backwards induction strategy to solve the game indicates that the equilibrium is {High, (High, High, Low)} as depicted in Fig. 4.8.

In this game, the Blue party (second mover) responds with Low advertising to High advertising from the Red party (first mover), and with High advertising to Low advertising, i.e., it responds choosing the opposite action than the first mover selects. Anticipating such a best response from the other party, the party acting as the first mover chooses a High level of advertising in order to induce the other party to respond with a Low level of advertisement.

Exercise 4—Trust and Reciprocity (Gift-Exchange Game)[B]

Consider the following game that has been used to experimentally test individuals' preferences for trust and reciprocity. The experimenter starts by giving player 1 ten dollars and player 2 zero dollars. The experimenter then asks player 1 how many dollars is he willing to give back to help player 2. If he chooses to give x dollars, the experimenter gives player 2 that amount tripled, i.e., $3x$ dollars. Subsequently, player 2 has the opportunity to give to any, all, or none of the money he has received from the experimenter to player 1. This game is often known as the "gift-exchange" game, since the trust that player 1 puts on player 2 (by giving him money at the beginning of the game), could be transformed into a large return to player 1, but only if player 2 reciprocates player's trust.

Part (a) Assuming that the two players are risk neutral and care only about their own payoff, find the subgame perfect equilibrium of this game. (The subgame perfect equilibrium, SPNE, does not need to coincide with what happens in experiments.)

Part (b) Does the game have a Nash equilibrium in which players receive higher payoffs?

Part (c) Several experiments show that players do not necessarily behave as predicted in the SPNE you found in part (a). In particular, the subject who is given the role of player 1 in the experiment usually provides a non-negligible amount to player 2 (e.g., $4), and the subject acting as player 2 responds by reciprocating player 1's kindness by giving him a positive monetary amount in return. An explanation for these experimental results is that the players may be altruistic. Show that the simplest representation of altruism—each player maximizing a weighted

Fig. 4.9 Gift-exchange
game modified

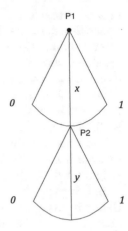

sum of his own dollar payoff and the other player's dollar payoff—cannot account
for this experimental regularity except for one very extreme choice of the weights.

Answer

Figure 4.9 depicts the extensive-form representation of the game. The amount
x represents the offer that player 1 gives to player 2, and y is the amount of money
player 2 gives to player 1 in return.

Note that the utility function of the first mover is $u_1(x, y) = 10 - x + y$, since he
retains $10 - x$ dollars after giving x dollars to player 2, and afterwards receives
y dollars from player 2 (and we are assuming no discounting). The utility function
of the second mover is $u_2(x, y) = 3x - y$, given that for every unit given up by
player 1, x, the experimenter adds up 2 units, and afterwards, player 2 gives y units
to player 1.

Part (a) *SPNE*: Working by backwards induction, we need to find the optimal
amount of y that player 2 will decide to give back to player 1 at the end of the game:

$$\max_{y \geq 0} \ u_2(x, y) = 3x - y$$

Obviously, since y enters negatively on player 2's utility function, the value of
y that maximizes $u_2(x, y)$ is $y^*(x) = 0$ for any amount of money, x, received from
player 1. Hence, $y^*(x) = 0$ can be interpreted as player 2's best response function
(although a very simple best response function, since giving zero dollars to player 1
is a strictly dominant strategy for player 2 once he is called on to move).

Player 1 can anticipate that, in the subgame originated after his decision, player 2
will respond with $y^*(x) = 0$. Hence, player 1 maximizes

$$\max_{x} u_1(x, y) = 10 - x + y$$
$$s.t. \ y^*(x) = 0$$

That is,

$$\max_{x} u_1(x, y) = 10 - x$$

By a similar argument, since x enters negatively into player 1's utility function, the value of x that maximizes $u_1(x,y)$ is $x^* = 0$. Therefore, the unique SPNE is $(x^*, y^*) = (0,0)$, which implies a SPNE payoff vector of $(u_1^*, u_2^*) = (10,0)$.

Part (b) *NE*. Note that player 2 should always respond with $y^*(x) = 0$ along the equilibrium path for any offer x chosen by player 1 in the first period, i.e., responding with $y = 0$ is a strictly dominant strategy for player 2. Hence, there is no credible threat that could induce player 1 to deviate from $x = 0$ in the first period. Therefore, the unique NE strategy profile (and outcome) is the same as in the SPNE described above.

Part (c) Operating by backwards induction, the altruistic player 2 chooses the value of y that maximizes the weighted sum of both players' utilities:

$$\max_y \; u_2(x,y) + \alpha_1 u_1(x,y)$$

where α_1 denotes player 2's concern for player 1's utility. That is:

$$\max_y (3x - y) + \alpha_1(10 - x + y) = 3x - \alpha_1 x - y + \alpha_1 y + 10\alpha_1$$

$$= 3x(1 - \alpha_1) + y(\alpha_1 - 1) + 10\alpha_1$$

Taking first-order conditions with respect to y, we obtain

$$\alpha_1 - 1 \leq 0 \quad \Leftrightarrow \quad \alpha_1 \leq 1$$

and, in an interior solution, where $y^* > 0$, we then have $\alpha_1 = 1$. (Otherwise, i.e., for all $\alpha_1 < 1$, we are at a corner solution where $y^* = 0$). Therefore, the only way to justify that player 2 would ever give some positive amount of y back to player 1 is if and only if $\alpha_1 \geq 1$. That is, if and only if player 2 cares about player 1's utility at least as much as he cares about his own.

Exercise 5—Stackelberg with Two Firms[A]

Consider a leader and a follower in a Stackelberg game of quantity competition. Firms face an inverse demand curve $p(Q) = 1 - Q$, where $Q = q_L + q_F$. denotes aggregate output. The leader faces a constant marginal cost $c_L > 0$ while the follower's constant marginal cost is $c_F > 0$, where $1 > c_F \geq c_L$, indicating that the leader has access to a more efficient technology than the follower.

Part (a) Find the follower's best response function, BRF_F, i.e., $q_F(q_L)$.

Part (b) Determine each firm's output strategy in the subgame perfect equilibrium (SPNE) of this sequential-move game.

Part (c) Under which conditions on c_L can you guarantee that both firms produce strictly positive output levels?

Part (d) Assuming cost symmetry, i.e., $c_L = c_F = c$, determine the aggregate equilibrium output level in the Stackelberg game, and its associated equilibrium price. Then compare them with those arising in (1) a monopoly; in (2) the Cournot game of quantity competition; and (3) in the Bertrand game of price competition (which, under cost symmetry, is analogous to a perfectly competitive industry). [*Hint*: For a more direct and visual comparison, depict the inverse demand curve p $(Q) = 1 - Q$, and locate each of these four aggregate output levels in the horizontal axis (with its associated prices in the vertical axis).]

Answer

Part (a) The follower observes the leader's output level, q_L, and chooses its own production, q_F, to solve:

$$\max_{q_F \geq 0}(1 - q_L - q_F)q_F - c_F q_F$$

Taking first order conditions with respect to q_F yields

$$1 - q_L - 2q_F - c_F = 0$$

and solving for q_F we obtain the follower's best response function

$$q_F(q_L) = \frac{1 - c_F}{2} - \frac{1}{2}q_L$$

which, as usual, is decreasing in the follower's costs, i.e., the vertical intercept decreases in c_F, indicating that, graphically, the best-response function experiences a downward shift as c_F increases. In addition, the follower's best-response function decreases in the leader's output decision (as indicated by the negative slope, $-\frac{1}{2}$).

Part (b) The leader anticipates that the follower will respond with best response function $q_F(q_L) = \frac{1-c_F}{2} - \frac{1}{2}q_L$, and plugs it into the leader's own profit maximization problem, as follows

$$\max_{q_L \geq 0}\left[1 - q_L - \underbrace{\left(\frac{1 - c_F}{2} - \frac{1}{2}q_L\right)}_{q_F}\right]q_L - c_L q_L$$

which simplifies into

$$\frac{1}{2}[(1 + c_F) - q_L]q_L - c_L q_L$$

Taking first order conditions with respect to q_L yields

$$\frac{1}{2}(1 - c_F) - q_L - c_L = 0$$

and solving for q_L we find the leader's equilibrium output level

$$q_L^* = \frac{1 + c_F - 2c_L}{2}$$

which, thus, implies a follower's equilibrium output of

$$q_F^* = q_F\left(\frac{1 + c_F - 2c_L}{2}\right) = \frac{1 - 3c_F + 2c_L}{4}$$

Note that the equilibrium output of every firm $i = \{L, F\}$ is decreasing in its own cost, c_i, and increasing in its rival's cost, c_j, where $j \neq i$. Hence, the SPNE of the game is

$$(q_L^*, q_F(q_L)) = \left(\frac{1 + c_F - 2c_L}{2}, \frac{1 - c_F}{2} - \frac{1}{2}q_L\right)$$

which allows the follower to optimally respond to both the equilibrium output level from the leader, q_L^*, but also to off-the-equilibrium production decisions $q_L \neq q_L^*$. **Part (c)** The follower (which operates under a cost disadvantage), produces a positive output level in equilibrium, i.e., $q_F^* > 0$, if and only if

$$\frac{1 - 3c_F + 2c_L}{4} > 0,$$

which, solving for c_L, yields

$$c_L > -\frac{1}{2} + \frac{3}{2}c_F \equiv C_A.$$

Similarly, the leader produces a positive output level, $q_L^* > 0$, if and only if

$$\frac{1 + c_F - 2c_L}{2} > 0$$

or, solving for c_L,

$$c_L < \frac{1}{2} + \frac{1}{2}c_F \equiv C_B.$$

Figure 4.10 depicts cutoffs C_A (for the follower) and C_B (for the leader), in the (c_F, c_L)—quadrant. (Note that we focus on points below the 45°-line, since the leader experiences a cost advantage relative to the follower, i.e., $c_L \leq c_F$.)

First, note that cutoff C_B is not binding since it lies above the 45°-line. Intuitively, the leader produces a positive output level for all (c_F, c_L)—pairs in the admissible region of cost pairs (below the 45°-line). However, cutoff C_A restricts the of cost pairs below the 45°-line to only that above cutoff C_A. Hence, in the shaded area of the figure, both firms produce a strictly positive output in equilibrium.

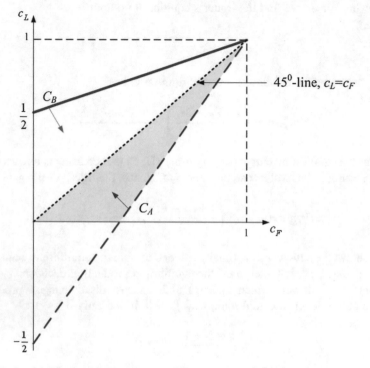

Fig. 4.10 Region of cost pairs for which both firms produce positive output

Part (d) *Stackelberg competition.* When firms are cost symmetric, $c_F = c_L = c$, the leader produces

$$q_L^* = \frac{1 + c - 2c}{2} = \frac{1 - c}{2}$$

while the follower responds with an equilibrium output of

$$q_F^* = \frac{1 - 3c + 2c}{4} = \frac{1 - c}{4}$$

thus producing half of the leader's output. Hence, aggregate equilibrium output in the Stackelberg game becomes

$$Q^{Stackel.} = q_L^* + q_F^* = \frac{1 - c}{2} + \frac{1 - c}{4} = \frac{3(1 - c)}{4}$$

which yields an associated price of

$$P\left(Q^{Stackel.}\right) = 1 - \frac{3(1 - c)}{4} = \frac{1 + 3c}{4}$$

Cournot competition. Under Cournot competition, each firm $i = \{1, 2\}$ simultaneously and independently solves

$$\max_{q_i}(1 - q_i - q_j)q_i - cq_i$$

which, taking first-order conditions with respect to q_i, and solving for q_i, yields a best response function of

$$q_i(q_j) = \frac{1 - c}{2} - \frac{1}{2}q_j$$

Simultaneously solving for q_i and q_j, we obtain a Cournot equilibrium output of $q_i^* = \frac{1-c}{3}$ for every firm $i = \{1, 2\}$.[1] Hence, aggregate output in Cournot is

$$Q^{Cournot} = q_1^* + q_2^* = \frac{2(1 - c)}{3}$$

which yields a market price of

$$P(Q^{Cournot}) = 1 - \frac{2(1 - c)}{3} = \frac{1 + c}{3}.$$

Bertrand competition. Under the Bertrand model of price competition and symmetric costs, the Nash equilibrium prescribes both firms to set a price that coincides with their common marginal cost, i.e., $p_1 = p_2 = c$. (This equilibrium is analyzed in Exercise 1 of Chap. 5, where we discuss other applications to Industrial Organization. However, since this equilibrium price coincides with that in perfectly competitive industries, we can at this point consider such a market structure in our subsequent comparisons.) Hence, aggregate output in this context coincides with that under a perfectly competitive industry,

$$c = 1 - Q, \text{ or } Q^{Bertrand} = 1 - c$$

Monopoly. A monopoly chooses the output level Q that solves

$$\max_{Q \geq 0}(1 - Q)Q - cQ.$$

Taking first order conditions with respect to Q yields

$$1 - 2Q - c = 0$$

and, solving for Q, we obtain a monopoly output of

---

[1]Recall from Chap. 2, that you can find this equilibrium output by inserting best response function $q_j(q_i)$ into $q_i(q_j)$, and then solving for q_i.

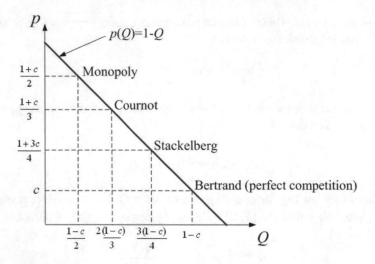

Fig. 4.11 Aggregate output and prices across different market structures

$$Q^{Monop} = \frac{1-c}{2}$$

which implies a monopoly price of

$$P(Q^{Monop}) = 1 - \frac{1-c}{2} = \frac{1+c}{2}.$$

Comparison. Figure 4.11 depicts the inverse linear demand $P(Q) = 1 - Q$, and locates the aggregate output levels found in the four market structures described above in the horizontal axis, afterwards mapping their corresponding prices in the vertical axis. Aggregate output is the lowest at a monopolistic market structure, larger under Cournot competition (when firms simultaneously and independently select output levels), larger under Stackelberg (when firms' competition is in quantities, but sequentially), and yet larger in a perfect competitive industry. The opposite ranking applies for equilibrium prices.

Exercise 6—First- and Second-Mover Advantage in Product Differentiation

Consider an industry with two firms, 1 and 2, competing in prices and no production costs. Firms sell a differentiated product, with the following direct demand functions

$$q_1 = 150 - 2p_1 + p_2 \text{ and } q_2 = 150 - 2p_2 + p_1$$

Intuitively, if firm i increases its own price p_i its sales decrease, while if its rival increases its price p_j firm i's sales increase. However, the own-price effect on sales is larger in absolute value than the cross-price effect, thus indicating that a marginal increase in p_i has a larger negative effect on q_i than the positive effect that an increase in p_j has (in this particular setting, the former is twice as big as the latter). **Part (a)** *Simultaneous-move game.* If firms simultaneously and independently choose their prices (competition a la Bertrand), find their best response functions, the equilibrium price pair (p_1^{Sim}, p_2^{Sim}), the equilibrium sales by each firm (q_1^{Sim}, q_2^{Sim}), and the equilibrium profits $(\pi_1^{Sim}, \pi_1^{Sim})$.

Answer
Firm 1 solves

$$\max_{p_1} \ (150 - 2p_1 + p_2)p_1$$

Taking first-order conditions with respect to p_1 yields

$$150 - 4p_1 + p_2 = 0.$$

And solving for p_1, we obtain firm 1's best response function

$$p_1(p_2) = \frac{150}{4} + \frac{1}{4}p_2$$

Similarly, firm 2 solves

$$\max_{p_2} \ (150 - 2p_2 + p_1)p_2.$$

Taking first-order conditions with respect to p_2 yields

$$150 - 4p_2 + p_1 = 0.$$

And solving for p_2, we obtain firm 2's best response function

$$p_2(p_1) = \frac{150}{4} + \frac{1}{4}p_1.$$

Note that both firms' best response functions are positively sloped, indicating that if firm j increases its price, firm i can respond by optimally increasing its own (i.e., prices are strategic complements). Inserting $p_2(p_1)$ into $p_1(p_2)$, we obtain the point where both best response functions cross each other,

$$(p_i, p_j) = (50, 50)$$

which constitutes the Nash equilibrium of this simultaneous-move game of price competition. Note that such equilibrium is symmetric, as both firms set the same price (i.e., graphically, best response functions cross at the 45-degree line). Using the direct demand functions, we find that sales by each firm are

$$q_1 = 150 - 2p_1 + p_2 = 100 \text{ and}$$
$$q_2 = 150 - 2p_2 + p_1 = 100.$$

Finally, equilibrium profits are

$$\pi_1 = 5000 \text{ and } \pi_2 = 5000.$$

Part (b) *Sequential-move game.* Let us now consider the sequential version of the above game of price competition. In particular, assume that firm 1 chooses its price first (leader) and that firm 2, observing firm 1's price, responds with its own price (follower). Find the equilibrium price pair $\left(p_1^{Seq}, p_2^{Seq}\right)$, the equilibrium sales by each firm $\left(q_1^{Seq}, q_2^{Seq}\right)$, and the equilibrium profits $\left(\pi_1^{Seq}, \pi_2^{Seq}\right)$. Which firm obtains the highest profit, the leader or the follower?

Answer
The follower (firm 2) observes firm 1's price, p_1, and best responds to it by using the best response function identified in part (a), i.e., $p_2(p_1) = \frac{150+p_1}{4}$. Since firm 1 (the leader) can anticipate the best response function that firm 2 will subsequently use in the second stage of the game, firm 1's profit maximization problem becomes

$$\max_{p_1} \left(150 - 2p_1 + \underbrace{\frac{150 + p_1}{4}}_{p_2} \right) * p_1$$

which depends on p_1 alone. Taking first-order conditions with respect to p_1, we obtain

$$150 - 4p_1 + \frac{150 + 2p_1}{4} = 0$$

Solving for p_1, yields an equilibrium price for the leader of $p_1 = \frac{750}{14} \approx 53.6$. Using the direct demand functions, we find that the sales of each firm are

$$q_1 = 150 - 2p_1 + \frac{150 + p_1}{4} \approx 93.8 \text{ and}$$

$$q_2 = 150 - 2p_2 + p_1 \approx 101.8.$$

And equilibrium profits in this sequential version of the game become

$$\pi_1 = 5027.7 \text{ and } \pi_2 = 5181.62.$$

Finally, note that the leader obtains a lower profit than the follower when firms compete in prices; as opposed to the profit ranking when firms sequentially compete in quantities (standard Stackelberg competition where the leader obtains a larger profit than the follower).

Exercise 7—Stackelberg Game with Three Firms Acting Sequentially[A]

Consider an industry consisting of three firms. Each firm $i = \{1, 2, 3\}$ has the same cost structure, given by cost function $C(q_i) = 5 + 2q_i$, where q_i denotes individual output. For every firm i, industry demand is given by the inverse demand function:

$$P(Q) = 18 - Q$$

where Q denotes aggregate output, i.e., $Q = q_1 + q_2 + q_3$. The production timing is as follows: Firm 1 produces its output first. Observing firm 1's output, q_1, firm 2 chooses its own output, q_2. Finally, observing both firm 1's and firm 2's output, firm 3 then produces its own output q_3. This timing of production is common knowledge among all three firms. The industry demand and cost functions are also known to each firm. Find values of output level q_1, q_2 and q_3 in the SPNE of the game.

Answer
Firm 3. Using backward induction, we first analyze the production decision of the last mover firm 3, which maximizes profits by solving:

$$\max_{q_3}(18 - (q_1 + q_2 + q_3))q_3 - (5 + 2q_3)$$

Taking first order conditions with respect to q_3, we obtain

$$18 - q_1 - q_2 - 2q_3 - 2 = 0.$$

Hence, solving for q_3 we obtain firm 3's best response function (which decreases in both q_1 and q_2)

$$q_3(q_1, q_2) = \frac{16 - q_1 - q_2}{2}.$$

Firm 2. Given this $BRF_3(q_1, q_2)$, we can now examine firm 2's production decision (the second mover in the game), which chooses an output level q_2 to solve

$$\max_{q_2}\left(18-\left(q_1+q_2+\underbrace{\frac{16-q_1-q_2}{2}}_{q_3}\right)\right)q_2-(5+2q_2).$$

Where we inserted firm 3's best response function, since firm 2 can anticipate firm 3's optimal response at the subsequent stage of the game. Simplifying this profit, we find

$$\max_{q_2}\left(10-\frac{1}{2}q_1-\frac{1}{2}q_2\right)q_2-5-2q_2.$$

Taking first order conditions with respect to q_2, yields

$$10-\frac{1}{2}q_1-q_2-2=0.$$

And solving for q_2 we find firm 2's best response function (which only depends on the production that occurs before its decision, i.e., q_1),

$$q_2(q_1)=8-\frac{1}{2}q_1.$$

Firm 1. We can finally analyze firm 1's production decision (the first mover). Taking into account the output level with which firm 2 and 3 will respond, as described by BRF_2 and BRF_3, respectively, firm 1 maximizes

$$\max_{q_1}\left(18-\left(q_1+\underbrace{\left(8-\frac{1}{2}q_1\right)}_{q_2}+\underbrace{\frac{16-q_1-\left(8-\frac{1}{2}q_1\right)}{2}}_{q_3}\right)\right)q_1-(5+2q_1)$$

which simplifies to

$$\max_{q_1}\left(6-\frac{1}{4}q_1\right)q_1-5-2q_1.$$

Taking first order conditions with respect to q_1, we obtain

$$6-\frac{1}{2}q_1-2=0.$$

Solving for q_1 we find the equilibrium production level for the leader (firm 1), $q_1^s=8$. Therefore, we can plug $q_1^s=8$ into firm 2's and 3's best response function to obtain

$$q_2^s = q_2(8) = 8 - \frac{1}{2}(8) = 4 \text{ units}$$

for firm 2 and

$$q_3^s = q_3(8, 4) = \frac{16 - 8 - 4}{2} = 2 \text{ units}$$

for firm 3. Hence, the output levels that arise in equilibrium are $(8, 4, 2)$. However, the SPNE of this Stackelberg game must specify optimal actions for all players, both along the equilibrium path (for instance, after firm 2 observes firm 1 producing $q_1^s = 8$ units) and off-the-equilibrium path (for instance, when firm 2 observes firm 1 producing $q_1 \neq 8$ units). We can accurately represent this optimal action at every point in the game in the following SPNE:

$$\left(q_1^s, q_2(q_1), q_3(q_1, q_2)\right) = \left(8, 8 - \frac{1}{2}q_2, \frac{16 - q_1 - q_2}{2}\right).$$

Exercise 8—Two-Period Bilateral Bargaining Game[A]

Alice and Bob are trying to split \$100. As depicted in the game tree of Fig. 4.12, in the first round of bargaining, Alice makes an offer at cost c (to herself), proposing to keep x_A and give the remaining $x_B = 100 - x_A$ to Bob. Bob either accepts her offer (ending the game) or rejects it. In round 2, Bob makes an offer of (y_A, y_B), at a cost of 10 to himself, which Alice accepts or rejects. If Alice accepts the offer, the game ends; but if she rejects it, the game proceeds to the third round, in which Alice makes an offer (z_A, z_B), at a cost c to herself. If Bob accepts her offer in the third round, the game ends and payoffs are accrued to each player; whereas if he rejects it, the money is lost. Assume that players are risk-neutral (utility is equal to money obtained minus any costs), and there is no discounting. If $c = 0$, what is the subgame-perfect equilibrium outcome?

Answer
Third period. Operating by backward induction, we start in the last round of the negotiation (third round). In $t = 3$, Alice will offer herself the maximum split z_A that still guarantees that it is accepted by Bob:

$$u_B(Accept_3) = u_B(Reject_3)$$
$$100 - z_A - 10 = -10, \text{ i.e., } z_A = 100.$$

That is, Alice offers $(z_A, z_B) = (100, 0)$ in $t = 3$.
Second period. In $t = 2$, Bob needs to offer Alice a split that makes her indifferent between accepting in $t = 2$ and rejecting in order to offer herself the entire pie in

Fig. 4.12 Two-period bilateral bargaining game

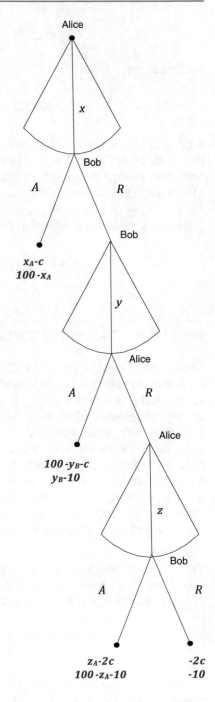

$t = 3$ (which we showed to be her equilibrium behavior in the last round of play). In particular, Alice is indifferent between accepting and rejecting if

$$u_A(Accept_2) = u_A(Reject_2)$$
$$100 - y_B = 100, \text{that is } y_B = 0$$

That is, Bob offers $(y_A, y_B) = (100, 0)$ in $t = 2$. (This is an extreme offer, whereby Bob cannot enjoy any negotiation power from making offers, and can only offer the entire pie to Alice. Intuitively, this offer emerges as equilibrium behavior in $t = 2$ in this exercise because players do not discount their future payoffs and they do not incur any cost in making offers, $c = 0$. Otherwise, the offers could not be so extreme).

First period. In $t = 1$, Alice needs to offer Bob a split that makes him indifferent between accepting in $t = 1$ and rejecting (which entails that he will offer himself $y_B = 0$ in the subsequent period, $t = 2$):

$$u_B(Accept_1) = u_B(Reject_1)$$
$$100 - x_A - 10 = 0 - 10, \text{that is } x_A = 100$$

That is, Alice offers $(x_A, x_B) = (100, 0)$ in $t = 1$. So the agreement is reached immediately and Alice gets the entire pie.

Exercise 9—Alternating Bargaining with a Twist[B]

Consider a two-player alternating offer bargaining game with 4 periods, and where player 1 is the first one to make offers. Rather than assuming discounting of future payoffs, let us now assume that the initial surplus shrinks by one unit if players do not reach an agreement in that period, i.e., surplus at period t becomes $S = 6 - t$ for all $t = \{1, 2, 3, 4\}$. As in similar bargaining games, assume that every player accepts an offer when being indifferent between accept and reject. Find the SPNE of the game.

Answer
Let us solve this sequential-move game by applying backward induction:
Period 4. In period 4, player 2 makes an offer to player 1, who accepts any offer $x_4 \geq 0$. Player 2, anticipating such a decision rule, offers $x_4 = 0$ in the 4th period, leaving player 2 with the remaining surplus $6 - 4 = 2$.
Period 3. In the previous stage (period 3), player 1 makes an offer to player 2, anticipating that, for player 2 to accept it, he must be indifferent between the offer and the payoff of \$2 he will obtain in the subsequent period (when player 2 becomes the proposer). Hence, the offer in period 3 must satisfy $x_3 \geq 2$, making player 1 an offer of exactly $x_3 = 2$, which leaves player 1 with the remaining surplus of $(6 - 3) - 2 = \$1$.

Period 2. In period 2, player 2 makes offers to player 1 anticipating that the latter will only obtain a payoff of \$1 if the game were to continue one more period. Hence, player 2's offer in period 2 must satisfy $x_2 \geq 1$, implying that player 2 offers exactly $x_2 = 1$ to player 1 (the minimum to guarantee acceptance), yielding him a remaining surplus of $(6 - 2) - 1 = \$3$.

Period 1. In the first period, player 1 is the individual making offers and player 2 is the responder. Player 1 anticipates that player 2's payoff will be \$3 if the game progresses towards the next stage (when player 2 becomes the proposer). His offer must then satisfy $x_1 \geq 3$, implying that player 1 offers $x_1 = \$3$ to player 2 to guarantee acceptance. Interestingly, player 1 offers a fair offer (half of the surplus) to player 2, which guarantees acceptance in the first stage of the game, implying that the surplus does not decrease over time.

Exercise 10—Backward Induction in Wage Negotiations[A]

Consider the following bargaining game between a labor union and a firm. In the first stage, the labor union chooses the wage level, w, that all workers will receive. In the second stage, the firm responds to the wage w by choosing the number of workers it hires, $h \in [0, 1]$. The labor union's payoff function is $u_L(w, h) = w \cdot h$, thus being increasing in wages and in the number of workers hired. The firm's profit is

$$\pi(w, h) = \left(h - \frac{h^2}{2} \right) - wh$$

Intuitively, revenue is increasing in the number of workers hired, h, but at a decreasing rate (i.e., $h - \frac{h^2}{2}$ is concave in h), reaching a maximum when the firm hires all workers, i.e., $h = 1$ where revenue becomes $1/2$.

Part (a) Applying backward induction, analyze the optimal strategy of the last mover (firm). Find the firm's best response function $h(w)$.

Part (b) Let us now move on to the first mover in the game (labor union). Anticipating the best response function of the firm that you found in part (a), $h(w)$, determine the labor union's optimal wage, w^*.

Answer

Part (a) The firm solves the following profit maximization problem:

$$\max_{h \in [0,1]} \left(h - \frac{h^2}{2} \right) - wh$$

Taking the first-order conditions with respect to h, we obtain:

$$1 - h - w \leq 0$$

Solving for h, yields a best response function of:

$$h(w) = 1 - w$$

Intuitively, the firm hires all workers when the wage is zero, but hires fewer workers as w increases. If the salary is the highest, $w = 1$, the firm responds hiring no workers. [Note that the above best response function $h(w)$ defines a maximum since second-order conditions yield $\frac{\partial^2 \pi(w,h)}{\partial h^2} = -1$, thus guaranteeing concavity in the profit function.]

Part (b) The labor union solves:

$$\max_{w \geq 0} u_L(w, h) = w \cdot h$$

$$\text{subject to } h(w) = 1 - w$$

The constraint reflects the fact that the labor union anticipates the subsequent hiring strategy of the firm. We can express this problem more compactly by inserting the (equality) constraint into the objective function, as follows:

$$\max_{w \geq 0} u_L(w(h), h) = w(1 - w) = w - w^2$$

Taking first-order conditions with respect to w yields:

$$1 - 2w \leq 0$$

and, solving for w, we obtain an optimal wage of $w^* = \frac{1}{2}$. Hence, the SPNE of the game is:

$$\{w^*, h(w)\} = \left\{ \frac{1}{2}, 1 - w \right\}$$

and, in equilibrium, salary $w^* = \frac{1}{2}$ is responded with $h\left(\frac{1}{2}\right) = 1 - \frac{1}{2} = \frac{1}{2}$ workers hired, i.e., half of workers are hired. In addition, equilibrium payoffs become

$$u_L^* = \frac{1}{2} \cdot \frac{1}{2} = \frac{1}{4} \text{ for the labor union}$$

$$\pi^* = \left(\frac{1}{2} - \frac{\left(\frac{1}{2}\right)^2}{2} \right) - \frac{1}{2} \cdot \frac{1}{2} = \frac{1}{8} \text{ for the firm.}$$

Exercise 11—Backward Induction-I[B]

Solve the game tree depicted in Fig. 4.13 using backward induction.

Answer

Starting from the terminal nodes, the smallest proper subgame we can identify is depicted in Fig. 4.14 (which initiates after player 1 chooses In). Recall that a subgame is the portion of a game tree that you can circle around without breaking any information sets. In this exercise, if we were to circle the portion of the game tree initiated after player 2 chooses A or B, we would be breaking player 3's information set. Hence, we need to keep enlarging the circle (area of the tree) until

Fig. 4.13 Extensive-form game

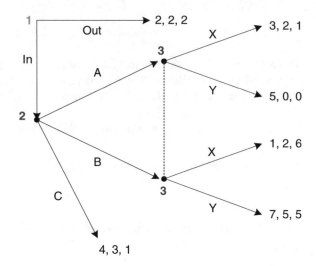

Fig. 4.14 Smallest proper subgame

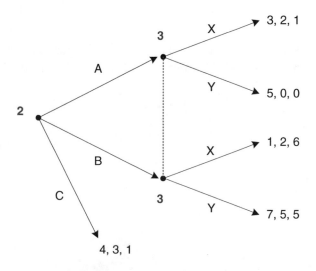

		Player 3	
		X	Y
Player 2	A	3,2,1	5,0,0
	B	1,2,6	7,5,5
	C	4,3,1	4,3,1

Fig. 4.15 Smallest proper subgame (in its normal-form)

Fig. 4.16 Extensive-form game (first stage)

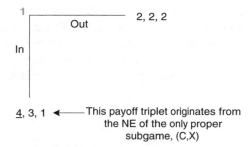

This payoff triplet originates from the NE of the only proper subgame, (C,X)

we do not break information sets. This happens when we circle the portion of the tree initiated after player 1 chooses In; as depicted in Fig. 4.14.

In this subgame in which only two players interact (players 2 and 3), player 3 chooses his action without observing player 2's choice. As a consequence, the interaction between players 2 and 3 in this subgame can be modeled as a simultaneous-move game. In order to find the Nash equilibrium of the subgame depicted in Fig. 4.14, we must first represent it in its normal (matrix) form, as depicted in Fig. 4.15.

For completeness, Fig. 4.15 also underlines the payoffs that arise when each player selects his best responses.[2] In outcome (C, X) the payoffs of player 2 and 3 are underlined, thus indicating that they are playing a mutual best response to each other's strategies. Furthermore, we do not need to examine player 1, since only player 2 and 3 are called on to move in the subgame. Hence, the Nash equilibrium of this subgame predicts that players 2 and 3 choose strategy profile (C, X). We can now plug the payoff triple resulting from the Nash equilibrium of this subgame, (4, 3, 1), at the end of the branch indicating that player 1 chooses In (recall that this was the node initiating the smallest proper subgame depicted in Fig. 4.14), as we illustrate in Fig. 4.16.

By inspecting the above game tree in which player 1 chooses In or Out, we can see that his payoff from In (4) is larger than from Out (2). Then, the SPNE of this game is (In, C, X).

---

[2]See exercises in Chap. 2 for more examples of this underlining process.

Exercise 12—Backward Induction-II[B]

Consider the sequential-move game depicted in Fig. 4.17. The game describes Apple's decision to develop the new iPhone (whatever the new name is, 5s, 6, ∞, etc.) with radically new software which allows for faster and better applications (apps). These apps are, however, still not developed by app developers (such as Rovio, the Finnish company that introduced Angry Birds). If Apple does not develop the new iPhone, then all companies make zero profit in this emerging market. If, instead, the new iPhone is introduced, then company 1 (the leader in the app industry) gets to decide whether to develop apps that are compatible with the new iPhone's software. Upon observing company 1's decision, the followers (firm 2 and 3) simultaneously decide whether to develop apps (D) or not develop (ND). Find all subgame perfect Nash equilibria in this sequential-move game.

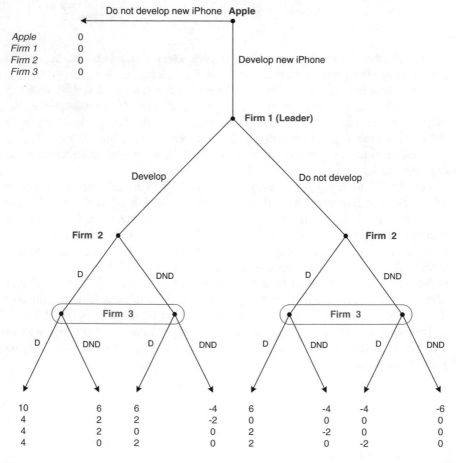

Fig. 4.17 Developing iPhone and apps game

		Firm 3	
		Develop	Do not develop
Firm 2	Develop	4, 4	2, 0
	Do not develop	0, 2	0, 0

Fig. 4.18 Smallest proper subgame (after firm 1 develops)

Answer
Company 1 develops. Consider the subgame between firms 2 and 3 which initiates after Apple develops the new iPhone and company 1 develops an application (in the left-hand side of the game tree of Fig. 4.17).

Since that subgame describes that firm 2 and 3 simultaneously choose whether or not to develop apps, we must represent it using its normal form in order to find the NEs of this subgame; as we do in the payoff matrix of Fig. 4.18.

We can identify the best responses for each player (as usual, the payoffs associated to those best responses are underlined in the payoff matrix of Fig. 4.18). In particular, $BR_2(D, D) = D$ and $BR_2(D, ND) = D$ for firm 2; and similarly for firm 3, $BR_3(D, D) = D$ and $BR_3(D, ND) = D$. Hence, *Develop* is a dominant strategy for each company, so there is a unique Nash equilibrium (*Develop, Develop*) in this subgame.

Company 1 does not develop. Next, consider the subgame associated with Apple having developed the new iPhone but company 1 not developing an application (depicted in the right-hand side of the game tree in Fig. 4.17). Since the subgame played between firms 2 and 3 is simultaneous, we represent it using its normal form in Fig. 4.19.

This subgame has two Nash equilibria: (*Develop, Develop*) and (*Do not develop, Do not develop*). (Note that, in our following discussion, we will have to separately analyze the case in which outcome (D, D) emerges as the NE of this subgame, and that in which (ND, ND) arises.)

Company 1—Case I. Let us move up the tree to the subgame initiated by Apple having developed the new iPhone. At this point, company 1 (the industry leader) has to decide whether or not to develop an application. Suppose that the Nash equilibrium for the subgame in which company 1 does not develop an application is (*Develop, Develop*). Replacing the two final subgames with the Nash equilibrium payoffs we found in our previous discussion, the situation is as depicted in the tree of Fig. 4.20. In particular, if firm 1 develops an application, then outcome (D, D) which entails payoffs (10, 4, 4, 4), as depicted in the terminal node at the bottom left-hand side of Fig. 4.20. If, in contrast, firm 1 does not develop an application for

		Firm 3	
		Develop	Do not develop
Firm 2	Develop	2, 2	-2, 0
	Do not develop	0, -2	0, 0

Fig. 4.19 Smallest proper subgame (after firm 1 does not develop)

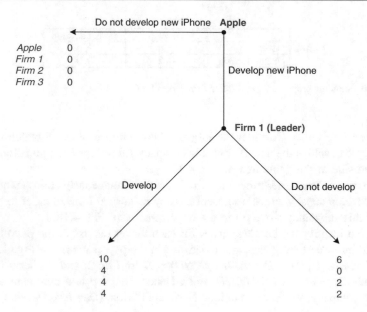

Fig. 4.20 Extensive-form subgame (Case I)

the new iPhone, then outcome (D, D) arises, yielding payoffs of (6, 0, 2, 2), as depicted in the bottom right–hand corner of Fig. 4.20.

We can now analyze firm 1's decision. If company 1 develops an application, then its payoff is 4, while its payoff is only 0 (since it anticipates the followers developing apps) from not doing so. Hence, company 1 chooses *Develop*.

Company 1—Case II. Now suppose the Nash equilibrium of the game that arises after firm 1 does not develop an app has neither firm 2 or 3 developing an app. Replacing the two final subgames with the Nash equilibrium payoffs we found in our previous discussion, the situation is as depicted in Fig. 4.21. Specifically, if firm 1 develops, outcome (D, D) emerges, which entails payoffs (10, 4, 4, 4); while if firm 1 does not develop firm 2 and 3 respond not developing apps either, ultimately yielding a payoff vector of (−6, 0, 0, 0). In this setting, if firm 1 develops an application, its payoff is 4; while its payoff is only 0 from not doing so. Hence, firm 1 chooses *Develop*.

Thus, regardless of which Nash equilibrium is used in the subgame initiated after firm 1 chooses *Do not develop* (in the right-hand side of the game in Figs. 4.20 and 4.21), firm 1 (the leader) optimally chooses to *Develop*.

First mover (Apple). Operating by backward induction, we now consider the first mover in this game (Apple). If Apple chooses to develop the new iPhone, then, as previously derived, firm 1 develops an application and this induces all followers 2 and 3 to do so as well. Hence, Apple's payoff is 10 from introducing the new iPhone. It is then optimal for Apple to develop the new iPhone, since its payoffs from so doing, 10, is larger than from not developing it, 0. Intuitively, since Apple anticipates all app developers will react introducing new apps, it finds the initial introduction of

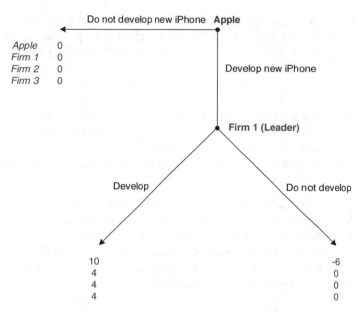

Fig. 4.21 Extensive-form subgame (Case II)

the iPhone to be very profitable. We can then identify two subgame perfect Nash equilibria (where a strategy for firm 2, as well as for firm 3, specifies a response to firm 1 choosing *Develop* and a response to company 1 choosing *Do not develop*[3]):

(*Develop iPhone, Develop, Develop/Develop, Develop/Develop*), and

(*Develop iPhone, Develop, Develop/Do not develop, Develop/Do not develop*).

Note that both SPNE result in the same equilibrium path, whereby, first, Apple introduces the new iPhone, the industry leader (firm 1) subsequently chooses to develop applications for the new iPhone, and finally firms 2 and 3 (observing firm 1's apps development) simultaneously decide to develop apps as well.

Exercise 13—Moral Hazard in the Workplace[B]

Consider the following moral hazard game between a firm and a worker. A firm offers either Contract 1 to a worker, which guarantees him a wage of $w = 26$ regardless of the outcome of the project that he develops in the firm, or Contract 2, which gives him a salary of $36 when the outcome of the project is good (G) but only $9 if the outcome is bad (B), i.e., $w_G = 36$ and $w_B = 9$. The worker can exert two levels of effort, either high (e_H) or low (e_L). The probability that, given a high effort level, the outcome of the project is good is $f(G|e_H) = 0.9$ and, therefore, the

---

[3]For instance, a strategy *Develop/Develop* for firm 2 reflects that this company chooses to develop apps, both after observing that firm 1 develops apps and after observing that firm 1 does not develop.

probability that, given a high effort, the outcome is bad is only $f(B|e_H) = 0.1$. If, instead, the worker exerts a low effort level, the good and bad outcome are equally likely, i.e., $f(G|e_L) = 0.5$ and $f(B|e_L) = 0.5$. The worker's utility function is

$$U_w(w, e) = \sqrt{w} - l(e),$$

where $\sqrt{w}$ reflects the utility from the salary he receives, and $l(e)$ represents the disutility from effort which, in particular, is $l(e_H) = 1$ when he exerts a high effort but $l(e_L) = 0$ when his effort is low (he shirks). The payoff function for the firm is $90 - w$ when the outcome of the project is good, but decreases to $30 - w$ when the outcome is bad.

Part (a) Depict the game tree of this sequential-move game. Be specific in your description of players' payoffs in each of the eight terminal nodes of the game tree.
Part (b) Find the subgame perfect equilibrium of the game.
Part (c) Consider now the existence of a social security payment that guarantees a payoff of x to the worker. Determine for which monetary amount of x the worker chooses to exert a low effort level in equilibrium.

Answer

Part (a) Fig. 4.22 depicts the game tree of this moral hazard game: First, the firm manager offers either contract 1 or 2 to the worker. Observing the contract offer, the worker then accepts or reject the contract. If he rejects the contract, both players obtain a payoff of zero. However, if the worker accepts it, he chooses to exert a high or low effort level. Finally, for every effort level, nature randomly determines whether such effort will result in a good or bad outcome, i.e., profitable or unprofitable project for the firm. As a summary, Fig. 4.23 depicts the time structure of the game.

In order to better understand the construction of the payoffs in each terminal node, consider for instance the case in which the worker accepts Contract 2 (in the right-hand side of the tree), responds exerting a low effort level and, afterwards, the outcome of the project is good (surprisingly!). In this setting, his payoff becomes $\sqrt{36} - 0 = 6$, since he receives a wage of $36 and exerts no effort, while the firm's payoff is $90 - 36 = 54$, given that the outcome was good and, as a result of the monetary incentives in Contract 2, the firm has to pay a salary of $36 to the worker when the outcome is good. If the firm offers instead Contract 1, the worker's salary remains constant regardless of the outcome of the project, but decreases in his effort (as indicated in the four terminal nodes in the left-hand side of the game tree).
Part (b) In order to find the SPNE, we must first notice that in this game we can define 4 proper subgames plus the game as a whole (as marked by the large circles in the game tree of Fig. 4.24).

Let us next separately analyze the Nash equilibrium of every proper subgame: *Subgame (1)*, initiated after the worker accepts Contract 1. Let's first find the expected utility the worker obtains from exerting a high effort level.

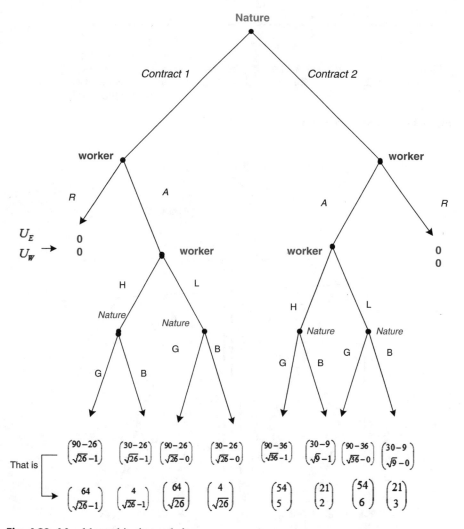

Fig. 4.22 Moral hazard in the workplace

$$\text{EU}_w(H|Contract\,1) = f(G|e_H)(\sqrt{26}-1) + f(B|e_H)(\sqrt{26}-1)$$
$$= 0.9(\sqrt{26}-1) + 0.1(\sqrt{26}-1) = \sqrt{26}-1$$

And the expected utility of exerting a low effort level under Contract 1 is

$$\text{EU}_w(L|Contract\,1) = 0.5\sqrt{26} + 0.5\sqrt{26} - 0 = \sqrt{26}$$

Hence, the worker exerts a low effort level, e_L, since

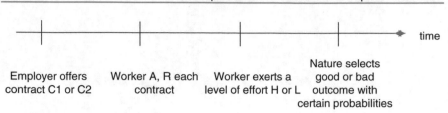

Fig. 4.23 Time structure of the moral hazard game

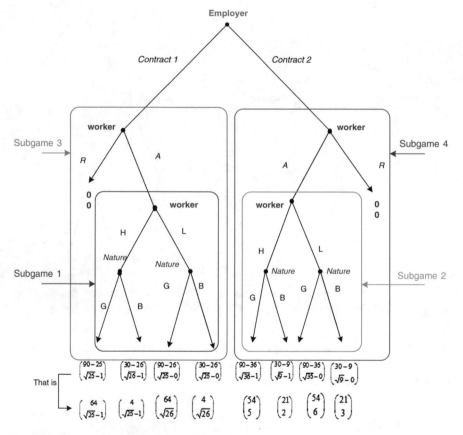

Fig. 4.24 Proper subgames in the moral hazard game

$$\mathrm{EU}_w(H|Contract\,1) < \mathrm{EU}_w(L|Contract\,1) \leftrightarrow \sqrt{26}-1 < \sqrt{26}$$

And the worker chooses to exert e_L after accepting Contract 1. Intuitively, Contract 1 offers the same salary regardless of the outcome. Thus, the worker does not have incentives to exert a high effort.

Fig. 4.25 Subgame 3 in the
moral hazard game

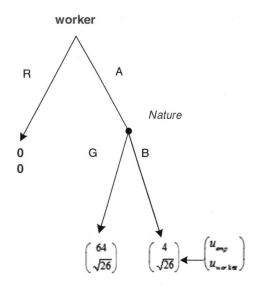

Subgame (2), initiated after the worker accepts Contract 2. In this case, the expected utility from exerting a high or low effort level is:

$$EU_w(H|Contract\,2) = 0.9 \cdot 5 + 0.1 \cdot 2 = 4.7, \text{ and}$$
$$EU_w(L|Contract\,2) = 0.5 \cdot 6 + 0.5 \cdot 3 - 0 = 4.5$$

Hence, the worker exerts a high effort level, e_H, after accepting Contract 2. _Subgame (3)_, initiated after receiving Contract 1. Operating by backwards induction, the worker can anticipate that, upon accepting Contract 1, he will exert a low effort (this was, indeed, the Nash equilibrium of subgame 1, as described above). Hence, the worker faces the reduced-form game in Fig. 4.25 when analyzing subgame (3).

As a consequence, the worker accepts Contract 1, since his expected payoff from accepting the contract, $0.5\sqrt{26} + 0.5\sqrt{26} = \sqrt{26}$, exceeds his payoff from rejecting it (zero). That is, anticipating a low effort level upon the acceptance of Contract 1,

$$EU_w(A|Contract\,1, e_L) \geq EU_w(R|Contract\,1, e_L) \leftrightarrow 0.5\sqrt{26} + 0.5\sqrt{26} \geq 0$$

Subgame (4), initiated after the worker receives Contract 2. Operating by backwards induction from subgame (2), the worker anticipates that he will exert a high effort level if he accepts Contract 2. Hence, the worker faces the reduced-form game in Fig. 4.26.

Therefore, the worker accepts Contract 2, given that:

$$EU_w(A|Contract\,2, e_H) \geq EU_w(R|Contract\,2, e_H)$$
$$0.9 \cdot 6 + 0.1 \cdot 3 - 1 \geq 0 \leftrightarrow 4.7 > 0$$

Fig. 4.26 Subgame 4 in the
moral hazard game

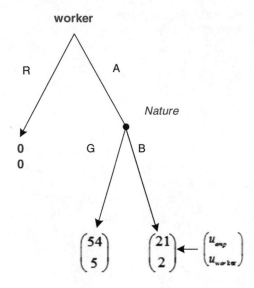

Game as a whole: Operating by backward induction, we obtain the reduced-form
game in Fig. 4.27.

When receiving Contract 1, the worker anticipates that he will accept it to
subsequently exert a low effort; while if he receives Contract 2 he will also accept it
but in this case exert a high effort.

Fig. 4.27 Reduced-form
game

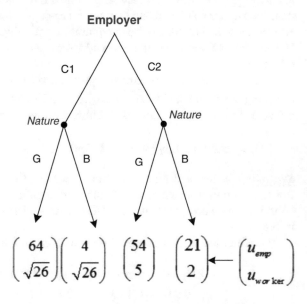

Analyzing the above reduced-form game, we can conclude that the employer prefers to offer Contract 2, since it yields a larger expected payoff. In particular,

$$E\pi(Contract\ 1|A, e_L) < E\pi(Contract\ 2|A, e_H), \text{ that is}$$
$$0.5 \cdot 64 + 0.5 \cdot 4 < 0.9 \cdot 54 + 0.1 \cdot 21 \leftrightarrow 34 < 50.7$$

Hence, in the SPNE the employer will offer Contract 2 to the worker paying him the wage scheme $(w_G, w_B) = (36, 9)$, which induces the worker to exert a high effort level in Contract 2 (in equilibrium) and a low effort in Contract 1 (off-the-equilibrium).

Part (c) In part (b) we obtained that, in equilibrium, the employer offers Contract 2 and the worker exerts high effort. Since the social security payment needs to achieve that the worker exerts a low effort, the easiest way to guarantee this is by making the worker reject Contract 2. Anticipating such rejection, the employer offers Contract 1, which implies an associated low level of effort. That is, a social security payment of x induces the worker to reject Contract 2 as long as x satisfies

$$EU_w(A|Contract\ 2, e_H) < EU_w(R|Contract\ 2, e_H)$$
$$0.9 \cdot \sqrt{36} + 0.1 \cdot \sqrt{9} - 1 < x$$

which simplifies to $x > 4.7$. That is, the social security payment must be relatively generous. In order to guarantee that the worker still accepts Contract 1, we need that

$$EU_w(A|Contract\ 1, e_L) > EU_w(R|Contract\ 1, e_L)$$
$$0.5 \cdot \sqrt{26} + 0.5 \cdot \sqrt{26} > x$$

which reduces to $x < \sqrt{26} \cong 5.09$, thus implying that the social security payment cannot be too generous since otherwise the worker would reject both types of contracts. Hence, when the social security payment is intermediate, i.e., it lies in the interval $x \in [4.7, \sqrt{26}]$, we can guarantee that the worker rejects Contract 2, but accepts Contract 1 (where he exerts a low effort level).

Applications to Industrial Organization

5

Introduction

This chapter helps us apply many of the concepts of previous chapters, dealing with simultaneous- and sequential-move games under complete information, to common industrial organization problems. In particular, we start with a systematic search for pure and mixed strategy equilibria in the Bertrand game of price competition between two symmetric firms, where we use several figures to illustrate our discussion. We then extend our explanation to settings in which firms are allowed to exhibit different costs.

We also explore the effects of cost asymmetry in the Cournot game of output competition. We afterwards move to a Cournot duopoly game in which one of the firms is publicly owned and managed. While private companies seek to maximize profits, public firms seek to maximize a combination of profits and social welfare, thus affecting its incentives and equilibrium output levels. At the end of the chapter, we examine applications in which firms can choose to commit to a certain strategy (such as spending in advertising, or competing in prices or quantities) before they start their competition with other firms. Interestingly, these pre-commitment strategies affect firm's competitiveness, and the position of their best-response functions, ultimately impacting subsequent equilibrium output levels and prices.

We also analyze an incumbent's incentives to invest in a more advanced technology that helps him to reduce its production costs in order to deter potential entrants from the industry; even if it leaves the incumbent with overcapacity (idle or unused capacity). We end the chapter with two more applications: one about a firm's tradeoff between directly selling to its customers or using a retailer; and another in which we analyze firms' incentives to merge, which depend on the proportion of firms in the industry that join the merger.

The original version of the chapter was revised: The erratum to the chapter is available at:
10.1007/978-3-319-32963-5_11

F. Munoz-Garcia and D. Toro-Gonzalez, *Strategy and Game Theory*,
Springer Texts in Business and Economics, DOI 10.1007/978-3-319-32963-5_5

Exercise 1—Bertrand Model of Price Competition[A]

Consider a Bertrand model of price competition between two firms facing an inverse demand function $p(q) = a - bq$, and symmetric marginal production costs, $c_1 = c_2 = c$, where $c < a$.

Part (a) Show that in the unique Nash equilibrium (NE) of this game, both firms set a price that coincides with their common marginal cost, i.e., $p_1^* = p_2^* = c$.

Part (b) Show that $p_1^* = p_2^* = c$ is not only the unique NE in pure strategies, but also the unique equilibrium in mixed strategies.

Answer

Part (a) We will analyze that $p_1^* = p_2^* = c$ is indeed a Nash equilibrium (NE). Notice that at this NE both firms make zero profits. Now let us check for profitable deviations:

- If firm i tries to deviate by setting a higher price $p_i > c$ then firm j will capture the entire market, and firm i will still make zero profits. Hence, such a deviation is not profitable.
- If firm i tries to deviate by setting a lower price $p_i < c$ then it sells to the whole market but at a loss for each unit sold given that $p_i < c$. Hence, such a deviation is not profitable either.

As a consequence, once firms charge a price that coincides with their common marginal cost, i.e., $p_1^* = p_2^* = c$, there does not exist any profitable deviation for any of the firms. Therefore, this pricing strategy is a NE.

Let us now show that, moreover, the NE strategies $p_1^* = p_2^* = c$ are the *unique* NE strategies in the Bertrand duopoly model with symmetric costs $c_1 = c_2 = c$.

- First, let us suppose that $\min\{p_1^*, p_2^*\} < c$. In this case, the firm setting the lowest price $p_i = \min\{p_1^*, p_2^*\}$ is incurring losses; as depicted in Fig. 5.1. If this firm raises its prices beyond the marginal cost, i.e., $p_i > c$, the worst that can happen is that its price is higher than that of its rival, $p_j < p_i$, and as a consequence firm i loses all its customers thus making zero profits (otherwise firm i will capture the entire market and make profits, which occurs when $p_j > p_i > c$). Then, the price choices given by $\min\{p_1^*, p_2^*\} < c$ cannot constitute a NE because there exist profitable deviations. [This argument is true for any firm $i = \{1, 2\}$ which initially gets the lowest price.]
- Second, let us suppose that $p_j = c$ and $p_k > c$ where $k \neq j$. Then, firm j captures the entire market, since it is charging the lowest price, but makes zero profits.

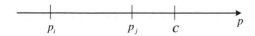

Fig. 5.1 $p_i < p_j < c$

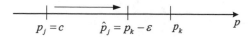

Fig. 5.2 Firm j has incentives to increase its price

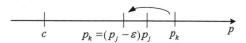

Fig. 5.3 Firm k has incentives to undercut firm j's price

Firm j has then incentives to raise its price to $\hat{p}_j = p_k - \varepsilon$, slightly undercutting its rival's price; as depicted in Fig. 5.2.

In particular, note that by deviating, firm j will still capture all the market (it is still the firm with the lowest price), but at a positive margin. Then, the initially considered price choices $p_j = c$ and $p_k > c$ where $k \neq j$ cannot constitute a NE, since there are profitable deviations for firm j.

- Third, let us suppose that both firms set prices above the common marginal cost, i.e., $p_j > c$ and $p_k > c$, but consider that such a price pair (p_j, p_k) is asymmetric, so that we can identify the lowest price $p_j \leq p_k$, whereby firm j captures the entire market. Hence, firm k earns zero profit, and has incentives to lower its price. If it undercuts firm j's price, i.e., $p_k = p_j - \varepsilon$, setting a price lower than that of its rival but higher than the common marginal cost (as illustrated in Fig. 5.3), then firm k captures all the market, raising its profits to $(p_k - c) \cdot q(p_k)$. Hence, we found a profitable deviation for firm k, and therefore, $p_k > p_j > c$ cannot be sustained as a NE.

Therefore, we have shown that any price configuration different from $p_1^* = p_2^* = c$ cannot constitute a NE. Since at least one firm has incentives to deviate from its strategy in such price profile. Therefore, the unique NE of the Bertrand duopoly model is $p_1^* = p_2^* = c$.

As a remark, we next describe how to represent firms' best-response functions in this game. Figure 5.4 depicts, for every price firm 2 sets, p_2, the profit maximizing price of firm 1. First, if $p_2 < c$, firm 2 is capturing all the market (at a loss), and firm 1 responds setting $p_1 = c$; as illustrated in the horizontal segment for all $p_2 < c$. Note that if firm 1 were to match firm 2's price, despite sharing sales, it would guarantee losses for each unit sold. Hence, it is optimal to respond with a price of $p_1 = c$. Second, if firm 2 increases its price to $c < p_2 < p^m$, where p^m denotes the monopoly price that firm 2 would set if being alone in the market, then firm 1 responds slightly undercutting its rival's price, i.e., $p_1 = p_2 - \varepsilon$; as depicted in the portion of firm 1's best response function that lies ε-below the 45-degree line. Finally, if firm 2 sets an extremely high price, i.e., $p_2 > p^m$, then firm 1 responds with a price $p_1 = p^m$. Such a monopoly price is optimal for firm 1, since it is lower

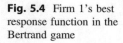
Fig. 5.4 Firm 1's best
response function in the
Bertrand game

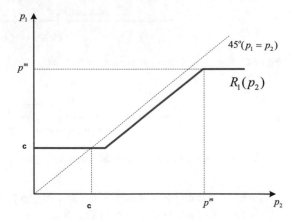

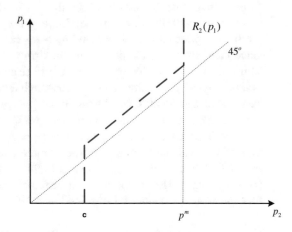
Fig. 5.5 Firm 2's best
response function in the
Bertrand game

than that of its rival (thus allowing firm 1 to capture all the market) and maximizes
firm 1's profits; as illustrated in the horizontal segment at the right hand of the
figure.

A similar analysis is applicable to firm's 2 best response function, as depicted
Fig. 5.5. Note that, in order to superimpose both firms' best response functions in
the same figure, we use the same axis as in the previous figure.

We can now superimpose both firms' best response functions, immediately
obtaining that the only point where they both cross each other is $p_1^* = p_2^* = c$, as
shown in our above discussion (Fig. 5.6).

Part (b) Let us work by contradiction, so let us assume that there exists a mixed
strategy Nash Equilibrium (MSNE). It has to be characterized by firms randomizing
between at least two different prices with strictly positive probability. Firm 1 must
randomize as follows:

$$\text{Prob.} \rightarrow q_1^1 \qquad\qquad q_1^2 = 1 - q_1^1$$
$$\text{Prices} \rightarrow p_1^1 \qquad\qquad p_1^2$$

Fig. 5.6 Equilibrium pricing
in the Bertrand model

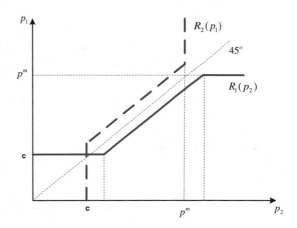

That is, firm 1 mixes setting a price p_1,

$$p_1 = \begin{cases} p_1^1 & \text{with} \quad \text{prob} \quad q_1^1 \\ p_1^2 & \text{with} \quad \text{prob} \quad q_1^2 = 1 - q_1^1 \end{cases}$$

Where, for generality, we allow the prices and their corresponding probabilities to take any values. Firm 2 must be similarly mixing between two prices, as follows:

$$\text{Prob.} \rightarrow q_2^1 \qquad\qquad q_2^1 = 1 - q_2^1$$

$$\text{Prices} \rightarrow p_2^1 \qquad\qquad p_2^2$$

That is, firm 2 sets a price p_2,

$$p_2 = \begin{cases} p_2^1 & \text{with} \quad \text{prob} \quad q_2^1 \\ p_2^2 & \text{with} \quad \text{prob} \quad q_2^2 = 1 - q_2^1 \end{cases}$$

For simplicity, let us first examine randomizations that cannot be profitable for either firm:

- First, note that in any MSNE we cannot have both firms losing money: $p_i^1 < c$ and $p_i^2 < c$ for all firm $i = \{1, 2\}$. In other words, we cannot have that both the lower and the upper bound of the randomizing interval lie *below* the marginal cost c, for both firms; as illustrated in Fig. 5.7.
- Second, if both $p_i^1 > c$ and $p_i^2 > c$ for all firm $i = \{1, 2\}$, both the lower and the upper bounds of the randomizing intervals of both firms lie *above* their common marginal cost; as depicted in Fig. 5.8. Then one of the firms, for instance firm j, can gain by deviating to a pure strategy that puts all probability weight on a single price, i.e., $p_j^1 = p_j^2 = p_j$, which slightly undercuts its rival's price. In particular, firm j can benefit from setting a price slightly below the lower bound of its rival's randomization interval, i.e., $p_j = p_k^1 - \varepsilon$. Hence we cannot have

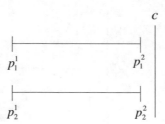

Fig. 5.7 Randomizing intervals below c

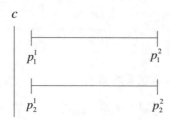

Fig. 5.8 Randomizing intervals above c

$p_i^1 > c$ and $p_i^2 > c$ being part of a MSNE, because there are profitable deviations for at least one firm.

- Finally, let us consider that for both firms $i = \{1, 2\}$ we have that $p_i^1 > c$ and $p_i^2 < c$, as depicted in Fig. 5.9. That is, the lower bound of randomization, p_i^1, lies below the common marginal cost, c, while the upper bound, p_i^2, lies above this cost. In this setting, every firm j could win the entire market by setting a price in which, rather than randomizing, uses a degenerate probability distribution (i.e., a pure strategy) by setting a price $p_j^1 = p_j^2 = p_j$, which slightly undercuts its rival's lower bound of price randomization, i.e., $p_j = p_k^1 - \varepsilon$. Therefore, having for both firms $p_i^1 < c$ and $p_i^2 > c$ cannot be a MSNE given that there are profitable deviations for every firm j.

Hence, we have reached a contradiction, since no randomizing pricing profile can be sustained as a MSNE of the Bertrand game. Thus, no MSNE exists. As a consequence, the unique NE implies that both firms play degenerated (pure) strategies, namely, $p_1^* = p_2^* = c$.

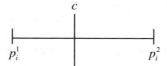

Fig. 5.9 Randomizing interval around c

Exercise 2—Bertrand Competition with Asymmetric Costs[B]

Consider a Bertrand model of price competition where firms 1 and 2 face different marginal production costs, i.e., $c_1 < c_2$, and an inverse demand function $p(Q) = a - bQ$, where Q denotes aggregate production, and $a > c_2 > c_1$.

Part (a) Find all equilibria in pure strategies: First, for the case in which prices are a continuous variable; and, second, when prices are a discrete variable (e.g., prices must be expressed in dollars and cents).

Part (b) Repeat your analysis for mixed strategies (for simplicity, you can focus on the case of continuous pricing).

Answer

Part (a)

Case 1 Pricing in terms of continuous monetary units is allowed.

Claim *In a Bertrand duopoly model with heterogeneous costs $c_1 < c_2 < a$ and continuous pricing there does not exist a Nash equilibrium in which both firms use degenerated (pure) strategies.*

Proof In order to prove this claim we will show that, for any possible pricing profile, we can find a profitable deviation. All the possible pricing profiles are:

(1) $p_i \leq p_j < c_1 < c_2$. In this case, both firms are making negative profits, since they are both setting a price below their corresponding marginal cost. Hence, they both have incentives to increase their prices.

(2) $p_i \leq c_1 < p_j < c_2$. Firm i is making negative (or zero) profits. Note that this is true regardless of its identity, either firm 1 or firm 2. Hence, this firm then has incentives to charge a price $p_i = p_j - \varepsilon$ for a small $\varepsilon > 0$. When firm i's identity is firm 1, or to a price $p_i = c_2$ when its identity is firm 2. Either way, firm i has incentives to deviate from the original configuration of prices.

(3) $c_1 < p_i \leq p_j < c_2$. Firm i serves all the market. So it is profitable for firm i to increase its price to $p_i = p_j - \varepsilon$, i.e., undercutting its rival's price by ε.

(4) $c_1 < p_1 \leq c_2 < p_2$. Both firms are making positive (or zero) profits, and firm 1 is capturing all the market. In this setting, firm 1 has incentives to set a price marginally close to its rival's cost, $p_1 = c_2 - \varepsilon$, which allows firm 1 to further increase its per-unit profits.

(5) $c_1 < p_1 \leq c_2 = p_2$. If $p_i = p_j = c_2$, then firms are splitting the market. In this context, the very efficient firm 1 has an incentive to set a price $p_1 = c_2 - \varepsilon$, which undercuts the lowest price that its (less efficient) competitor can profitably charge, helping firm 1 to capture all the market.

(6) $c_1 < c_2 \leq p_i \leq p_j$. In this context, firm i serves all the market if $p_i < p_j$, or firms share the market, if $p_i = p_j$. However, firm j has an incentive to undercut firm i's price by setting $p_j = p_i - \varepsilon$.

(7) $p_1 \leq c_1 < c_2 \leq p_2$. In this setting, firm 1 serves all the market and makes negative (zero) profits if $p_1 < c_1$ ($p_1 = c_1$, respectively). Firm 1 has incentives to set $p_1 = c_2 - \varepsilon$ for small ε, which implies that this firm captures all the market and makes a positive margin on every unit sold.

(8) $p_2 \leq c_1 < c_2 \leq p_1$. This case is analog to the previous one. Since firm 1 is making zero profits (no sales), this firm has incentives to deviate to a lower price, i.e., $p_1 = c_2 - \varepsilon$, allowing it to capture all the market.

Hence, we have checked that, for any profile of prices, some (or both) firms have profitable deviations. Therefore, none of these strategy profiles constitutes a pure strategy Nash equilibria. Hence, in the Bertrand duopoly model with heterogeneous costs, $c_1 < c_2$, and continuous pricing there does not exist pure strategy Nash equilibria.

Case 2 Discrete pricing

Claim *If $c_2 - c_1 > \varepsilon$ (i.e., $c_2 > c_1 + \varepsilon$) the unique Bertrand equilibrium is for the least competitive firm (firm 2 in this case) to set a price that coincides with its marginal cost, $p_2 = c_2$, while the more competitive firm (company 1) sets a price slightly below that of its rival, i.e., $p_1 = c_2 - \varepsilon$.*

Note that this pricing profile yields an equilibrium output of $q_2 = 0$ for firm 2 and $q_1 = \frac{a - (c_2 - \varepsilon)}{b}$ for firm 1. In this context, firm 2 does not have incentives to deviate to a lower price. In particular, while a lower price captures some consumers, it would be at a loss for every unit sold. Similarly, firm 1 does not have incentives to lower prices, given that doing so would reduce its profit per unit. Firm 1 cannot increase prices either (further approaching them to c_2) since, unlike in the previous part of the exercise (where such convergence can be done infinitely), in this case prices are discrete. Intuitively, firm 1 is charging 1 cent less than firm 2's marginal cost c_2.

Part (b)

Claim *In the Bertrand duopoly model with heterogeneous costs $c_1 < c_2 < a$ the following strategy constitutes a mixed strategy Nash equilibrium (MSNE) of the game:*

- *The most competitive firm (Firm 1 in this case) sets a price that coincides with its rival's marginal cost, i.e. $p_1 = c_2$*
- *The least competitive firm (Firm 2) sets a price p_2 by continuously randomizing over the interval $[c_2, c_2 + \varepsilon]$ for any $\varepsilon > 0$, with a cumulative distribution function $F(p; \varepsilon)$, with associated density $f(p; \varepsilon) > 0$ at all p; as depicted in Fig. 5.10.*

Proof In order to check that this strategy profile constitutes a MSNE we need to check that each firm i is playing its best response, BR_i.

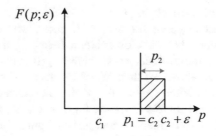

Fig. 5.10 MSNE in the Bertrand game with asymmetric firms

- Let us fix firm 2 randomizing over $[c_2, c_2 + \varepsilon]$, and check if firm 1 is playing its BR_1 by setting $p_1 = c_2$ (as prescribed). First, note that firm 1 (the most competitive company) sets the lowest price of all firms in equilibrium, except in the event that firm 2 sets a price exactly equal to $p_2 = c_2$. This event, however, occurs with a zero-probability measure, since firm 2 continuously randomizes over the interval $[c_2, c_2 + \varepsilon]$. Hence, firm 1 captures all the market demand, except in a zero-probability measure event.

  If firm 1 deviates towards a lower price, i.e., setting $p_1 < c_2$, then it lowers its profits for every unit sold relative to those in $p_1 = c_2$. Hence, $p_1 < c_2$ is not a profitable deviation for firm 1.

  If firm 1 sets instead a higher price, $p_1 > c_2$, then two effects arise: on one hand, firm 1 makes a larger margin per unit (as long as its price, p_1, still lies below that of firm 2) but, on the other hand, firm 1 gives rise to a negative effect, since it increases its probability of ending up with no sales (if the realization of p_2 is $p_2 < p_1$) at a rate of $f(p; \varepsilon)$. In order to show that the negative effect of setting a price $p_1 > c_2$, i.e., a price in the interval $[c_2, c_2 + \varepsilon]$, dominates its positive effect (and thus firm 1 does not have incentives to deviate), let us next examine firm 1's profits

$$\pi_1(p) = \underbrace{(1 - F(p; \varepsilon))}_{prob\{p_1 < p_2\}} \overbrace{D(p)}^{\substack{Amount\ sold \\ at\ price\ p}} \underbrace{(p - c_1)}_{per\ unit\ profits}$$

In this context, a marginal increase in its price yields the first order condition:

$$\frac{\partial \pi_1(p)}{\partial p} = -f(p; \varepsilon)D(p)(p - c_1) + (1 - F(p; \varepsilon))[D'(p)(p - c_1) + D(p)] < 0$$

Hence, firm 1 doesn't have incentives to set a price above $p_1 = c_2$.

- Let us now fix firm 1's strategy at $p_1 = c_2$, and check if firm 2 randomizing over $[c_2, c_2 + \varepsilon]$ is a best response for firm 2. If firm 2 sets a price below c_2 (or randomizing continuously with a lower bound below c_2) then firm 2 makes zero (or negative) profits. If firm 2 sets a price above c_2 (or randomizing continuously with an upper bound above c_2) then firm 2 makes zero profits as well, since it captures no customers willing to buy the product at such a high price. Therefore, there does not exist a profitable deviation for firm 2, and the strategy profile can be sustained as a MSNE of the Bertrand game of price competition when firms exhibit asymmetric production cost.[1]

Exercise 3—Duopoly Game with A Public Firm[B]

Consider a market (e.g. oil and natural gas in several countries) with one public firm (firm 0), and one private firm, (firm 1). Both firms produce a homogenous good with identical and constant marginal cost $c > 0$ per unit of output, and face the same inverse linear demand function $p(X) = a - b X$ with aggregate output $X = x_0 + x_1$, and where $a > c$. The private firm maximizes its profit

$$\pi_1 = p(X)x_1 - cx_1,$$

And the public firm maximizes a combination of social welfare and profits

$$V_0 = \theta W + (1 - \theta)\pi_0$$

where social welfare (W) is given by $W = \int_0^X p(y)dy - c(x_0 + x_1)$, and its profits are $\pi_0 = p(X)x_0 - cx_0$. Intuitively, parameter θ represents the weight that the manager of the public firm assigns to social welfare, while $(1 - \theta)$ is the weight that he assigns to its profits. Both firms simultaneously and independently choose output (as in the Cournot model of quantity competition).

Part (a) In order to better understand the incentives of these firms, and compare them with two private firms competing in a standard Cournot model, first find the best-response functions of the private firm, $x_1(x_0)$, and of the public firm, $x_0(x_1)$.

Part (b) Depict the best response functions. For simplicity, you can restrict your analysis to $\theta = 0, \theta = 0.5, \theta = 1$. Intuitively explain the rotation in the public firm's best response function as θ increases. Connect your results with those of a standard Cournot model where both firm's only care about profits, i.e., $\theta = 0$.

Part (c) Calculate the equilibrium quantities for the private and public firms. Find the aggregate output in equilibrium.

Part (d) Calculate the socially optimal output level and compare it with the equilibrium outcome you obtained in part (c).

---

[1]For more details on this MSNE, see Andreas Blume (2003) "Bertrand without fudge" *Economic Letters*, 7, pp. 167–68.

Answer

Part (a) The private firm maximizes

$$\pi_1 = [a - b(x_0 + x_1)]x_1 - cx_1,$$

Taking the first-order conditions with respect to x_1, we find the best-response function of the private firm:

$$x_1(x_0) = \frac{a - c}{2b} - \frac{x_0}{2}.$$

The public firm maximizes $V_0 = \theta W + (1 - \theta)\pi_0$. That is,

$$V = (1 - \theta)([a - b(x_0 + x_1)]x_0 - cx_0) + \theta\left[\int_0^{x_0 + x_1} (a - bx)dx - c(x_0 + x_1)\right],$$

Taking the first-order conditions with respect to x_0, we have

$$\frac{\partial V}{\partial x_0} = (1 - \theta)(a - bx_1 - c - 2bx_0) + \theta[a - b(x_0 + x_1) - c)] = 0,$$

And solving for x_0, we find the best-response function of the public firm:

$$x_0 = \frac{a - c}{2b\left(1 - \frac{\theta}{2}\right)} - \frac{x_1}{2\left(1 - \frac{\theta}{2}\right)}.$$

Part (b) In order to better understand the effect of θ on the public firm's best response function and, as a consequence, on equilibrium behavior, let us briefly examine some comparative statistics of parameter θ. In particular, when θ increases from 0 to 1 the best-response function of the public firm pivots outward, with its vertical intercept being unaffected. Let us analyze some extreme cases (Fig. 5.11 depicts the public firm's best response function evaluated at different values of θ):

- At $\theta = 0$, the best response function of the public firm becomes analog to that of private firm: the public firm behaves exactly as the private firm and the equilibrium is symmetric, with both firms producing the same amount of output. Intuitively, this is not surprising: when $\theta = 0$ the manager of the public company assigns no importance to social welfare, and only cares about profits.
- As θ increases, the slope the best-response function of the public firm, $2\left(1 - \frac{\theta}{2}\right)$, becomes flatter, and the public firm produces a larger share of industry output, i.e. the crossing point between both best response functions happens more to the southeast.

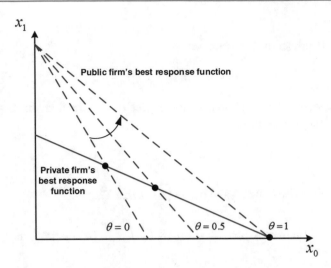

Fig. 5.11 Best response function of the private firm (*solid line*) and public firm (*dashed line*)

- At $\theta = 1$, only the public company produces in equilibrium, while the private firm does not produce (corner solution).

Part (c) The equilibrium quantities solve the system of two equations:

$$x_1 = \frac{a-c}{2b} - \frac{x_0}{2}, \text{ and}$$
$$x_0 = \frac{a-c}{2b\left(1 - \frac{\theta}{2}\right)} - \frac{x_1}{2\left(1 - \frac{\theta}{2}\right)}.$$

Simultaneously solving for x_0 and x_1, we find the individual output levels:

$$x_0 = \frac{1}{3 - 2\theta}\frac{a-c}{b}, \quad \text{and} \quad x_1 = \frac{1-\theta}{3 - 2\theta}\frac{a-c}{b}.$$

Note that for $\theta = 0$ the outcome is the same as that for a Cournot duopoly, with both firms producing $x_0 = x_1 = \frac{a-c}{3b}$, and for $\theta = 1$ the public firm produces the competitive outcome, $x_0 = \frac{a-c}{b}$, and the private firm produces nothing, $x_1 = 0$. (This result was already anticipated in our discussion of the pivoting effect in the best response function in the previous question.)

The aggregate output as a function of θ is, therefore,

$$X(\theta) = x_0 + x_1 = \frac{2-\theta}{3 - 2\theta}\frac{a-c}{b}.$$

Finally, note that differentiating the aggregate output with respect to θ yields

$$\frac{dX(\theta)}{d\theta} = \frac{1}{(3-2\theta)^2} \frac{a-c}{b} > 0.$$

Hence, an increase in θ results in an increase in output, and an increase in social welfare.

Part (d) The socially optimal output level corresponds to $p(X) = c$, which implies $a - bX = c$, i.e., $X = \frac{a-c}{b}$. The equilibrium output of the mixed duopoly, $\frac{2-\theta}{3-2\theta}\frac{a-c}{b}$, is below the socially optimal level, $\frac{a-c}{b}$, for any θ satisfying $\frac{2-\theta}{3-2\theta} < 1$. That is, $2 - \theta < 3 - 2\theta$ or $\theta < 1$. However, it exactly attains this output level at $\theta = 1$.

Exercise 4—Cournot Competition with Asymmetric Costs[A]

Consider a duopoly game in which firms compete a la Cournot, i.e., simultaneously and independently selecting their output levels. Assume that firm 1 and 2's constant marginal costs of production differ, i.e., $c_1 > c_2$. Assume also that the inverse demand function is $p(q) = a - bq$, with $a > c_1$. Aggregate output is $q = q_1 + q_2$.

Part (a) Find the pure strategy Nash equilibrium of this game. Under what conditions does it yield corner solutions (only one firm producing in equilibrium)?

Part (b) In interior equilibria (both firms producing positive amounts), examine how do equilibrium outputs vary when firm 1's costs change.

Answer

Part (a) In a Nash equilibrium (q_1^*, q_2^*), firm 1 maximizes its profits by selecting the output level q_1 that solves

$$\max_{q_1 \geq 0, q_2^*} \pi_1(q_1, q_2^*) = \left(a - b\left(q_1 + q_2^*\right)\right)q_1 - c_1 q_1 = aq_1 - bq_1^2 - bq_2^* q_1 - c_1 q_1$$

where firm 1 takes the equilibrium output of firm 2, q_2^*, as given. Similarly for firm 2, which maximizes

$$\max_{q_2 \geq 0, q_1^*} \pi_2(q_1^*, q_2) = \left(a - b\left(q_1^* + q_2\right)\right)q_2 - c_2 q_2 = aq_2 - bq_1^* q_2 - bq_2^2 - c_2 q_2$$

which takes the equilibrium output of firm 1, q_1^*, as given. Taking first order conditions with respect to q_1 and q_2 in the above profit-maximization problem, we obtain:

$$\frac{\partial \pi_1(q_1, q_2^*)}{\partial q_1} = a - 2bq_1 - bq_2^* - c_1 \tag{5.1}$$

$$\frac{\partial \pi_2(q_1^*, q_2)}{\partial q_2} = a - bq_1^* - 2bq_2 - c_2. \tag{5.2}$$

Solving for q_1 in (5.1) we obtain firm 1's best response function:

$$q_1(q_2) = \frac{a - bq_2 - c_1}{2b} = \frac{a - c_1}{2b} - \frac{q_2}{2}$$

(Note that, as usual, we rearranged the expression of the best response function to obtain two parts: the vertical intercept, $\frac{a-c_1}{2b}$, and the slope, $-\frac{1}{2}$). Similarly, solving for q_2 in (5.2) we find firm 2's best response function

$$q_2(q_1) = \frac{a - bq_1 - c_2}{2b} = \frac{a - c_2}{2b} - \frac{q_1}{2}.$$

Plugging firm 1's best response function into that of firm 2, we obtain:

$$q_2^* = \frac{a - c_2}{2b} - \frac{1}{2}\left(\frac{a - c_1}{2b} - \frac{q_2}{2}\right)$$

Since this expression only depends on q_2, we can now solve for q_2 to obtain the equilibrium output level for firm 2, $q_2^* = \frac{a + c_1 - 2c_2}{3b}$. Plugging the output we found q_2^* into firm 1's best response function $q_1 = \frac{a - 2bq_2 - c_2}{b}$, we obtain firm 1's equilibrium output level:

$$q_1^* = \frac{a - 2b\left[\frac{a+c_1-2c_2}{3b}\right] - c_2}{b} = \frac{a + c_2 - 2c_1}{3b}$$

Hence, for $q_1^* = \frac{a+c_2-2c_1}{3b} \leq 0$ we need firm 1's costs to be sufficiently high, i.e., solving for c_1 we obtain $\frac{a+c_2}{2} \leq c_1$. A symmetric condition holds for firm 2's output. In particular, $q_2^* = \frac{a+c_1-2c_2}{3b} \leq 0$ arises if $\frac{a+c_1}{2} \leq c_2$. Therefore, we can identify three different cases (two giving rise to corner solutions, and one providing an interior solution):

(1) if $c_i \geq \frac{a+c_j}{2}$ for all firm $i = \{1, 2\}$, then both firms costs are so high that no firm produces a positive output level in equilibrium;
(2) if $c_i \geq \frac{a+c_j}{2}$ but $c_j < \frac{a+c_i}{2}$, then only firm j produces a positive output (intuitively, firm j would be the most efficient company in this setting, thus leading it to be the only producer); and
(3) if $c_i \leq \frac{a+c_j}{2}$ for all firm $i = \{1, 2\}$, then both firms produce positive output levels.

Fig. 5.12 Cournot game—
Cost pairs, and production
decisions

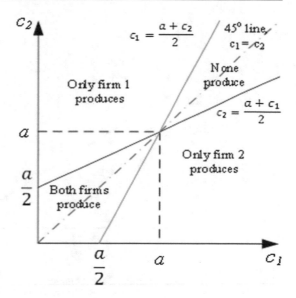

Figure 5.12 summarizes these equilibrium results as a function of c_1 and c_2. In addition, the figure depicts what occurs when firms are cost symmetric, $c_1 = c_2 = c$ along the 45°-line, whereby either both firms produce positive output levels, i.e., if $c < \frac{a+c}{2}$ or $c < a$; or none of them does, i.e., if $c > a$.

For instance, when $c_2 > \frac{a+c_1}{2}$, only firm 1 produces, as depicted in Fig. 5.13, where the best response functions of firms 1 and 2 only cross at the vertical axis where $q_1^* > 0$ and $q_2^* = 0$.

Part (b) In order to know how the (q_1^*, q_2^*) varies when (c_1, c_2) change we separately differentiate each (interior) equilibrium output level with respect to c_1 and c_2, as follows:

$$\frac{\partial q_1^*}{\partial c_1} = -\frac{2}{3b} < 0 \text{ and } \frac{\partial q_1^*}{\partial c_2} = \frac{1}{3b} > 0$$

Hence, firm 1's equilibrium output decreases as its own costs go up, but increases as its rival's costs increase. A similar intuition applies to firm 2's equilibrium output:

$$\frac{\partial q_2^*}{\partial c_2} = -\frac{2}{3b} < 0 \text{ and } \frac{\partial q_2^*}{\partial c_1} = \frac{1}{3b} > 0.$$

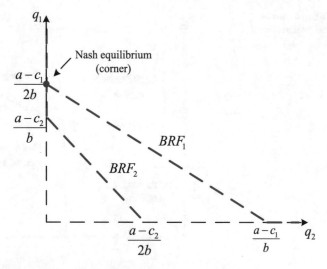

Fig. 5.13 A corner solution in the Cournot game

Exercise 5—Strategic Advertising and Product Differentiation[C]

Consider a two stage model. At the first stage, each firm selects how much money to spend on advertising. The amount selected by firm $i = \{1,2\}$ is denoted by A_i. After the firm selects its level of advertising, it faces an inverse demand for its product given by:

$$p_i(q_i, q_j) = \left[a - bq_i - d_iq_j\right]$$

where $i \neq j$, $b, d_i > 0$ and $0 < d_i < b$, where parameter d_i is inversely related to A_i. Hence, if firm i advertises more its product is perceived to be more differentiated relative to that of firm j (i.e., $\partial d_i / \partial A_i < 0$).

At the second stage, each firm decides how much to produce. Find the SPNE of the two-stage game by solving it backwards. In particular, for given A_1 and A_2, derive the equilibrium quantities of the second stage ($q_1(A_1,A_2)$, $q_2\,(A_1,A_2)$). Afterwards, given the derivation of the second stage, find the equilibrium spending on advertising. (For simplicity, assume no production costs.)

Answer

Using backwards induction we start at the second stage:
Second Stage Game: For given levels of advertising, every firm i chooses the output level q_i that maximizes its profits, taking the production level of its rival, q_j, as given.

$$\max_{q_i}[a - bq_i - d_iq_j]q_i$$

Taking first order conditions with respect to q_i, we obtain $a - 2bq_i - d_iq_j = 0$ and solving for q_i, we obtain firm i's best response function: $q_i(q_j) = \frac{a}{2b} - \frac{d_i}{2b}q_j$. Plugging firm j's best response function into firm i's, we find

$$q_i = \frac{a}{2b} - \frac{d_i\left(\frac{a}{2b} - \frac{d_jq_i}{2b}\right)}{2b}$$

which yields an equilibrium output level of:

$$q_i^* = \frac{(2b - d_i)a}{4b^2 - d_id_j} \tag{5.3}$$

First Stage Game: Given the equilibrium output levels, q_i^*, found in expression (5.3) above, we can now move to the first stage game, where every firm independently and simultaneously selects an advertisement expenditure, A_i, that maximizes its profit function.

$$\max\left[a - bq_i^* - d_iq_j^*\right]q_i^* - A_i$$

where we plug the equilibrium value of q_i^* in the second stage of the game. Taking first order conditions with respect to A_i, we find

$$\frac{d\pi_i}{dA_i} = \underbrace{\frac{\partial\pi_i}{\partial q_i}\frac{\partial q_i^*}{\partial A_i}}_{0} + \underbrace{\frac{\partial\pi_i}{\partial q_j}\frac{\partial q_j^*}{\partial A_i}}_{\substack{\text{Indirect} \\ \text{effect}}} - \underbrace{\frac{\partial\pi_i}{\partial A_i}}_{\substack{\text{Direct} \\ \text{effect}}}$$

where the first term on the right-hand side is zero since $\frac{\partial q_i^*}{\partial A_i} = 0$ because of the envelope theorem. The third term represents the direct effect that advertising entails on firm i's profits (the cost of spending more dollars on advertising). The second term, however, reflects the indirect (or strategic) effect from advertising, capturing the fact that a change in a firm's advertising expenditure affects the equilibrium production of its rival, q_j^*, which ultimately affects firm i's profits. Importantly, this suggest that, for A_i to be effective, it must be irreversible and perfectly observable by its rivals, since otherwise firm i would not be able to affect its rival's output decision in the second period game.

Hence, expanding the above first-order condition, we obtain:

$$\frac{d\pi_i}{dA_i} = \frac{\partial\pi_i}{\partial q_j}\frac{\partial q_j^*}{\partial A_i} - \frac{\partial\pi_i}{\partial A_i}$$

$$= \frac{\partial\pi_i}{\partial q_j}\frac{\partial q_j^*}{\partial d_i}\frac{\partial d_i}{\partial A_i} - 1$$

$$= \left(-d_i q_i^*\right)\frac{\partial q_j^*}{\partial d_i}\frac{\partial d_i}{\partial A_i} - 1$$

where in particular, $\frac{\partial q_j^*}{\partial d_i}$ is found by using the best response function, $q_j^* = \frac{(2b-d_j)a}{4b^2-d_j d_i}$ as follows

$$\frac{\partial q_j^*}{\partial d_i} = \frac{(2b - d_j)ad_j}{(4b^2 - d_i d_j)^2} = \frac{q_j d_j}{(4b^2 - d_i d_j)}$$

Substituting in the above expression, we obtain

$$\frac{d\pi_i}{dA_i} = \left(-d_i q_i^*\right)\left[\frac{q_j^* d_j}{(4b^2 - d_i d_j)}\right]\frac{\partial d_i}{\partial A_i} - 1$$

$$= -q_i^* q_j^*\frac{\partial d_i}{\partial A_i}\left[\frac{d_i d_j}{(4b^2 - d_i d_j)}\right] - 1 = 0.$$

Therefore, solving for $\frac{\partial d_i}{\partial A_i}$, we find

$$\frac{\partial d_i}{\partial A_i} = -\frac{4b^2 - d_i d_j}{d_i d_j} \cdot \frac{1}{q_i^* q_j^*}. \tag{5.4}$$

Hence, the optimal level of advertising by firm i in the first stage, A_i, is the level of A_i such that the advertising effect on firm i's product solves the above expression.

As a remark, note that if, instead, advertising was selected simultaneously with output, equilibrium advertising would be determined as follows:

$$\frac{d\pi_i}{dA_i} = -q_i q_j\frac{\partial d_i}{\partial A_i} - 1 = 0$$

Where now term $\frac{\partial q_j^*}{\partial d_i}$ is absent. Intuitively, advertising A_i does not affect firm j's optimal output level in this case. Solving for $\frac{\partial d_i}{\partial A_i}$, we find

$$\frac{\partial d_i}{\partial A_i} = -\frac{1}{q_i q_j}. \tag{5.5}$$

Comparing the right-hand side of expressions (5.4) and (5.5), we can conclude that a given increase in advertising expenditures produces a larger decrease in d_i (more product differentiation), i.e. $\frac{\partial d_i}{\partial A_i}$ is larger (in absolute value), when firms determine their level of advertising sequentially (before choosing their output level) than when they choose it simultaneously with their output level.

Exercise 6—Cournot Oligopoly^B

Compute the Cournot equilibrium with N firms when the firms face the inverse demand function $P(Q) = Q^{-1/\varepsilon}$, where $\varepsilon > 1$, and have identical constant marginal cost c.

Answer
Every firm i simultaneously and independently chooses its output level q_i in order to maximize profits

$$\max_{q_i}\left(Q^{-\frac{1}{\varepsilon}} - c\right)q_i.$$

Taking first order conditions with respect to q_i, we obtain,

$$Q^{-\frac{1}{\varepsilon}} - c - \frac{1}{\varepsilon}Q^{-\frac{1}{\varepsilon}-1} \cdot q_i = 0.$$

At a symmetric equilibrium, $Q = N \cdot q_i$, and hence $q_i = Q/N$. Rearranging the above first-order conditions,

$$Q^{-\frac{1}{\varepsilon}} - c - \frac{1}{\varepsilon}Q^{-\frac{1}{\varepsilon}-1} \cdot \frac{Q}{N} = 0 \Leftrightarrow Q^{-\frac{1}{\varepsilon}} - \frac{1}{\varepsilon}Q^{-\frac{1}{\varepsilon}-1} \cdot \frac{Q}{N} = c$$

$$\Leftrightarrow Q^{-\frac{1}{\varepsilon}}\left[\frac{\varepsilon N - 1}{\varepsilon N}\right] = c.$$

And solving for Q we find, $Q^* = \left[\frac{\varepsilon N - 1}{c\varepsilon N}\right]^\varepsilon$. In the limit when $N \to \infty$ we obtain $\lim_{N\to\infty} Q^* = \left(\frac{1}{c}\right)^\varepsilon$, which coincides with the outcome for a perfectly competitive market. To see this take the demand function $P(Q) = Q^{-1/\varepsilon}$, and solve for Q,

$$Q = (P)^{-\varepsilon} = (1/P)^\varepsilon$$

In a perfectly competitive market it must be true that $p(Q) = c$, hence

$$Q^* = (1/c)^\varepsilon$$

Exercice 7—Commitment in Prices or Quantities?[B]

This exercise is based on Singh and Vives (1984).[2] Consider a market that consists of two firms offering differentiated products. The inverse demand that each firm faces is:

$$p_i = a - bq_i - dq_j \qquad 0 \le d \le b.$$

Consider the following two stage game. In the first stage each firm can commit either to supply a certain quantity $\bar{q}_i$ or to set a price $\bar{p}_i$. In the second stage, the remaining variables that were not chosen in the first stage are determined in order to clear the market (prices, if quantities were chosen in the first stage; or quantities, if prices were selected in the first stage). Show that a commitment to quantity in the first stage is a dominant strategy for each firm.

Answer
Operating by backward induction, let us first analyze the second stage of the game.
Second Stage
Part (a) If both firms commit to quantities in the first stage, then firm i's market demand is

$$p_i(q_i, q_j) = a - bq_i - dq_j.$$

Part (b) If both firms commit to prices in the first stage, then market demand is found by solving for q_i in $p_i(q_i, q_j)$, as follows

$$q_i = \frac{a(b-d) - bp_i - dq_j}{(b-d)(b+d)}.$$

Part (c) If one firm commits to prices, $\bar{p}_i$, and the other to quantities, $\bar{q}_j$, then market demand for firm i is

$$\bar{p}_i = a - bq_i - d\bar{q}_j \rightarrow q_i = \frac{a - d\bar{q}_j - \bar{p}_i}{b}$$

and that for firm j is

$$\bar{q}_j = \frac{a(b-d) - b\bar{p}_i - dp_j}{(b-d)(b+d)} \rightarrow p_j = a - b\bar{q}_j - d\left(\frac{a - d\bar{q}_j - \bar{p}_i}{b}\right).$$

---

[2]Singh, N., & Vives, X. (1984). "Price and Quantity Competition in a Differentiated Duopoly". *The Rand Journal of Economics*, 15, 4, 546–554.

First Stage

If both firms commit to quantities, then from the demand function in point (a), we obtain the Cournot outcome. In particular, taking first order conditions with respect to q_i in

$$\max_{q_i} \left(a - bq_i - dq_j \right) q_i$$

we obtain

$$a - 2bq_i - dq_j = 0$$

which yields a best response function of $q_i(q_j) = \frac{a}{2b} - \frac{d}{2b} q_j$. Simultaneously solving for q_i and q_j, we find equilibrium outputs of $q_i = q_j = \frac{(2b-d)a}{4b^2-d^2}$, thus entailing equilibrium profits of $\pi_i^C \equiv \frac{ba^2}{(b+d)^2}$ for every firm $i = \{1,2\}$.

If both firms commit to prices, from point (b), we obtain the Bertrand outcome. In particular, every firm i chooses its price level p_i that maximizes

$$\max_{p_i} p_i \left(\frac{a(b-d) - bp_i - dq_j}{(b-d)(b+d)} \right).$$

Taking first order conditions with respect to p_i yields

$$\frac{a(b-d) - 2bp_i - dq_j}{b^2 - d^2} = 0$$

and solving for p_i, we obtain firm i's price

$$p_i = \frac{a(b-d) - dq_j}{2b}$$

The price of firm j, p_j, is symmetric. Hence, plugging these results into firm i's profit function, we obtain firm i's equilibrium profits $\pi_i^B \equiv \frac{(b-d)a^2}{(b+d)(2b-d)^2}$ for every firm $i = \{1,2\}$, where $\pi_i^B < \pi_i^C$ for all parameter values.

From (c), we obtain a hybrid outcome, which can be found by maximizing firm i's profits:

$$\max_{\bar{p}_i} \pi_i = \left(\frac{a - d\bar{q}_j - \bar{p}_i}{b} \right) \bar{p}_i$$

Taking first-order conditions with respect to p_i, since firm i commits in using prices, we obtain:

$$\frac{a - d\bar{q}_j - 2\bar{p}_i}{b} = 0 \rightarrow \bar{p}_i(\bar{q}_j) = \frac{a}{2} - \frac{d}{2}\bar{q}_j.$$

Similarly, firm j, which commits to using quantities, maximizes profits

$$\max_{\bar{q}_j} \left[a - b\bar{q}_j - d\left(\frac{a - d\bar{q}_j - \bar{p}_i}{b}\right) \right] \bar{q}_j$$

Taking first-order conditions with respect to $\bar{q}_j$, we find

$$a - 2b\bar{q}_j - d\left(\frac{a - 2d\bar{q}_j - \bar{p}_i}{b}\right) = 0 \rightarrow \bar{q}_j(\bar{p}_i) = \frac{a(b-d) + d\bar{p}_i}{2(b^2 - d^2)}.$$

Substituting the expression we found for $\bar{p}_i$ into $\bar{q}_j$, we have

$$\bar{q}_j = \frac{a(b-d) + d\left(\frac{a - d\bar{q}_j}{2}\right)}{2(b^2 - d^2)} \Rightarrow \bar{q}_j = \frac{2ab - ad}{4b^2 - 3d^2}.$$

Plugging the equilibrium quantity of firm j, $\bar{q}_j$, into $\bar{p}_i(\bar{q}_j) = \frac{a}{2} - \frac{d}{2}\bar{q}_j$, we obtain

$$\bar{p}_i = \frac{a}{2} - \frac{d}{2}\overbrace{\left(\frac{2ab - ad}{4b^2 - 3d^2}\right)}^{\bar{q}_j} = \frac{a(b-d)(2b+d)}{4b^2 - 3d^2}.$$

We can now find q_i by using the demand function firm i faces, $\bar{p}_i = a - bq_i - d\bar{q}_j$, and the expressions for $\bar{p}_i$ and $\bar{q}_j$ we found above. In particular,

$$\underbrace{\frac{a(b-d)(2b+d)}{4b^2 - 3d^2}}_{\bar{p}_i} = a - bq_i - d\underbrace{\frac{2ab - ad}{4b^2 - 3d^2}}_{\bar{q}_j}$$

Solving for q_i, we obtain an equilibrium output of $q_i = \frac{a(b-d)(2b+d)}{b(4b^2 - 3d^2)}$. Hence, the profits of the firm that committed to prices (firm i) are

$$\pi_i^p = \bar{p}_i \cdot q_i = \frac{a(b-d)(2b+d)}{4b^2 - 3d^2} \cdot \frac{a(b-d)(2b+d)}{b(4b^2 - 3d^2)} = \frac{a^2(b-d)^2(2b+d)^2}{b(4b^2 - 3d^2)^2}.$$

Similarly, regarding firm j, we can obtain its equilibrium price, p_j, by using the demand function firm j faces, $p_j = a - b\bar{q}_j - dq_i$, and the expressions for $\bar{q}_j$ and q_i found above. Specifically,

$$p_j = a - b \underbrace{\left[\frac{a(2b-d)}{4b^2 - 3d^2}\right]}_{\bar{q}_j} - d \underbrace{\left[\frac{a(b-d)(2b+d)}{b(4b^2 - 3d^2)}\right]}_{q_i} = \frac{a(b-d)(2b-d)(b+d)}{b(4b^2 - 3d^2)}$$

Thus, the profits of the firm that commits to quantities (firm j) are

$$\pi_j^Q = p_j\bar{q}_j = \frac{a(b-d)(2b-d)(b+d)}{b(4b^2 - 3d^2)} \cdot \frac{2ab - ad}{4b^2 - 3d^2} = \frac{a^2(b-d)(2b - d^2)(b+d)}{b(4b^2 - 3d^2)^2}$$

Summarizing, the payoff matrix that firms face in the first period game is:

Firm j

		Prices	Quantities
Firm i	Prices	π_i^B, π_j^B	π_i^P, π_j^Q
	Quantities	π_i^Q, π_j^P	π_i^C, π_j^C

If firm j chooses prices (fixing our attention on the left column), firm i's best response is to select quantities since $\pi_i^Q > \pi_i^B$ given that $d > 0$ by assumption. Similarly, if firm j chooses quantities (in the right column), firm i's best responds with quantities since $\pi_i^C > \pi_i^P$ given that $d > 0$. Since payoffs are symmetric, a similar argument applies to firm j. Therefore, committing to quantities is a strictly dominant strategy for both firms, and the unique Nash equilibrium is for both firms to commit in quantities.

Exercise 8—Fixed Investment as a Pre-Commitment Strategy[B]

Consider a market that consists of two firms producing a homogeneous product. Firms face an inverse demand of $p_i = a - bq_i - dq_j$. The cost function facing each firm is given by

$$TC_i(c_i) = F(c_i) + c_i q_i \ , \quad \text{where } F'(c_i) < 0, F''(c_i) \geq 0.$$

Hence, the fixed costs of production decline as the unit variable cost increases. Alternatively, as fixed costs increase, the unit variable costs decline. Consider the following two-stage game: in the first stage, each firm chooses its technology, which determines its unit variable cost c_i (to obtain a lower unit variable cost c_i the firm needs to invest in improved automation, which increases the fixed cost $F(c_i)$). In the second stage, the firms compete as Cournot oligopolists.

Part (a) Find the equilibrium of the two-stage game, and discuss the properties of the "strategic effect." Which strategic terminology applies in this case?

Part (b) Assume that $F(c_i) = \alpha - \beta + 0.5c_i^2$. Derive the closed form solution of the equilibrium of the two stage game. What restrictions on parameters α and β are necessary to support an interior equilibrium?

Answer

Part (a) In the first stage, firm i chooses a level of c_i that maximizes its profits, given its equilibrium output in the second stage of the Cournot oligopoly game. That is,

$$\max_{c_i} \pi_i(c_i, q_i(c_i), q_j(c_i))$$

Taking first-order conditions with respect to c_i, we find

$$\frac{d\pi_i}{dc_i} = \underbrace{\frac{\partial \pi_i}{\partial c_i}}_{(1)} + \underbrace{\frac{\partial \pi_i}{\partial q_i}\frac{\partial q_i}{\partial c_i}}_{(2)} + \underbrace{\frac{\partial \pi_i}{\partial q_j}\frac{\partial q_j}{\partial c_i}}_{(3)}.$$

Let us separately interpret these three terms:

(1) *Direct Effect*: an increase in c_i produces a change in costs of $\frac{\partial \pi_i}{\partial c_i}$.
(2) This effect is zero since in the second stage $q_i(c_i)$ is chosen in equilibrium (i.e., by the envelope theorem).
(3) *Strategic Effect*. $\frac{\partial \pi_i}{\partial q_j}\frac{\partial q_j}{\partial c_i}$ can be expressed as:

$$\underbrace{\frac{\partial \pi_i}{\partial q_j}}_{-} \ \underbrace{\frac{\partial q_j}{\partial q_i}}_{-} \ \underbrace{\frac{\partial q_i}{\partial c_i}}_{-} < 0$$

Intuitively, the strategic effect captures the fact that an increase in c_i (in the third component) reduces firm i's production, which increases firm j's output, ultimately reducing firms i's profits. Hence, the overall strategic effect is negative. Therefore, the firm seeks to invest in increased automation, for unit variable costs to be lower than in the absence of competition. Firms, hence, want to become more competitive in the second period game, i.e., a "top dog" type of commitment. Let us analyze this first-order condition in more detail.

Second stage. For a given c_i and c_j chosen in the first stage, in the second stage firm i chooses q_i to maximize the following profits

$$\max_{q_i} \pi_i = (a - bq_i - bq_j - c_i)q_i - F(c_i)$$

Taking first order conditions with respect to q_i we obtain $a - 2bq_i - bq_j - c_i = 0$, which yields the best response function $q_i(q_j) = \frac{a-c}{2b} - \frac{1}{2}q_j$, ultimately entailing a Nash equilibrium output level of $q_i = \frac{a-2c_i+c_j}{3b}$. (This is a familiar result in Cournot duopolies with asymmetric marginal costs.) Hence, first stage equilibrium profits are

$$\pi_i = \frac{(a - 2c_i + c_j)^2}{9b} - F(c_i).$$

First stage. In the first stage, firm i chooses c_i to maximize its profits, yielding first order conditions:

$$\frac{\partial \pi_i}{\partial c_i} = -\frac{4(a - 2c_i + c_j)}{9b} - F'(c_i) = 0$$

and second order conditions:

$$\frac{\partial^2 \pi_i}{\partial c_i^2} = \frac{8}{9b} - F''(c_i) \leq 0$$

Hence, to guarantee that second order conditions hold, we need $F''(c_i) \geq 8/9b$. In addition, to guarantee that firms choose a positive output at the symmetric equilibrium, we need that the first order condition satisfies $\left.\frac{\partial \pi_i}{\partial c_i}\right|_{c_1=c_2=0} > 0$. Hence, $-\frac{4a}{9b} - F'(0) > 0$, or $F'(0) > \frac{4a}{9b}$.

Note that we do not solve for c_i in the first order condition $\frac{\partial \pi_i}{\partial c_i}$ since we would not be able to obtain a closed form solution for the best response function $c_i(c_j)$ of firm i until we do not have more precise information about the parametric function of $F(c_i)$. We will do that in the next section of the exercise, where we are informed about the functional form of $F(c_i)$.

Assessing Both Effects. To assess the strategic effect, note that in the first stage:

$$\frac{\partial \pi_i}{\partial c_i} = \frac{\partial \pi_i}{\partial c_i} + \frac{\partial \pi_i}{\partial c_j}\frac{\partial q_j}{\partial c_i},$$

where $\pi_i = (a - bq_i - bq_j - c_i)q_i - F(c_i)$, implying that the first term (direct effect) is hence:

$$\frac{\partial \pi_i}{\partial c_i} = -q_i - F'(c_i)$$

While the second term (strategic effect) is given by

$$\frac{\partial \pi_i}{\partial q_j} \frac{\partial q_j}{\partial c_i} = (-bq_i)\frac{1}{3b} < 0$$

Part (b) We now solve the above exercise assuming that the functional relationship between fixed and variable costs is $F(c_i) = \alpha - \beta c_i + 0.5c_i^2$. In this context, $F'(c_i) = -\beta + c_i$, and $F''(c_i) = 1$. Substituting them into the first order condition of the first stage, yields:

$$\frac{\partial \pi_i}{\partial c_i} = -\frac{4(a - 2c_i + c_j)}{9b} + \beta - c_i$$

While the second order condition becomes

$$\frac{\partial^2 \pi_i}{\partial c_i^2} = \frac{8}{9b} - 1 \leq 0.$$

Therefore, in order for the second-order condition to hold, we need $b \geq 8/9$. At the symmetric equilibrium $c_i = c_j = c$, we thus have

$$-\frac{4(a - c)}{9b} + \beta - c = 0.$$

Solving for c, we obtain $c^* = \frac{4a - 9b\beta}{4 - 9b}$. Hence, we need $-F'(0) > \frac{4a}{9b}$, i.e., $\beta > \frac{4a}{9b}$ to guarantee an interior solution $c^* > 0$.

Exercise 9—Entry Deterring Investment[B]

Consider an industry with two firms, each firm $i = \{1,2\}$ with demand function $q_i = 1 - 2p_i + p_j$. Firm 2's (the entrant's) marginal cost is 0. Firm 1's (the incumbent's) marginal cost is initially ½. By investing $I = 0.205$, the incumbent can buy a new technology and reduces its marginal cost to 0 as well.

Part (a) Consider the following time structure: The incumbent chooses whether to invest, then the entrant observes the incumbent's investment decision; and subsequently the firms compete in prices. Show that in the subgame perfect equilibrium the incumbent does not invest.

Part (b) Show that if the investment decision is *not* observed by the entrant, the incumbent's investing becomes part of the SPNE. Comment.

Answer

Part (a) *Second period*. Recall that in the first period the incumbent decides to invest/not invest, while in the second period firms compete in a Bertrand duopoly. Operating by backward induction, let us next separately analyze firms' pricing decisions (during the second stage) in each of the two possible investment profiles that emerge from the first period (either the incumbent invests, or it does not).
The Incumbent Invests. In this setting, both firms' costs are zero, i.e., $c_1 = c_2 = 0$. Then, each firm $i \neq j$ chooses the price p_i that maximizes

$$\max_{p_i} \; (1 - 2p_i + p_j)p_i.$$

Taking first order conditions with respect to p_i, we have $1 - 4p_i + p_j = 0$. Solving for p_i, we obtain the firm i's best response function $p_i(p_j) = \frac{1+p_j}{4}$. (Note that second order conditions also hold, since they are $-4 < 0$, guarantee that such a price level maximizes firm i's profit function.)

By symmetry, firm j's best response function is $p_j(p_i) = \frac{1+p_i}{4}$. Plugging $p_j(p_i)$ into $p_i(p_j)$, we obtain $p_i = \frac{1 + \frac{1+p_i}{4}}{4} = \frac{5+p_i}{16}$ and thus equilibrium prices are $p_i = \frac{1}{3}$ and $p_j = \frac{1}{3}$. These prices yield equilibrium profits of:

$$\frac{1}{3}\left(1 - \frac{2}{3} + \frac{1}{3}\right) = \frac{2}{9} \text{ for every firm } i$$

Nonetheless, the profits that the incumbent earns after including the investment in the cost-reducing technology are

$$\pi_i = \frac{2}{9} - 0.205 = 0.017.$$

The Incumbent does not Invest. In this case, the incumbent's costs are higher than the entrant's, i.e., $c_1 = \frac{1}{2}$ and $c_2 = 0$. First, the incumbent's profit maximization problem is

$$\max_{p_1} \; (1 - 2p_1 + p_2)(p_1 - 0.5) = p_1 - 2p_1^2 + p_2p_1 - 0.5 + p_1 - 0.5p_2$$

Taking the first order conditions with respect to p_1, we find

$$1 - 4p_1 + p_2 + 1 = 0.$$

And solving for p_1, we obtain firm 1's (the incumbent's) best response function

$$p_1(p_2) = \frac{2+p_2}{4} \qquad \text{(A)}$$

Second, the entrant's profit-maximization problem is

$$\max_{p_2} \ (1 - 2p_2 + p_1)p_2.$$

Taking first order conditions with respect to p_2,

$$1 - 4p_2 + p_1 = 0.$$

And solving for p_2, we find firm 2's (the entrant's) best response function

$$p_2(p_1) = \frac{1+p_1}{4} \qquad \text{(B)}$$

Solving simultaneously for (p_1, p_2) in expressions (A) and (B), we obtain the equilibrium price for the incumbent

$$p_1 = \frac{2 + \frac{1+p_1}{4}}{4} = \frac{9+p_1}{16} \rightarrow p_1 = \frac{3}{5}.$$

And the equilibrium price for the entrant

$$p_2 = \frac{1 + \frac{3}{5}}{4} = \frac{2}{5}.$$

Entailing associated profits of

$$\pi_1 = \left(1 - 2 \cdot \frac{3}{5} + \frac{2}{5}\right)\left(\frac{3}{5} - \frac{1}{2}\right) = 0.02 \text{ for the incumbent}$$

$$\pi_2 = \left(1 - 2 \cdot \frac{2}{5} + \frac{3}{5}\right)\frac{2}{5} = 0.32 \text{ for the entrant.}$$

First stage. After analyzing the second stage of the game, we next examine the first stage, whereby the incumbent must decide whether or not to invest in cost-reducing technologies. In particular, the incumbent decides not to invest in cost-reducing technologies since its profits when it does not invest, 0.02, are larger than when it does, 0.017. Intuitively, the incumbent prefers to compete with a more efficient entrant than having to incur a large investment cost in order to achieve the same cost efficiency as the entrant, i.e., the investment is too costly.

Part (b) Now the entrant only chooses a single price p_2, i.e., a price that is not conditional on whether the incumbent invested or not. We know from part (a) that the entrant's best response is to set a price:

$$p_2 = \begin{cases} \frac{1}{3} & \text{If the incumbent chooses Invest} \\ \frac{2}{5} & \text{If the incumbent chooses Not Invest} \end{cases}$$

(1) If $p_2 = \frac{1}{3}$, then the incumbent profits depends on whether it invests or not. After *investment*, the incumbent maximizes:

$$\max_{p_1} \left(1 - 2p_1 + \frac{1}{3}\right)p_1$$

Taking first order conditions with respect to p_1, we find $1 - 4p_1 + \frac{1}{3} = 0$, and solving for p_1 we obtain $p_1 = \frac{1}{3}$. Hence the incumbent's profits in this case are

$$\pi_1 = \left(1 - \frac{2}{3} + \frac{1}{3}\right)\frac{1}{3} - 0.205 = 0.017.$$

After *no investment*, the incumbent maximizes

$$\max_{p_1} \left(1 - 2p_1 + \frac{1}{3}\right)\left(p_1 - \frac{1}{2}\right)$$

Taking first order conditions with respect to p_1, we find $1 - 4p_1 + \frac{1}{3} + 1 = 0$, and solving for p_1 we obtain $p_1 = \frac{7}{12}$. Therefore the incumbent's profits are

$$\pi_1 = \left(1 - \frac{14}{12} + \frac{1}{3}\right)\left(\frac{7}{12} - \frac{1}{2}\right) = 0.014.$$

As a consequence, $\pi_1^{Invest} > \pi_1^{NotInvest}$ if $p_2 = \frac{1}{3}$

(2) If $p_2 = \frac{2}{5}$, then the Incumbent profits are different from above. In particular upon *investment*, the incumbent maximizes

$$\max_{p_1} \left(1 - 2p_1 + \frac{2}{5}\right)p_1$$

Taking first order conditions with respect to p_1, we find $1 - 4p_1 + \frac{2}{5} = 0$, and solving for p_1 we obtain $p_1 = \frac{7}{20}$. Thus, the incumbent's profits are

$$\pi_1 = \left(1 - \frac{14}{20} + \frac{2}{5}\right)\frac{7}{20} - 0.205 = 0.04.$$

Upon *no investment*, the incumbent maximizes

$$\max_{p_1} \left(1 + 2p_1 + \frac{2}{5}\right)\left(p_1 - \frac{1}{2}\right).$$

Taking first order conditions with respect to p_1, we find $1 - 4p_1 + \frac{2}{5} + 1 = 0$, and solving for p_1 we obtain $p_1 = \frac{3}{5}$, thus yielding profits of

$$\pi_1 = \left(1 - \frac{6}{5} + \frac{2}{5}\right)\left(\frac{3}{5} - \frac{1}{2}\right) = 0.02.$$

Therefore, $\pi_1^{Invest} > \pi_1^{NotInvest}$ if $p_2 = \frac{2}{5}$.

Hence, incumbent invests in pure strategies, both when $p_2 = \frac{1}{3}$ and when $p_2 = \frac{2}{5}$, and the unique SPNE of the entry game is (Invest, $p_1 = p_2 = \frac{1}{3}$).

Exercise 10—Direct Sales or Using A Retailer?[C]

Assume two firms competing in prices where the demand that firm i faces is given by

$$q_i = 1 - 2p_i + p_j \text{ for every firm } i, j = 1, 2 \text{ and } j \neq i$$

There are no costs of production. Each firm has the option between selling the product directly to the consumers, or contracting with a retailer that will sell the product of the firm. If a firm contracts with the retailer, the contract is a franchise fee contract specifying: (a) the wholesale price (p_w^i) of each unit of the product that is transferred between the firm and the retailer; and (b) the fixed franchise fee (F_i) for the right to sell the product.

The game proceeds in the following two stages: At the first stage, each firm decides whether to contract with a retailer or to sell the product directly to the consumers. The firm that sells through a retailer, signs the contract at this stage. At the second stage, the firm or its retailer sets the retail price (p_i) for the product.

Solve for the equilibrium of this two stage game. How does it compare with the regular Bertrand equilibrium? Explain.

Answer

Let us separately find which are the firm's profits in each of the possible selling profiles that can arise in the second stage of the game: (1) both firms sell directly their good to consumers; (2) both sell their product through retailers; and (3) only one firm i sells the product through a retailer. We will afterwards compare firms' profits in each case.

Both Firms Sell Directly to Consumers: Then every firm i selects a price p_i that solves

$$\max_{p_i} \ (1 - 2p_i + p_j)p_i$$

Taking first order conditions with respect to p_i, $1 - 4p_i + p_j = 0$, and solving for p_i, we obtain $p_i(p_j) = \frac{1+p_j}{4}$. Similarly for firm j and p_j, where firm j's best response function is $p_j(p_i) = \frac{1+p_i}{4}$. Plugging $p_j(p_i)$ into $p_i(p_j)$, we obtain the equilibrium price $p_i = \frac{1}{3}$, with associated profits of

$$\pi_i = \left(1 - \frac{2}{3} + \frac{1}{3}\right)\frac{1}{3} = \frac{2}{9}$$

Both Firms Use Retailers: Let us first analyze the retailer's profit-maximization problem, where it chooses a price p_i for the product, but has to pay a franchise fee F_i to firm i in order to become a franchisee, and a wholesale price p_w^i to firm i for every unit sold.

$$\max_{p_i} \ (1 - 2p_i + p_j)(p_i - p_w^i) - F_i$$

Intuitively, $p_i - p_w^i$ denotes the margin that the retailer makes per unit when it sells the product of firm i. Taking first order conditions with respect to p_i, $1 - 4p_i + p_j + 2p_w^i = 0$, and solving for p_i, we obtain the retailer's best response function, $p_i(p_w^i, p_j) = \frac{1+p_j+2p_w^i}{4}$.

By symmetry, the franchisee of firm j has a similar best-response function, i.e., $p_j(p_w^j, p_i)$. Hence, plugging $p_j(p_w^j, p_i)$ into $p_i(p_w^i, p_j)$, we find $p_i(p_w^i, p_j) = \frac{1 + 2p_w^i + \left(\frac{1+p_i+2p_w^j}{4}\right)}{4}$ and solving for p_i, we obtain the optimal price that the franchisee of firm i (retailer) charges for its products to customers

$$p_i = \frac{5 + 8p_w^i + 2p_w^j}{15} \quad \text{for every firm } i = \{1, 2\} \text{ and } j \neq i$$

Let us now study the firm's decision of what wholesale price to charge to the franchisee, p_w^i, and what franchisee fee to charge, F_i, in order to maximize its profits. In particular, firm i solves:

$$\max_{p_w^i, F_i} p_w^i q_i + F_i$$

And given that $F_i = (p_i - p_w^i)q_i$ (no profits for the retailer), then firm i's profit-maximization problem can be simplified to

$$\max_{p_w^i} p_w^i q_i + p_i q_i - p_w^i q_i$$

(which contains a single choice variable, p_w^i). Taking first-order conditions with respect to p_w^i,

$$\frac{\partial \pi_i}{\partial p_i} \frac{\partial p_i}{\partial p_w^i} + \frac{\partial \pi_i}{\partial p_j} \frac{\partial p_j}{\partial p_w^j} = 0$$

In addition, since firm i acts before the franchisee, it can anticipate the latter's behavior, who charges a price of $p_i = \frac{5 + 8p_w^i + 2p_w^j}{15}$. This simplifies our above first order condition to $(1 - 4p_i + p_j)\frac{8}{15} + p_i \frac{2}{15} = 0$, or $8 - 30p_i + 8p_j = 0$. By symmetry $8 - 30p_j + 8p_i = 0$ for firm j, which allows us to simultaneously solve both equations for p_i, obtaining $p_i = \frac{4}{11}$ for all $i = \{1,2\}$.

As a consequence, sales of every good at these prices are

$$q_i = 1 - 2 \cdot \frac{4}{11} + \frac{4}{11} = \frac{7}{11}.$$

And the associated profits that firm i obtains from selling goods through a franchisee are

$$\pi_i = \frac{4}{11} \cdot \frac{7}{11} = \frac{28}{121} \simeq 0.231.$$

Only one Firm Uses Retailers: If firm j sells directly to customers, it sets a price p_j to maximize

$$\max_{p_j} p_j(1 - 2p_j + p_i)$$

Taking first order conditions with respect to p_j, we obtain $1 - 4p_j + p_i = 0$, which yields firm j's best response function $p_j(p_i) = \frac{1 + p_i}{4}$.

If instead firm i uses a retailer, then the retailer's profit-maximization problem is

$$\max_{p_i} \left(1 - 2p_i + p_j\right)\left(p_i - p_w^i\right) - F_i$$

Taking first order conditions with respect to p_i (the only choice variable for the retailer), we obtain $1 - 4p_i + p_j + 2p_w^i = 0$, which yields $p_i(p_j, p_w^i) = \frac{1 + p_j + 2p_w^i}{4}$.

Combining firm j's best response function and the retailer's best response function, we find equilibrium prices of

$$p_j = \frac{1 + \frac{1 + p_j + 2p_w^i}{4}}{4} \rightarrow p_j = \frac{5 + 2p_w^i}{15}, \text{ and}$$

$$p_i = \frac{1 + \frac{5 + 2p_w^i}{15} + 2p_w^i}{4} \rightarrow p_i = \frac{5 + 8p_w^i}{15}.$$

We can now move to firm i's profit-maximization problem (recall that firm i uses a franchisee),

$$\max_{p_w^i, F_i} p_w^i q_i + F_i$$

And since $F_i = (p_i - p_w^i)q_i$, the profit maximization problem becomes

$$\max_{p_w^i} p_i q_i + p_i q_i - p_w^i q_i$$

Taking first order conditions with respect to the wholesale price that firm i charges to the franchisee, p_w^i, we obtain

$$\frac{\partial \pi_i}{\partial p_i}\frac{\partial p_i}{\partial p_w^i} + \frac{\partial \pi_i}{\partial p_j}\frac{\partial p_j}{\partial p_w^i} = 0.$$

That is,

$$\left(1 - 4p_i + p_j\right)\frac{8}{15} + p_i\frac{2}{15} = 0$$

And solving for p_i yields

$$p_i = \frac{4}{15} + \frac{4}{15}p_j \text{ for all } i = \{1, 2\}.$$

Since $p_i = \frac{5+8p_w^i}{15}$ and $p_j = \frac{5+2p_w^i}{15}$, from our above results, the previous expression becomes

$$\frac{5+8p_w^i}{15} = \frac{4+4\cdot\frac{5+2p_w^i}{15}}{15}$$

And solving for the wholesale price, p_w^i, we obtain $p_w^i = \frac{5}{112}$. Therefore, the selling prices are

$$p_i = \frac{5+8\cdot\frac{5}{112}}{15} = \frac{5}{14} \text{ and } p_j = \frac{5+2\cdot\frac{5}{112}}{15} = \frac{19}{56},$$

while the amounts sold by each firm are

$$q_i = \left(1 - 2\cdot\frac{5}{14} + \frac{19}{56}\right) = \frac{5}{8} \text{ and } q_j = \left(1 - 2\cdot\frac{19}{56} + \frac{5}{14}\right) = \frac{19}{28},$$

and their associated profits are $\pi_i = \frac{5}{8}\cdot\frac{5}{14} = \frac{25}{112} \simeq 0.22$ and $\pi_j = \frac{19}{56}\cdot\frac{19}{28} = \frac{361}{1568} \simeq 0.23$.

Let us summarize the profits firms earn in each of the previous three cases with the following normal-form game.

		Firm j	
		Use Retailer	Sell Directly
Firm i	Use Retailer	28/121, 28/121	25/121, 361/1568
	Sell Directly	361/1568, 25/121	2/9, 2/9

If firm j uses a retailer (in the left column), firm i's best response is to also use a retailer; while if firm j sells directly (in the right-hand column), firm i's best response is to use a retailer. Hence, using a retailer is a strictly dominant strategy for firm i, i.e., it prefers to use a retailer regardless of firm j's decision. Since firm j's payoffs are symmetric, the unique Nash equilibrium of the game has both firms using a retailer.

Interestingly, in this equilibrium both firms obtain higher profits franchising their sales (using a retailer), i.e., $\pi_i = \pi_j = \frac{28}{121} \simeq 0.231$, than when they do not use a retailer (in which case their profits are $\pi_i = \pi_j = \frac{2}{9} = 0.22$).

Exercise 11—Profitable and Unprofitable Mergers^A

Consider an industry with n identical firms competing à la Cournot. Suppose that the inverse demand function is $P(Q) = a - bQ$, where Q is total industry output, and $a, b > 0$. Each firm has a marginal costs, c, where $c < a$, and no fixed costs.

Part (a) *No merger*. Find the equilibrium output that each firm produces at the symmetric Cournot equilibrium. What is the aggregate output and the equilibrium price? What are the profits that every firm obtains in the Cournot equilibrium? What is the equilibrium social welfare?

Part (b) *Merger*. Now let m out of n firms merge. Show that the merger is profitable if and only if it involves a sufficiently large number of firms.

Part (c) Are the profits of the nonmerged firms larger when m of their competitors merge than when they do not?

Part (d) Show that the merger reduces consumer welfare.

Answer

Part (a) At the symmetric equilibrium with n firms, we have that each firm i solves

$$\max_{q_i}(a - bQ_{-i} - bq_i)q_i - cq_i$$

where $Q_{-i} \equiv \sum_{j \neq i} q_j$ denotes the aggregate production of all other $j \neq i$ firms. Taking first-order conditions with respect to q_i, we obtain

$$a - bQ_{-i} - 2bq_i^* - c = 0.$$

And at the symmetric equilibrium $Q_{-i} = (n-1)q_i^*$. Hence, the above first order conditions become $a - b(n-1)q_i^* - 2bq_i^* - c = 0$ or $a - c = b(n+1)q_i^*$. Solving for q_i^*, we find that the individual output level in equilibrium is

$$q_i^* = \frac{a-c}{(n+1)b}.$$

Hence, aggregate output in equilibrium is $Q^* = nq_i^* = \frac{n}{(n+1)}\frac{a-c}{b}$; while the equilibrium price is

$$p^* = a - bQ^* = a - b\frac{n}{(n+1)}\frac{a-c}{b} = \frac{a-cn}{n+1}$$

And equilibrium profits that every firm i obtains are

$$\pi_i^* = (p^* - c)q_i^* = \frac{(a-c)^2}{(n+1)^2 b}.$$

Fig. 5.14 Profitable mergers
satisfy $n < m + \sqrt{m} - 1$

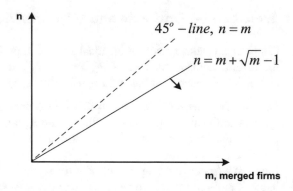

The level of social welfare with n firms, W_n, is defined by $W_n = \int_0^Q [a - bQ]dQ - cQ$.

Calculating the integral, we obtain $aQ - \left(\frac{b}{2}\right)(Q^2) - cQ$. Substituting for $Q = \frac{n}{(n+1)}\frac{a-c}{b}$ yields

$$W_n = \frac{n(n+2)(a-c)^2}{2n(n+1)^2}.$$

Part (b) Assume that m out of n firms merge. While before the merger there are n firms in this industry, after the merger there are $n - m + 1$. [For instance, if $n = 10$ and $m = 7$, the industry now has $n - m + 1 = 4$ firms, i.e., 3 nonmerged firms and the merged firm.] In order to examine whether the merger is profitable for the m merged firms, we need to show that the profit after the merger, π_{n-m+1}, satisfies $\pi_{n-m+1} \geq m\pi_n$,

$$\frac{(a-c)^2}{(n-m+2)^2} \geq m\frac{(a-c)^2}{(n+1)^2}$$

Solving for n, we obtain that $n < m + \sqrt{m} - 1$, as depicted in the (n,m)-pairs below the line $m + \sqrt{m} - 1$ in Fig. 5.14.

And note that this condition is compatible with the fact that the merger must involve a subset of all firms, i.e., $m \leq n$; as depicted in the points below the 45-degree line. For instance, in an industry with $n = 100$ firms, the condition we found determines that this merger is only profitable if at least 89 firms merge. This result is due to Salant, Switzer and Reynolds (1983) and, since the function $m + \sqrt{m} - 1$ can be approximated by a straight line with a slope of 0.8, it is usually referred as the "80 % rule," intuitively indicating that the market share of the merged firms must be at least 80 % for their merger to be profitable.

Part (c) The output produced by the merged firms decreases (relative to their output before the merger), implying that each of the *nonmerged firms* earns larger profits after the merger because of $\pi_{n-m+1} \geq \pi_n$. This condition holds for mergers of any size, i.e., both when condition $n < m + \sqrt{m} - 1$ holds and otherwise. This surprising result is often referred to as the "merger paradox."

Part (d) Since the equilibrium price p^* is decreasing in n, and the merger reduces the number of firms in the industry, it increases the equilibrium price. Similarly, since W_n is increasing in n, equilibrium social welfare is reduced as a consequence of the merger.

Repeated Games and Correlated Equilibria

6

Introduction

In this chapter we explore agents' incentives to cooperate when they interact in infinite repetitions of a stage game, such as the Prisoner's Dilemma game or the Cournot oligopoly game. Repeated interactions between the same group of individuals, or repeated competition between the same group of firms in a given industry, are fairly common. As a consequence, this chapter analysis helps at characterizing strategic behavior in several real-life settings in which agents expect to interact with one another for several time periods. In these settings, we evaluate:

1. the discounted stream of payoffs that agents can obtain if they recurrently cooperate;
2. how to find a player's optimal deviation, and the associated payoff that he would obtain from deviating; and
3. under which conditions the payoff in (1) is larger than the payoff in (2) thus motivating players to cooperate.

We accompany our discussion with several figures in order to highlight the trade-off between the current benefits that players obtain from deviating versus the future losses they would sustain if they deviate today (i.e., the cooperative profits that a firm gives up if it breaks the cooperative agreement today). To emphasize these intuitions, we explore the case in which punishments from cooperation last two, three, or infinite periods. We then apply this framework to study firms' incentives to cooperate in a cartel: first, when the industry consists of only two firms; and second, extend it to settings with more than two firms competing in either quantities or prices, with perfect or imperfect monitoring technologies (i.e., when

The original version of the chapter was revised: The erratum to the chapter is available at: 10.1007/978-3-319-32963-5_11

F. Munoz-Garcia and D. Toro-Gonzalez, *Strategy and Game Theory*,
Springer Texts in Business and Economics, DOI 10.1007/978-3-319-32963-5_6

competing firms can immediately detect a defection from other cartel participants or, instead, need a few periods before noticing such defection). For completeness, we also investigate equilibrium behavior in repeated games where players can choose among three possible strategies in the stage game, which allows for richer deviation profiles. We then apply our analysis to bargaining games among three players, and study how equilibrium offers are affected by players' time preferences.

We end the chapter with exercises explaining how to graphically represent the set of feasible payoffs in infinitely repeated games, and how to further restrict such area to payoffs which are also individually rational for all players. This graphical representation of feasible and individually rational payoffs helps us present the so-called "Folk theorem", as we next describe.

Folk Theorem. For every Nash equilibrium in the unrepeated version of the game with expected payoff vector $\bar{v} = (\bar{v}_1, \bar{v}_2, \ldots, \bar{v}_N)$, where $\bar{v}_i \in \mathbb{R}$, there is a sufficiently high discount factor $\delta \geq \bar{\delta}$, where $\delta \in [0, 1]$ for which any feasible and individually rational payoff vector $v = (v_1, v_2, \ldots, v_N)$ can be sustained as the SPNE of the infinitely repeated version of the game.

Intuitively, if players care enough about their future payoffs (high discount factor), they can identify an equilibrium strategy profile in the infinitely repeated game that provides them with higher average per-period payoff than the payoff they obtain in the Nash equilibrium of the unrepeated version of the game.

Exercise 1—Infinitely Repeated Prisoner's Dilemma Game[A]

Consider the infinitely repeated version of the following prisoner's dilemma game, where C denotes confess and NC represents not confess (Table 6.1):

Part (a) Can players support the cooperative outcome (NC, NC) as a SPNE by using tit-for-tat strategies, whereby players punish deviations from NC in a past period by reverting to the stage-game Nash equilibrium where they both confess (C, C) for just one period, and then returning to cooperation?

Part (b) Consider now that players play tit-for-tat, but reverting to the stage-game Nash equilibrium where they both confess (C, C) for two periods, and then players return to cooperation. Can the cooperative outcome (NC, NC) be supported as a SPNE?

Table 6.1 Prisoner's dilemma game

		Player 2	
		C	NC
Player 1	C	2, 2	6, 0
	NC	0, 6	3, 3

Part (c) Suppose that players use strategies that punish deviation from cooperation by reverting to the stage-game Nash equilibrium for ten periods before returning to cooperation. Compute the threshold discount factor, δ, above which cooperation can be sustained as a SPNE of the infinitely repeated game.

Answer

Part (a) At any period t, when cooperating in the (NC, NC) outcome, every player $i = \{1, 2\}$ obtains a payoff stream of:

$$3 + 3\delta + 3\delta^2 + \cdots$$

where δ is the discount factor. If, instead, player i deviates to C, he obtains a payoff of 6 today. However, in the following period, he is punished by the (C, C) outcome, which yields a payoff of 2, and then both players return to the competitive outcome (NC, NC), which yields a payoff of 3. Hence, the payoff stream that a deviating player obtains is:

$$\underbrace{6}_{gain} + \underbrace{2\delta}_{punishment} + \underbrace{3\delta^2 + \cdots}_{back\ to\ cooperation}$$

As a remark, note that once player $j \neq i$ detects that player i is defecting (i.e., playing NC), he finds that NC is a best response to player i playing NC. Hence, the punishment is credible, since player j has incentives to select NC, thus giving rise to the (NC, NC) outcome.

Hence, cooperation can be sustained if:

$$3 + 3\delta + 3\delta^2 + \cdots \geq 6 + 2\delta + 3\delta^2 + \cdots$$

and rearranging,

$$3\delta - 2\delta \geq 6 - 3$$

which yields $\delta \geq 3$, a condition that cannot hold since delta must satisfy $\delta \in [0, 1]$.

Hence, no cooperation can be sustained with temporary reversion to the Nash Equilibrium of the stage game for only one period. Intuitively, the punishment for deviating is too small to induce players to stick to cooperation over time. Figure 6.1 represents the stream of payoffs that a given player obtains by cooperating, and compares it with his stream of payoffs from cheating. Graphically, the instantaneous gain the cheating player obtains today, $6 - 3 = 3$, offsets the future loss he would suffer from being punished during only one period tomorrow, $3 - 2 = 1$.

Part (b) At any period t, when cooperating by choosing NC, players get the same payoff stream we identified in part (a):

$$3 + 3\delta + 3\delta^2 + 3\delta^3 + \cdots$$

If, instead, a player $i = \{1, 2\}$ defects to confess, C, he will now suffer a punishment during two periods, thus yielding a payoff stream of:

Fig. 6.1 Payoff stream when defection is punished for one period

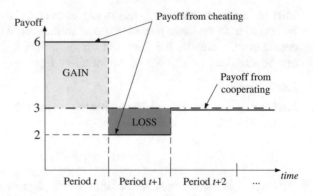

$$\underbrace{6}_{gain} + \underbrace{2\delta + 2\delta^2}_{punishment} + \underbrace{3\delta^3 + \cdots}_{back\ to\ cooperation}$$

Hence, cooperation can be sustained if:

$$3 + 3\delta + 3\delta^2 + 3\delta^3 + \cdots \geq 6 + 2\delta + 2\delta^2 + 3\delta^3 + \cdots$$

and rearranging,

$$3 + 3\delta + 3\delta^2 \geq 6 + 2\delta + 2\delta^2$$
$$\delta^2 + \delta - 3 \geq 0$$

Solving for δ we have:

$$\delta = \frac{-1 \pm \sqrt{1 + (4 \times 1 \times 3)}}{2 \times 1} = \frac{-1 \pm \sqrt{13}}{2}$$

which yields,

$$\delta_1 = -2.3 < 0 \quad \text{and} \quad \delta_2 = 1.3 > 1$$

Indeed, depicting $\delta^2 + \delta - 3 = 0$, we can see that the function lies in the negative quadrant, for all $\delta \in [0, 1]$, as Fig. 6.2 illustrates. Hence, $\delta^2 + \delta - 3 \geq 0$ cannot be satisfied for any $\delta \in [0, 1]$.

Therefore, cooperation cannot be supported in this case either, when we use a temporary punishment of only two periods. Graphically, the instantaneous gain from cheating (which increases a player's payoff from 3 to 6 in Fig. 6.3) still offsets the future loss from being punished, since the punishment only lasts for two periods.

Part (c) At any period t, when a player $i = \{1, 2\}$ chooses to cooperate, choosing NC, while his opponent cooperates, his payoff stream becomes:

Fig. 6.2 Function $\delta^2 + \delta - 3$ δ, Discount factor

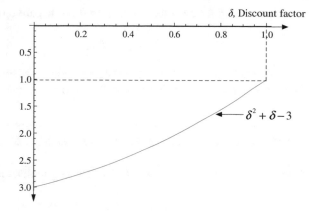

$$3 + 3\delta + 3\delta^2 + 3\delta^3 + \cdots + 3\delta^{10} + 3\delta^{11} + \cdots$$

If instead, player i deviates, he obtains a payoff of 6 today, but he is subsequently punished for 10 periods, obtaining a payoff of only 2 in each period, before returning to the cooperation, where his payoff is 3. Hence, his payoff stream is:

$$6 + 2\delta + 2\delta^2 + 2\delta^3 \cdots + 2\delta^{10} + 3\delta^{11} + \cdots$$

Hence, cooperation can be sustained if:

$$3 + 3\delta + 3\delta^2 + 3\delta^3 + \cdots + 3\delta^{10} + 3\delta^{11} + \cdots \geq 6 + 2\delta + 2\delta^2 + 2\delta^3 \cdots + 2\delta^{10} + 3\delta^{11} + \cdots$$

Rearranging:

$$(3-2)\delta + (3-2)\delta^2 + (3-2)\delta^3 + \cdots + (3-2)\delta^{10} \geq 6 - 3$$
$$\delta + \delta^2 + \delta^3 + \cdots + \delta^{10} \geq 3$$
$$\delta\left(1 + \delta + \delta^2 + \cdots + \delta^9\right) \geq 3$$

Fig. 6.3 Payoff stream when defection is punished for two periods

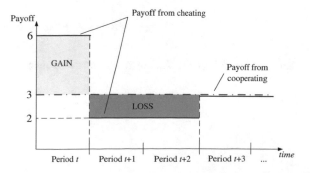

After discarding all solutions of δ that are negative or larger than 1, we get:

$$\delta \geq 0.76$$

Hence, a long punishment of ten periods allows players to sustain cooperation if they assign a sufficiently high importance to future payoffs, i.e., δ is close to 1.

Exercise 2—Collusion when firms compete in quantities[A]

Consider two firms competing as Cournot oligopolists in a market with demand:

$$p(q_1, q_2) = a - bq_1 - bq_2$$

Both firms have total costs, $TC(q_i) = cq_i$ where $c > 0$ is the marginal cost of production, and $a > c$.

Part (a) Considering that firms only interact once (playing an unrepeated Cournot game), find the equilibrium output, the market price, and the equilibrium profits for every firm.

Part (b) Now assume that they could form a cartel. Which is the output that every firm should produce in order to maximize the profits of the cartel? Find the market price, and profits of every firm. Are their profits higher when they form a cartel than when they compete as Cournot oligopolists?

Part (c) Can the cartel agreement be supported as the (cooperative) outcome of the infinitely repeated game?

Answer

Part (a) Firm 1 chooses q_1 to maximize its profits

$$\max_{q_1} \pi_1 = (a - bq_1 - bq_2)q_1 - cq_1$$

$$= aq_1 - bq_1^2 - bq_2q_1 - cq_1$$

Taking first order conditions with respect to q_1 we find:

$$\frac{\partial \pi_1}{\partial q_1} = a - bq_1 - bq_2 - c = 0$$

and solving for q_1 we find firm 1's best response function

$$q_1(q_2) = \frac{a - c}{2b} - \frac{1}{2}q_2$$

We can obtain a similar expression for firm 2's best response function, $q_2(q_1)$, since firms are symmetric

$$q_2(q_1) = \frac{a-c}{2b} - \frac{1}{2}q_1$$

Plugging $q_2(q_1)$ into $q_1(q_2)$, we can find firm 1's equilibrium output:

$$q_1 = \frac{a-c}{2b} - \frac{1}{2}\left(\frac{a-c}{2b} - \frac{1}{2}q_1\right) \Rightarrow q_1 = \frac{a-c}{4b} + \frac{1}{4}q_1$$

$$\Rightarrow q_1 = \frac{a-c}{3b}$$

and similarly, firm 2's equilibrium output is

$$q_2 = \frac{a-c}{2b} - \frac{1}{2}\left(\frac{a-c}{3b}\right) = \frac{a-c}{2b} - \frac{a-c}{6b} = \frac{a-c}{3b}$$

Therefore, the market price is

$$p(q_1, q_2) = a - bq_1 - bq_2 = a - b\left(\frac{a-c}{3b}\right) - b\left(\frac{a-c}{3b}\right)$$

$$= \frac{3a - a + c - a + c}{3} = \frac{a + 2c}{3}$$

and the equilibrium profits of every firm $i = \{1, 2\}$ are

$$\pi_1^{cournot} = \pi_2^{cournot} = p(q_1, q_2) \times q_i - c \times q_i$$

$$= \left(\frac{a+2c}{3}\right)\left(\frac{a-c}{3b}\right) - c\left(\frac{a-c}{3b}\right) = \frac{(a-c)^2}{9b}$$

Part (b) Since the cartel seeks to maximize the firms' joint profits, they simultaneously choose q_1 and q_2 that solve

$$\max_{q_1, q_2} \pi_1 + \pi_2 = (a - bq_1 - bq_2)q_1 - cq_1 + (a - bq_1 - bq_2)q_2 - cq_2$$

which simplifies to

$$\max_{q_1, q_2}(a - bq_1 - bq_2)(q_1 + q_2) - c(q_1 + q_2)$$

Notice, however, that this profits maximization problem can be further simplified to

$$\max_Q(a - bQ)Q - cQ$$

since $Q = q_1 + q_2$. This maximization problem coincides with that of a regular monopolist. In other words, the overall production of the cartel of two firms which are symmetric in costs coincides with that of a standard monopoly. Indeed, taking first order conditions with respect to Q, we obtain

$$a - 2bQ - c \leq 0$$

Solving for Q, we find $Q = \frac{a-c}{2b}$, which is an interior solution given that $a > c$ by definition. Therefore, each firm's output in the cartel is

$$q_1 = q_2 = \frac{\frac{a-c}{2b}}{2} = \frac{a-c}{4b}$$

and the market price is:

$$p = a - bQ = a - b\left(\frac{a-c}{2b}\right) = \frac{a+c}{2}$$

Therefore, each firm's profit in the cartel is:

$$\pi_1 = pq_1 - TC(q_1) = \left(\frac{a+c}{2}\right)\left(\frac{a-c}{4b}\right) - c\left(\frac{a-c}{4b}\right) = \frac{(a-c)^2}{8b}$$
$$\pi_1^{cartel} = \pi_2^{cartel} = \frac{(a-c)^2}{8b}$$

Comparing the profits that every firm earns in the cartel, $\frac{(a-c)^2}{8b}$, against those under Cournot competition, $\frac{(a-c)^2}{9b}$, we can thus conclude that both firms have incentive to collude in the cartel for all parameter values.

$$\pi_1^{cartel} = \pi_2^{cartel} > \pi_1^{cournot} = \pi_2^{cournot}$$

Part (c) *Cooperation*: If at any period t, a given firm i cooperates producing the cartel output, while all other firms produce the cartel output as well, its profit is $\frac{(a-c)^2}{8b}$. As a consequence, the discounted sum of the infinite stream of profits from cooperating in the cartel is

$$\frac{(a-c)^2}{8b} + \delta\frac{(a-c)^2}{8b} + \delta^2\frac{(a-c)^2}{8b} + \cdots = \frac{1}{1-\delta}\frac{(a-c)^2}{8b}.$$

(Recall that $1 + \delta + \delta^2 + \delta^3 + \cdots$ can be simplified to $\frac{1}{1-\delta}$.)

Deviation: Since firms are symmetric, we only consider one of the firms (firm 1). In particular, we need to find the optimal deviation that, conditional on firm 2 choosing the cartel output, maximizes firm 1's profits. That is, which is the output that maximizes firm 1's profits from deviating? Since firm 2 sticks to cooperation (i.e., produces the cartel output $q_2 = \frac{a-c}{4b}$), if firm 1 seeks to maximize its current profits, we only need to plug $q_2 = \frac{a-c}{4b}$ into firm 1's best response function, $q_1(q_2) = \frac{a-c}{2b} - \frac{1}{2}q_2$, as follows

$$q_1^{dev} = q_1\left(\frac{a-c}{4b}\right) = \frac{a-c}{2b} - \frac{1}{2}\frac{a-c}{4b} = \frac{3(a-c)}{8b}$$

This result provides us with firm 1's optimal deviation, given that firm 2 is still respecting the cartel agreement. In this context, firm 1's profit is

$$\pi_1 = \left[a - b\left(\frac{3(a-c)}{8b}\right) - b\left(\frac{a-c}{4b}\right) - c\right]\left(\frac{3(a-c)}{8b}\right)$$

$$= 3\left(\frac{a-c}{8}\right)\left(\frac{3(a-c)}{8b}\right) = \frac{9(a-c)^2}{64b}$$

while that of firm 2 is

$$\pi_2 = \left[a - b\left(\frac{3(a-c)}{8b}\right) - b\left(\frac{a-c}{4b}\right) - c\right]\left(\frac{a-c}{4b}\right)$$

$$= 3\left(\frac{a-c}{8}\right)\left(\frac{a-c}{4b}\right) = \frac{3(a-c)^2}{32b}$$

Hence, firm 1 has incentives to unilaterally deviate since its *current* profits are larger by deviating than by cooperating. That is,

$$\pi_1^{deviate} = \frac{9(a-c)^2}{64b} > \pi_1^{cartel} = \frac{(a-c)^2}{8b}$$

Incentives to cooperate: We can now compare the profits that firms obtain from cooperating in the cartel agreement against those from choosing an optimal deviation plus the profits they would obtain from being punished thereafter (discounted profits in the Cournot oligopoly). In particular, for cooperation to be sustained we need

$$\frac{1}{1-\delta}\frac{(a-c)^2}{8b} > \frac{9(a-c)^2}{64b} + \frac{\delta}{1-\delta}\frac{(a-c)^2}{9b}$$

Solving for δ in this inequality, we obtain

$$\frac{1}{8(1-\delta)} > \frac{9}{64} + \frac{\delta}{9(1-\delta)},$$

which yields

$$\delta > \frac{9}{17}$$

Hence, firms need to assign a sufficiently high importance to future payoffs, $\delta \in \left(\frac{9}{17}, 1\right)$, for the cartel agreement to be sustained.

Credible punishments: Finally, notice that firm 2 has incentives to carry out the punishment. Indeed, if it doesn't revert to the Nash equilibrium of the stage game (producing the Cournot equilibrium output), firm 2 obtains profits of $\frac{3(a-c)^2}{32b}$, since firm 1 keeps producing its optimal deviation $q_1^{dev} = \frac{3(a-c)}{8b}$ while firm 2 produces the cartel output $q_2^{Cartel} = \frac{a-c}{4b}$. If, instead, firm 2 practices the punishment, producing the Cournot output $\frac{a-c}{3b}$, its profits are $\frac{(a-c)^2}{9b}$, which exceed $\frac{3(a-c)^2}{32b}$ for all parameter values. Hence, upon observing that firm 1 deviates, firm 2 prefers to revert to the production of its Cournot output level than being the only firm that produces the cartel output.

Exercise 3—Collusion when *N* firms compete in quantities[B]

Consider n firms producing homogenous goods and choosing quantities in each period for an infinite number of periods. Demand in the industry is given by $p = 1 - Q$, Q being the sum of individual outputs. All firms in the industry are identical: they have the same constant marginal costs $c < 1$, and the same discount factor δ. Consider the following trigger strategy:

- Each firm sets the output q^m that maximizes joint profits at the beginning of the game, and continues to do so unless one or more firms deviate.
- After a deviation, each firm sets the quantity q, which is the Nash equilibrium of the unrepeated Cournot game.

Part (a) Find the condition on the discount factor that allows for collusion to be sustained in this industry.

Part (b) Indicate whether a larger number of firms in the industry facilitates or hinders the possibility of reaching a collusive outcome.

Answer

Part (a) *Cooperation*: First, we need to find the quantities that maximize joint profits $\pi = (1 - Q)Q - cQ$. This output level coincides with that under monopoly, i.e., $Q = \frac{1-c}{2}$, yielding aggregate profits of

$$\pi = \left(1 - \frac{1-c}{2}\right)\frac{1-c}{2} - c\frac{1-c}{2} = \frac{(1-c)^2}{4}$$

for the cartel. Therefore, at the symmetric equilibrium, individual quantities are $q^m = \frac{1}{n}Q = \frac{1}{n}\frac{1-c}{2}$ and individual profits under the collusive strategy are $\pi^m = \frac{(1-c)^2}{4n}$.

Deviation: As for the deviation profits, the optimal deviation by firm i, q^{dev}, is given by solving

$$q^d(q^m) = \text{argmax}_q [1 - (n-1)q^m - q - c]q.$$

Intuitively, the above expression represents that firm i selects the output level that maximizes its profits given that all other $(n-1)$ firms are still respecting the collusive agreement, thus producing q^m. In particular, the value of q that maximizes the above expression is obtained through the first order conditions:

$$[1 - (n-1)q^m - 2q - c] = 0$$

$$1 - (n-1)\left(\frac{1}{n}\frac{1-c}{2}\right) - 2q - c = 0$$

which, solving for q, yields,

$$q^d = (n+1)\frac{(1-c)}{4n}.$$

The profits that a firm obtains by deviating from the collusive output are hence

$$\pi^d = \left[1 - (n-1)q^m - q^d(q^m)\right] \times q^d(q^m) - c \times q^d(q^m)$$
$$= \left[1 - (n-1)\left(\frac{1}{n}\frac{1-c}{2}\right) - (n+1)\frac{1-c}{4n}\right](n+1)\frac{1-c}{4n} - c(n+1)\frac{1-c}{4n},$$

which simplifies to $\pi^d = \frac{(1-c)^2(n+1)^2}{16n^2}$.

Incentives to collude: Given the above profits from colluding and from deviating, every firm i chooses to collude as long as

$$\frac{1}{1-\delta}\pi^m \geq \pi^d + \frac{\delta}{1-\delta}\pi^c$$

which, solving for δ, yields

$$\delta \leq \frac{\pi^m - \pi^d}{\pi^c - \pi^d}$$

and multiplying both sides by -1, we obtain

$$\delta \geq \frac{\pi^d - \pi^m}{\pi^d - \pi^c}.$$

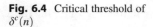

Fig. 6.4 Critical threshold of $\delta^c(n)$

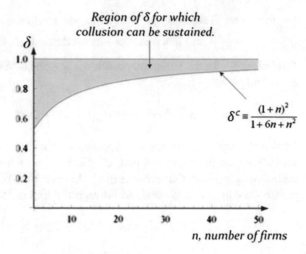

Intuitively, the numerator represents the profit gain that a firm experiences when it unilaterally deviates from the cartel agreement, $\pi^d - \pi^m$. When such profit gain increases, cartel becomes more difficult to sustain, i.e., it is only supported for higher discount factors. Plugging the expressions of profits π^d, π^m, and π^c we previously found, yields a cutoff discount factor of

$$\delta \geq \frac{(1+n)^2}{1+6n+n^2}$$

For compactness, we denote this ratio as

$$\frac{(1+n)^2}{1+6n+n^2} \equiv \delta^c$$

Hence, under punishment strategies that involve a reversion to Cournot equilibrium forever after a deviation takes place, tacit collusion arises if and only if firms are sufficiently patient. Figure 6.4 depicts cutoff δ^c, as a function of the number of firms, n, and shades the region of δ that exceeds such a cutoff.

Differentiating the critical threshold of the discount factor, δ^c, with respect to the number of firms, n, we find that

$$\frac{\partial \delta}{\partial n} = \frac{4(n^2 - 1)}{(1+6n+n^2)} > 0$$

Intuition: Other things being equal, as the number of firms in the agreement increases, the more difficult it is to reach and sustain collusion in the cartel agreement, i.e., as the market becomes less concentrated collusion becomes less likely.

Exercise 4—Collusion when *N* firms compete in prices^C

Consider a homogenous industry where n firms produce at zero cost and play the Bertrand game of price competition for an infinite number of periods. Assume that:

- When firms choose the same price, they earn a per-period profit $\pi(p) = p\alpha\frac{D(p)}{n}$, where parameter α represents the state of demand.
- When a firm i charges a price of p_i lower than the price of all of the other firms, it earns a profit $\pi(p_i) = p_i\alpha D(p_i)$, and all of the other firms obtain zero profits.

Imagine that in the current period demand is characterized by $\alpha = 1$, but starting from the following period demand will be characterized by $\alpha = \theta$ in each of the following periods. All the players know exactly the evolution of the demand state at the beginning of the game, and firms have the same common discount factor, δ.

Assume that $\theta > 1$ and consider the following trigger strategies. Each firm plays the monopoly price p_m in the first period of the game, and continues to charge such a price until a profit equal to zero is observed. When this occurs, each firm charges a price equal to zero forever. Under which conditions does this trigger strategy represent a SPNE? [*Hint*: In particular, show how θ and n affects such a condition, and give an economic intuition for this result.]

Answer

Cooperation: Let us denote the collusive price by $p^c \in (c, p_m]$. At time $t = 0$, parameter α takes the value of $\alpha = 1$, whereas at any subsequent time period $t = \{1,2,...\}$, parameter α becomes $\alpha = \theta$. Hence, by colluding, firm i obtains a discounted stream of profits of

$$\frac{\pi(p^c)}{n} + \delta\theta\frac{\pi(p^c)}{n} + \delta^2\theta\frac{\pi(p^c)}{n} + \delta^3\theta\frac{\pi(p^c)}{n} + \cdots$$
$$= \frac{\pi(p^c)}{n}(1 + \delta\theta + \delta^2\theta + \delta^3\theta + \cdots)$$

Deviation: When deviating, firm i charges a price marginally lower than the collusive price p^c and captures all the market, thus obtaining a profit of $\pi(p^c)$. However, after that deviation, all firms revert to the Nash equilibrium of the unrepeated Bertrand game, which yields a profit of zero thereafter. Therefore, the payoff stream from deviating is $\pi(p^c) + 0 + \delta 0 + \cdots = \pi(p^c)$.

Incentives to collude: Hence, every firm i colludes as long as

$$\frac{\pi(p^c)}{n}(1 + \delta\theta + \delta^2\theta + \delta^3\theta + \cdots) \geq \pi(p^c),$$

rearranging, we obtain

$$1 + \delta\theta + \delta^2\theta + \delta^2\theta + \cdots \geq n$$

Fig. 6.5 Critical threshold $\tilde{\delta}(n, \theta)$

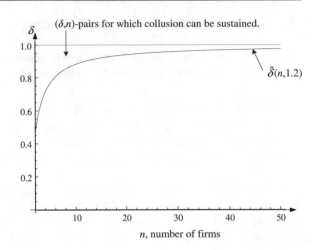

or equivalently, $\delta \geq \frac{n-1}{n-1+\theta}$. For compactness, we hereafter denote the previous ratio as $\frac{n-1}{n-1+\theta} \equiv \tilde{\delta}(n, \theta)$. Figure 6.5 depicts this critical threshold of the discount factor, evaluated at $\theta = 1.2$, i.e., demand increases 20 % after the first year.

Comparative statics: We can next examine how the critical discount factor $\tilde{\delta}(n, \theta)$ is affected by changes in demand, θ, and in the number of firms, n. In particular,

$$\frac{\partial \tilde{\delta}(n, \theta)}{\partial \theta} = \frac{-(n-1)}{(n-1+\theta)^2} < 0, \quad \text{whereas} \quad \frac{\partial \tilde{\delta}(n, \theta)}{\partial n} = \frac{\theta}{(n-1+\theta)^2} > 0.$$

In words, the higher the increments in demand, θ, the higher the present value of the stream of profits received from $t = 1$ onwards. That is, the opportunity cost of deviation increases as demand becomes stronger. Graphically, the critical discount factor $\tilde{\delta}(n, \theta)$ shifts downwards, thus expanding the region of (δ, n)-pairs for which collusion can be sustained. Figure 6.6 provides an example of this comparative statics result, whereby θ is evaluated at $\theta = 1.2$ and at $\theta = 1.8$. On the other hand, when n increases, the collusion is more difficult to sustain in equilibrium. That is, the region of (δ, n)-pairs for which collusion can be sustained shrinks as n increases; as depicted by rightward movements in Fig. 6.6.

Fig. 6.6 Cutoff $\tilde{\delta}(n, \theta)$ for $\theta = 1.2$ and $\theta = 1.8$

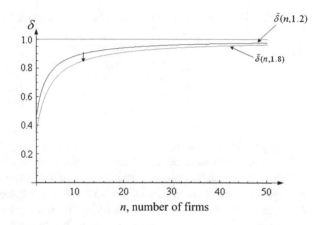

Exercise 5—Repeated games with three available strategies to each player in the stage game[A]

Consider the following simultaneous-move game between player 1 (in rows) and player 2 (in columns) (Table 6.2).

This stage game is played twice, with the outcome from the first stage observed before the second stage begins. There is no discounting i.e., the discount factor δ is $\delta = 1$. Can payoff (4, 4) be achieved in the first stage in a pure-strategy subgame perfect equilibrium? If so, describe a strategy profile that does so and prove that it is a subgame perfect Nash equilibrium. If not, prove why not.

Answer

 PSNE: For player 2, his set of best responses are the following: if player 1 plays T, then player 2's best response is $BR_2(T) = L$ if instead player 1 plays M, then $BR_2(M) = C$; and if player 1 plays B, $BR_2(B) = R$. For player 1, his best responses are the following: if player 2 plays L, then $BR_1(L) = T$; if he plays C, then $BR_1(C) = M$; and if player 2 plays R, then $BR_1(R) = B$. We can represent the payoffs in each best response with the bold underlined numbers in the next matrix (Table 6.3).

 Thus, we can identify two pure strategy Nash equilibria (PSNE): $\{(T, L), (M, C)\}$, with corresponding equilibrium payoffs pairs $u(T, L) = (3, 1)$, and $u(M, C) = (1, 2)$.

Table 6.2 Simultaneous-move game with three available actions to each player

	L	C	R
T	3, 1	0, 0	5, 0
M	2, 1	1, 2	3, 1
B	1, 2	0, 1	4, 4

Table 6.3 Underlining best response payoffs in the unrepeated game

		Player 2		
Player 1		L	C	R
	T	**3**, **1**	0, 0	**5**, 0
	M	2, 1	**1**, **2**	3, 1
	B	1, 2	0, 1	4, **4**

MSNE: Before starting our search of mixed strategy Nash equilibrium (MSNE) in this game, it would be convenient to eliminate those strategies that are strictly dominated for either player. Strictly dominated strategies do not receive any probability weight at the MSNE, implying that we can focus on the remaining (undominated) strategies, thus simplifying our calculations. From the above normal form game, there does not exist any pure strictly dominated strategies. Hence, we will analyze if there exists some mixed strategy that strictly dominates a pure strategy.

When player 1 randomizes between strategies T and M with associated probabilities $\frac{2}{3}$ and $\frac{1}{3}$ respectively, $r_1 = \{\frac{2}{3}T, \frac{1}{3}M, 0\}$, we can show that he obtains an expected utility that exceeds that from selecting the pure strategy B.[1] Indeed, for any possible strategy $s_2 \in S_2$ chosen by his opponents (player 2), we have that player 1's payoffs satisfy $u_1(r_1, s_2) > u_1(B, s_2)$. In particular, when player 2 selects L, in the left-hand column, $s_2 = L$, we find that player 1's expected payoff from selecting the mixed strategy r_1 exceeds that from selecting B,

$$u_1(r_1, L) > u_1(B, L)$$

$$\text{i.e.,}\ \frac{2}{3} \cdot 3 + \frac{1}{3} \cdot 2 > 1 \leftrightarrow 8 > 3$$

Similarly, when player 2 selects $s_2 = C$, we also obtain that player 1's payoff satisfies

$$u_1(r_1, C) > u_1(B, C)$$

$$\text{i.e.,}\ \frac{2}{3} \cdot 0 + \frac{1}{3} \cdot 1 > 0 \leftrightarrow \frac{1}{3} > 0$$

And finally, for the case in which player 2 selects $s_2 = R$,

$$u_1(r_1, R) > u_1(B, R)$$

---

[1]You can easily find many other probability weights on T and M that yield an expected utility exceeding the utility from playing strategy B, for any strategy selected by player 2 (i.e., for any column), ultimately allowing you to delete B as being strictly dominated.

Table 6.4 Surviving strategies in the unrepeated game

Player 1		Player 2		
		L	C	R
	T	3, 1	0, 0	5, 0
	M	2, 1	1, 2	3, 1

$$\text{i.e.,} \frac{2}{3} \cdot 5 + \frac{1}{3} \cdot 3 > 4 \leftrightarrow \frac{13}{3} > 4$$

Therefore, the payoffs obtained from the mixed strategy $r_1 = \{\frac{2}{3}T, \frac{1}{3}M, 0\}$ strictly dominate the payoffs obtained from the pure strategy B. Hence, we can eliminate the pure strategy B of player 1 because of being strictly dominated by r_1. In this context, the reduced-form matrix that emerges after deleting the row corresponding to B is shown in Table 6.4.

Given this reduced-form matrix, let us check if, similarly as we did for player 1, we can find a mixed strategy of player 2 that strictly dominates one of his pure strategies. In particular, if player 2 randomizes between $s_2 = L$ and $s_2 = C$ with equal probability $\frac{1}{2}$, his resulting expected utility is larger than that under pure strategy R, regardless of the specific strategy selected by player 1 (either T or M). In particular, if player 1 chooses $s_1 = T$ (in the top row), player 2's expected utility from randomizing with strategy r_2 yields an expected payoff above the payoff he would obtain from selecting R.[2]

$$u_2(r_2, T) > u_2(R, T), \text{i.e.,} \frac{1}{2} \cdot 1 + \frac{1}{2} \cdot 0 > 0 \leftrightarrow \frac{1}{2} > 0$$

And similarly, if player 1 chooses $s_1 = M$ (in the bottom row)

$$u_2(r_2, M) > u_2(R, M), \quad \text{i.e.,} \frac{1}{2} \cdot 1 + \frac{1}{2} \cdot 2 > 1 \leftrightarrow \frac{3}{2} > 1$$

Hence, we can conclude that player 2's pure strategy $s_2 = R$ is strictly dominated by the randomization $r_2 = \{\frac{1}{2}L, \frac{1}{2}C, 0\}$. Therefore, after deleting $s_2 = R$ from the payoff matrix, we end up with the following reduced-form game (Table 6.5).

At this point, however, we cannot delete any further strategies for player 1 or 2. Hence, when developing our analysis of the MSNE of the game, we will only consider $S_1 = \{T, M\}$ and $S_2 = \{L, C\}$ as the strategy sets of players 1 and 2, respectively.

---

[2]Note that we do not analyze the case of $u_2(r_2, B) > u_2(R, B)$ given that strategy $s_1 = B$ was already deleted from the strategy set of player 1.

Table 6.5 Surviving strategies in the unrepeated game (after two rounds of deletion)

Player 1		Player 2	
		L	C
	T	3, 1	0, 0
	M	2, 1	1, 2

We can now start to find the MSNE of the remaining 2×2 matrix. Player 2 wants to randomize over L and C with a probability q that makes player 1 indifferent between his pure strategies $s_1 = \{T, M\}$. That is,

$$EU_1(T) = EU_1(M)$$

$$3q + 0(1 - q) = 2q + 1 \cdot (1 - q), \quad \text{which yields} \quad q = \frac{1}{2}$$

And similarly, player 1 randomizes with a probability p that makes player 2 indifferent between her pure strategies $S_2 = \{L, C\}$. That is,

$$EU_2(L) = EU_2(C)$$

$1 \cdot p + 1 \cdot (1 - p) = 0 \cdot p + 2(1 - p)$, yielding $p = \frac{1}{2}$. Hence, the MSNE strategy profile is given by:

$$r^* = \left\{ \left(\frac{1}{2}T, \frac{1}{2}M \right), \left(\frac{1}{2}L, \frac{1}{2}C \right) \right\}$$

And the expected payoffs associated to this MSNE are:

$$EU_1 = \frac{1}{2} \left(3 \cdot \frac{1}{2} + 0 \cdot \frac{1}{2} \right) + \frac{1}{2} \left(2 \cdot \frac{1}{2} + 1 \cdot \frac{1}{2} \right) = \frac{3}{4} + \frac{3}{4} = \frac{6}{4} = \frac{3}{2}$$

$$EU_2 = \frac{1}{2} \left(1 \cdot \frac{1}{2} + 1 \cdot \frac{1}{2} \right) + \frac{1}{2} \left(0 \cdot \frac{1}{2} + 2 \cdot \frac{1}{2} \right) = \frac{1}{2} + \frac{1}{2} = 1$$

Hence, the expected utility pair in this MSNE is $\left(\frac{3}{2}, 1 \right)$.

Summarizing, if the game is played only once, we have three Nash-equilibria (two PSNEs and one MSNE), with the following associated payoffs:

$$\text{PSNEs} : u(T, L) = (3, 1) \quad \text{and} \quad u(M, C) = (1, 2)$$

$$\text{MSNE} : EU(r_1, r_2) = \left(\frac{3}{2}, 1 \right)$$

Repeated game: But the efficient payoff (4, 4) is not attainable in these equilibria of the unrepeated game.[3] However, in the two-stage game the following strategy profile (punishment scheme) constitutes a SPNE:

- Play (B, R) in the first stage.
- If the first-stage outcome is (B, R), then play (T, L) in the second stage.
- If the first-stage outcome is not (B, R), then play the MSNE r^* in the second stage.

Proof Let's analyze if the former punishment scheme played by every agent induces both players to not deviate from outcome (B,R), with associated payoff (4, 4), in the first stage of the game. To clarify our discussion, we separately examine the incentives to deviate by players 1 and 2,

Player 1. Let's take as given that player 2 adopts the previous punishment scheme:

- If player 2 sticks to (B, R), then player 1 obtains $u_1 = 4 + 3 = 7$, where 4 reflects his payoffs in the first stage of the game, when both cooperate in (B, R); and 3 represents his payoffs in the second stage, where outcome (T, L) arises according to the above punishment scheme.
- If, instead, player 1 deviates from (B, R), then his optimal deviation is to play (T, R) which yields a utility level of 5 in the first period, but a payoff of $\frac{3}{2}$ in the second period (the punishment implies the play of the MSNE in the second period). Therefore, player 1 does not have incentives to deviate since his payoff from selecting the cooperative outcome (B, R), 4 + 3=7, exceed his payoff from deviating $5 + \frac{3}{2} = \frac{13}{2}$.

Player 2. Taking as given that player 1 sticks to the punishment scheme:

- If player 2 plays (B, R) in the first stage, then she obtains an overall payoff of $u_2(B, R) = 4 + 1 = 5$. As we can see, player 2 doesn't have incentives to deviate, because his best response function is in fact $BR_2(B) = R$ when we take as given that player 1 is playing B.

Therefore, no player has incentives to deviate from (B, R) in the first stage of the game. As a consequence, the efficient payoff (4, 4) can be sustained in the first stage of the game in a pure strategy SPNE strategy profile.

---

[3]Note that payoff pair (4, 4) is efficient since a movement to another payoff pair, while it benefit one player, reduces the utility level of the other player.

Exercise 6—Infinite-Horizon Bargaining Game Between Three Players[C]

Consider a three-player variant of Rubinstein's infinite-horizon bargaining game, where players divide a surplus (pie) of size 1, and the set of possible agreements is

$$X = \{(x_1, x_2, x_3) | x_i \geq 0 \quad \text{for} \quad i = 1, 2, 3 \quad \text{and} \quad x_1 + x_2 + x_3 = 1\}.$$

Players take turns to offer agreements, starting with player 1 in round $t = 0$, and the game ends when *both* responders accept an offer. If agreement x is reached in round t, player i's utility is given by $\delta^t x_i$, where $\frac{1}{2} < \delta < 1$. Show that, if each player is restricted to using stationary strategies (in which she makes the same proposal every time she is the proposer, and uses the same acceptance rule whenever she is a responder), there is a unique subgame perfect equilibrium outcome. How much does each player receive in equilibrium?

Answer
First, note that every proposal is voted using the unanimity rule. For instance, if player 1 offers x_2 to player 2 and x_3 to player 3, then players 2 and 3 independently and simultaneously decide if they accept or reject player 1's proposal. If they both accept, players get $x = (x_1, x_2, x_3)$, while if either player rejects, player 1's offer is rejected (because of unanimity rule) and player 2 becomes the proposer in the following period, making an offer to players 1 and 3, that they choose whether to accept or reject.

Consider any player i, and let x^{prop} denote the offer the proposer at this time period makes to himself, x^{next} the offer that he makes to the player who will become the proposer in the next period, and x^{two} the offer he makes to the player who will become the proposer two periods from now. In addition, we know that the sum of all offers must be equal to the size of the pie, $x^{\text{prop}} + x^{\text{next}} + x^{\text{two}} = 1$.

We know that the offer the proposer makes must satisfy two main conditions:

1. First, his offer to the player who will become the proposer in the next period, x^{next}, must be weakly higher than the discounted value of the offer that such player would make to himself during the next period (when he becomes the proposer), δx^{prop}. That is, $x^{\text{next}} \geq \delta x^{\text{prop}}$. Moreover, since the proposer seeks to minimize the offers he makes, he wants to reduce x^{next} as much as possible, but still guarantee that his offer is accepted, i.e., $x^{\text{next}} = \delta x^{\text{prop}}$.

2. Second, the offer he makes to the player who will be making proposals two periods from now, x^{two}, must be weakly higher than the discounted value of the offer that such a player will make to himself two periods from now as a proposer, $\delta^2 x^{\text{prop}}$. That is, $x^{\text{two}} \geq \delta^2 x^{\text{prop}}$, and since the proposer seeks to minimize the offer x^{two}, he will choose a proposal that satisfies $x^{\text{two}} = \delta^2 x^{\text{prop}}$.

Using the above three equations, $x^{prop} + x^{next} + x^{two} = 1$, $x^{next} = \delta x^{prop}$, and $x^{two} = \delta^2 x^{prop}$, we can simultaneously solve for x^{prop}, x^{next} and x^{two}. In particular, plugging $x^{next} = \delta x^{prop}$ and $x^{two} = \delta^2 x^{prop}$ into the constraint $x^{prop} + x^{next} + x^{two} = 1$, we obtain

$$x^{prop} + \delta x^{prop} + \delta^2 x^{prop} = 1,$$

and solving for x^{prop} yields

$$x^{prop} = \frac{1}{1 + \delta + \delta^2}$$

Using this result into the other two equations, we obtain

$$x^{next} = \delta x^{prop} = \frac{\delta}{1 + \delta + \delta^2} \quad \text{and} \quad x^{two} = \delta^2 x^{prop} = \frac{\delta^2}{1 + \delta + \delta^2},$$

which constitute the equilibrium payoffs in this game. Intuitively, this implies that player 1 offers x^{prop} to himself in the first round of play, i.e., in period $t = 0$, x^{next} to player 2, and x^{two} to player 3, his offers are then accepted by players 2 and 3, and the game ends.

Comparative statics: Let us next check how these payoffs are affected by the common discount factor, δ, as depicted in Fig. 6.7. When players are very impatient, i.e., δ is close to zero, the player who makes the first proposal fares better than do others, and retains most of the pie, i.e., x^{prop} is close to one, while x^{next} and x^{two} are close to zero. However, when players become more patient, the proposer is forced to make more equal divisions of the pie, since otherwise his proposal would be rejected. In the extreme case of perfectly patient players, when $\delta = 1$, all players obtain a third of the pie.

Fig. 6.7 Equilibrium payoffs in the 3-player bargaining game

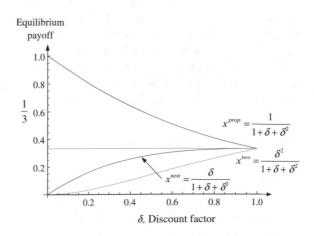

Exercise 7—Representing Feasible, Individually Rational Payoffs^C

Consider an infinitely-repeated game where the stage game is depicted Table 6.6.

Player 1 chooses rows and player 2 chooses columns. Players discount the future using a common discount factor δ.

Part (a) What outcomes in the stage-game are consistent with Nash equilibrium play?

Part (b) Let v_1 and v_2 be the repeated game payoffs to player 1 and player 2 respectively. Draw the set of feasible payoffs from the repeated game, explaining any normalization you use.

Part (c) Find the set of individually rational feasible set of payoffs.

Part (d) Find a Nash equilibrium in which the players obtain (9, 9) each period. What restrictions on δ are necessary?

Answer

We first need to find the set of Nash equilibrium strategy profiles for the stage game represented in Table 6.6: Let us start identifying best responses for each player. In particular, if player 1 plays U (in the top row), player 2's best response is $BR_2(U) = R$; while if player 1 plays D (in the bottom row), $BR_2(D) = R$, thus indicating that R is a strictly dominant strategy for player 2. Regarding player 1's best responses, we find that $BR_1(L) = D$ and $BR_1(R) = D$, showing that D is a strictly dominant strategy for player 1. Hence, (D, R) is the unique Nash equilibrium (NE) outcome of the stage game, in which players use strictly dominant strategies, and which yields an equilibrium payoff pair of (7, 7).

Figure 6.8 depicts the utility of player 1 and 2. For player 1, for instance, $u_1(D,q)$ represents his utility from selecting D, for any probability q with which his opponent selects Left, i.e., $10q + 7(1-q) = 7 + 3q$, and, similarly, $u_1(U,q)$ reflects player 1's utility from playing U, i.e., $9q + 1(1-q) = 1 + 8q$. A similar argument applies for player 2: when he selects L, his expected utility is $u_2(L,P) = 9p + 1(1-p) = 7 + 3p$. From Fig. 6.8 it becomes obvious that strategy U is strictly dominated for player 1, since it provides a strictly lower utility than D regardless of the precise value of q with which player 2 randomizes. Similarly, strategy L is strictly dominated for player 2.

From previous chapters, we know that every Nash equilibrium strategy profile must put weight only on strategies that are not strictly dominated. This explains why in this case the unique Nash equilibrium of the stage game does not put any

Table 6.6 Simultaneous-move game

Player 1		Player 2	
		L	R
	U	9, 9	1, 10
	D	10, 1	7, 7

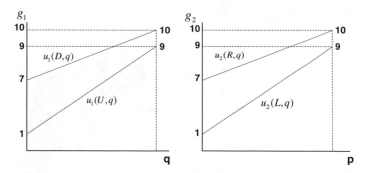

Fig. 6.8 Expected utilities for players 1 and 2

weight on strategy U for player 1 and strategy L for player 2, since they are both strictly dominated.

Anticipating that player 1 will choose D, since it gives an unambiguously larger payoff then U (as depicted in the figure), player 2 can minimize his opponent's payoff by selecting $q = 0$, which entails a minimax payoff of $v_1 = 7$ for player 1. Similarly, anticipating player 2 will choose R, since it provides an unambiguously larger payoff than L (as illustrated in the figure), player 1 can minimize his opponent's payoff by selecting $p = 0$, which yields a minimax payoff of $v_2 = 7$ for player 2. Therefore, the minimax payoff $(v_1, v_2) = (7, 7)$ is consistent with the Nash equilibrium payoff of (7, 7).

Part (b) We know that the set of feasible payoffs V is represented by the convex hull of all possible payoffs obtained in the stage game, as depicted in Fig. 6.9. (This set of feasible payoffs is often referred to in the literature as FP set.)

Fig. 6.9 Set of feasible payoffs

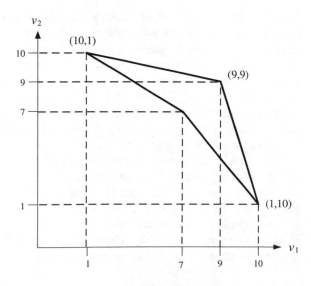

Fig. 6.10 Set of feasible
payoff after the normalization

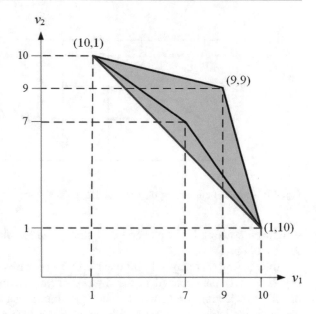

But obviously the above set of feasible payoffs in the repeated game is not convex. In order to address this non-convexity, we can convexify the set of feasible payoffs of the game by assuming that all players observe the outcome of a public randomization device at the start of each period. Using this normalization, the set of feasible payoffs can be depicted as the shaded area in Fig. 6.10.

Part (c) The minimax values for player i are given by

$$\underline{v}_i = \min_{\alpha_j} \left[\max_{\alpha_i} u_i(\alpha_i, \alpha_j) \right] \text{ where } \alpha_i, \alpha_j \in [0, 1] \, j \neq i.$$

Player 1. She maximizes her expected payoffs by playing D, a strictly dominant strategy, regardless of the probability used by player 2, q. Player 2 minimizes the payoff obtained from the previous maximization problem by choosing the minimax profile, implying that the minimax payoff for player 1 is $\underline{v}_1 = 7$.

Player 2. A similar argument applies to player 2, who maximizes her expected payoffs by playing R, a strictly dominant strategy, regardless of the probabilities used by player 1, p. Player 2 minimizes the payoff obtained from the previous maximization problem by choosing the minimax profile $p = 0$, thus entailing that the minimax payoff for player 2 is $\underline{v}_2 = 7$.

The feasible and individually rational set of payoffs (often referred to as set of FIR payoffs) is given by all those feasible payoffs $v_i \in V$ such that $v_i > \underline{v}_i$, that is, all feasible payoffs in which every player obtains strictly more than his payoffs from playing the minimax profiles, i.e., $v_i > 7$ for every player i in this game, as Fig. 6.11 depicts in the shaded area.

Part (d) For a sufficiently high discount factor, i.e., $\delta \in (\underline{\delta}, 1)$, we can design an appropriate punishment scheme that introduces incentives for every player not to

Fig. 6.11 Set of feasible, individually rational payoffs

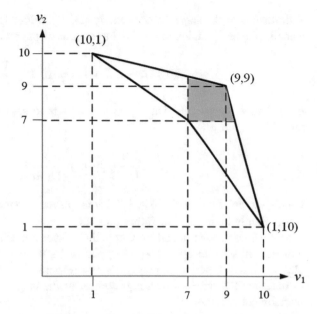

deviate from payoff (9, 9). In fact, this result is just a particular application of the Folk theorem, given that the proposed outcome (9, 9) is feasible and satisfies individual rationality for every player; i.e., $v_i > \underline{v}_i$ for all i.

Proof Let's take the payoff from outcome (U, L), 9, for every player $i = 1, 2$. Given symmetry, we can proceed all this proof by focusing on one of the players. Hence, without loss of generality let's analyze player 1's strategies. In particular, consider that player 1 uses the following grim-trigger strategy:

- Play U in round $t = 0$ and continue to do so as long as the realized action profile in previous periods was (U, L).
- Play D as soon as an outcome different from (U, L) is observed.

Hence, in order to show that strategy profile (U, L) is a Nash equilibrium of the infinitely repeated game we need to demonstrate that player 1 does not want to deviate from selecting U at every time period. To show that, we separately evaluate the payoff stream that player 1 obtains when cooperating and defecting.

Cooperation: If at any given time period t player 1 decides to play U forever after, his discounted stream of payoffs becomes

$$9 + \delta 9 + \delta^2 9 + \ldots = \frac{9}{1 - \delta}$$

Deviation: If, instead, player 1 deviates to D, his current payoff increases (from 9 to 10), but his deviation is observed (both by player 2 and by himself), triggering as a consequence an infinite punishment whereby both players revert to the Nash equilibrium of the stage game, namely (D, R), with associated payoff pair (7, 7). Hence, player 1 obtains the following stream of payoffs from deviating

$$10 + \delta 7 + \delta^2 7 + \cdots = 10 + \frac{\delta}{1 - \delta} 7$$

Incentives to cooperate: Thus, player 1 prefers to select U, i.e., cooperate, if and only if

$$\frac{9}{1 - \delta} \geq 10 + \frac{\delta}{1 - \delta} 7$$

which, solving for δ, yields $\delta \geq \frac{1}{3}$. That is, player 1 cooperates as long as he assigns a sufficiently high value to future payoffs.

Incentives to punish: In addition, note that player 2 has incentives to implement the punishment scheme upon observing player 1's deviation. In particular, given that player 1 has deviated to D, player 2's best response is $BR_2(D) = R$, thus giving rise to outcome (D, R), the Nash equilibrium of the stage game, as prescribed by the punishment scheme.

Hence, for any $\delta > \frac{1}{3}$ we can design an appropriate punishment scheme that supports the above specified strategy for player 1 as his optimal strategy in this infinitely repeated game. We can show the same results for player 2, given the symmetry of their payoffs. Hence, the outcome (U, L) can be supported as a subgame perfect equilibrium strategy profile in this infinitely repeated game.

Exercise 8—Collusion and Imperfect Monitoring[C]

Two firms produce a differentiated product, and every firm $i = \{1, 2\}$ faces the following linear demand:

$$p_i = a - bq_i - dq_j, \quad \text{where} \quad a, b, d > 0 \quad \text{and} \quad 0 < d < b$$

Intuitively, when $d \to 0$ firm j's output decisions do not affect firm i's sales (i.e., products are heterogeneous), while when $d \to b$, firm j's output decisions have a significant impact on firm i's sales, i.e., products are relatively homogeneous. The firms expect to compete against each other for infinite periods, that is, there is no entry or exit in the industry. [*Hint*: Assume that when a firm detects deviations from collusive behavior, it retaliates by reverting to the non-cooperative Cournot (Bertrand) production level (pricing, respectively), in all subsequent periods. Prior to detection, each firm assumes that its rival follows the collusive outcome.]

Part (a) Assume that firms compete as Cournot oligopolists. What is the minimum discount factor necessary in order to sustain the collusive outcome assuming that (i) Deviations are detected immediately, and (ii) Deviations are detected with a lag of one period.

Part (b) Repeat part (a) of the exercise under the assumption that firms compete as Bertrand oligopolists. (You can assume immediate detection)

Answer

Part (a)

(i) **Deviations are immediately detected.**
   Cournot outcome. Every firm i selects output level q_i by solving

$$\max_{q_i} \pi_i = (a - bq_i - dq_j)q_i$$

Taking first-order conditions with respect to q_i, we obtain $\frac{\partial \pi_i}{\partial q_i} = a - 2bq_i - dq_j = 0$, which yields the best response function $q_i(q_j) = \frac{a - dq_j}{2b}$ for every firm i and $j \neq i$.

Note that, as products become more heterogeneous, d close to zero indicating very differentiated products, firm i's best response function approaches the monopoly output $q_i = \frac{a}{2b}$.

Plugging firm j's best response function $q_j(q_i) = \frac{a - dq_i}{2b}$, into firm i's best response function yields

$$q_i = \frac{a - d\left(\frac{a - dq_i}{2b}\right)}{2b} = \frac{a}{2b} - \frac{d(a - dq_i)}{4b^2}$$

Solving for q_i in this equality, we obtain the (interior) Cournot equilibrium output when firms sell a differentiated product

$$q_i^c = \frac{a}{2b + d}$$

Therefore, firm i's equilibrium profits become (recall that production is costless):

$$\pi_i^c = \left(a - b \cdot \frac{a}{2b + d} - d \cdot \frac{a}{2b + d}\right)\frac{a}{2b + d} = \frac{ba^2}{(2b + d)^2}$$

Cooperative outcome (Monopoly). The merged firm would select the level of aggregate output, q, that solves

$$\max_q (a - bq - dq)q$$

Taking first-order conditions with respect to q, we obtain

$$\frac{\partial \pi}{\partial q} = a - 2(b+d)q = 0$$

and solving for q yields

$$q = \frac{a}{2(b+d)}$$

Hence, since firms are symmetric, every firm produces half of the collusive output, i.e., $q_i^m = q_j^m = \frac{a}{4(b+d)}$.

Thus, the profits that every firm i obtains if the collusive agreement is respected are

$$\pi_i^m = \left(a - b\frac{a}{2(b+d)} - d\frac{a}{2(b+d)}\right)\frac{a}{4(b+d)} = \frac{a^2}{8(b+d)}$$

Best deviation for firm i. Let us assume that firm $j \neq i$ produces the cooperative output level $q_j^m = \frac{a}{4(b+d)}$. In order to determine the optimal deviation for firm i, we just need to plug firm j's output q_j^m into firm i's best response function, as follows

$$q_i\left(q_j^m\right) = \frac{a - d\left(\frac{a}{4(b+d)}\right)}{2b} = \frac{a(4b+3d)}{8b(b+d)}$$

Therefore, the profits that firm i obtains from deviating are

$$\pi_i^D = \left(a - b\frac{a(4b+3d)}{8b(b+d)} - d\frac{a}{4(b+d)}\right)\frac{a(4b+3d)}{8b(b+d)} = \frac{a^2(4b+3d)^2}{64b(b+d)}$$

Incentives to collude: Thus, for firms to collude, we need that

$$\pi_i^m\left(1 + \delta + \delta^2 + \cdots\right) \geq \pi_i^D + \pi_i^C\left(\delta + \delta^2 + \cdots\right)$$

That is,

$$\pi_i^m\frac{1}{1-\delta} \geq \pi_i^D + \pi_i^C\frac{\delta}{1-\delta}$$

Plugging the expression of profits π_i^m, π_i^D and π_i^C we found in previous steps yields

$$\frac{a^2}{8(b+d)} \cdot \frac{1}{1-\delta} \geq \frac{a^2(4b+3d)^2}{64b(b+d)} + \frac{ba^2}{(2b+d)^2} \cdot \frac{\delta}{1-\delta}$$

Multiplying by $(1 - \delta)$ on both sides of the inequality, we obtain

$$\frac{a^2}{8(b+d)} \geq \frac{a^2(4b+3d)^2}{64b(b+d)}(1-\delta) + \delta\frac{ba^2}{(2b+d)^2}$$

and solving for δ yields the minimal discount factor that supports collusion in this industry,

$$\delta \geq \frac{4b^2 + 4bd + d^2}{8b^2 + 8bd + d^2} \equiv \delta_1$$

Importantly, note that δ_1 increases in the parameter reflecting product differentiation, d, since the derivative

$$\frac{\partial \delta_1}{\partial d} = \frac{4bd(2b+d)}{(8b^2 + 8bd + d^2)^2}$$

is positive for all parameter values $b, d > 0$. Intuitively, cooperation is more difficult to sustain when the product both firms sell is relatively homogeneous (i.e., d increases approaching b); but becomes easier to support as firms sell more differentiated products (i.e., d is close to zero), as if each firm operated as an independent monopolist.

(ii) **Deviations are detected with a lag of one period.**

In this case, collusion can be supported if

$$\pi_i^m(1 + \delta + \delta^2 + \cdots) \geq \pi_i^D + \delta\pi_i^D + \pi_i^C(\delta^2 + \delta^3 + \cdots)$$

where note that, if firm i deviates, it obtains the deviating profits of π_i^D for *two* consecutive periods. The above inequality can be alternatively expressed as

$$\pi_i^m\frac{1}{1-\delta} \geq \delta\pi_i^D + (1+\delta)\pi_i^D$$

Plugging the expression for profits π_i^m, π_i^D and π_i^C we obtained in previous steps of this exercise yields

$$\frac{a^2}{8(b+d)} \cdot \frac{1}{1-\delta} \geq \frac{a^2(4b+3d)^2}{64b(b+d)}(1+\delta) + \frac{ba^2}{(2b+d)^2} \cdot \frac{\delta^2}{1-\delta}$$

Multiplying both sides of the inequality by $(1-\delta)$ we obtain

$$\frac{a^2}{8(b+d)} \geq \frac{a^2(4b+3d)^2}{64b(b+d)}(1-\delta^2) + \delta^2 \frac{ba^2}{(2b+d)^2}$$

since $(1+\delta)(1-\delta) = (1-\delta^2)$. Solving for δ, we obtain the minimal discount factor sustained cooperation

$$\delta \geq \sqrt{\frac{(4b^2+4bd+d^2)}{(8b^2+8bd+d^2)}} \equiv \delta_2$$

Note that, similarly as for cutoff δ_1 (when deviations where immediately observed and punished), δ_2 is also increasing in d. However, $\delta_2 > \delta_1$ implying that, when deviations are only detected (and punished) after one-period lag, cooperation can be sustained under more restricting conditions than when deviations are immediately detected. Figure 6.12 depicts cutoffs δ_2 and δ_1 as a function of parameter d, both of them evaluated at $b = 1/2$. Graphically, the region of discount factors sustaining cooperation shrinks as firms detect deviations with a lag one one period.

Part (b) Let us now analyze cooperative outcomes if firms compete a la Bertrand. The inverse demand function every firm $i = \{1, 2\}$ faces is

$$p_i = a - bq_i - dq_j$$

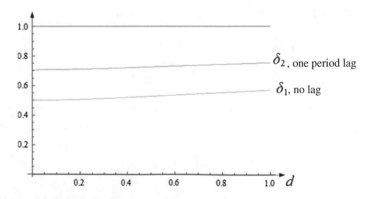

Fig. 6.12 Minimal discount factors δ_1 and δ_2

Hence, solving for q_i and q_j we find the direct demands,

$$q_i = \frac{a}{b+d} + \frac{dp_j - bp_i}{b^2 - d^2} \quad \text{for firm } i$$

$$q_j = \frac{a}{b+d} + \frac{dp_i - bp_j}{b^2 - d^2} \quad \text{for firm } j$$

Since $p_i = p_j = p$ in the collusive outcome, firms choose a common price p that maximizes their joint profits

$$\max_p \left(\frac{a}{b+d} + \frac{p(d-b)}{b^2 - d^2} \right) p = \left(\frac{a}{b+d} - \frac{p}{b+d} \right) p$$

Taking first-order conditions with respect to the common price p, we obtain

$$\frac{\partial \pi}{\partial p} = \frac{a}{b+d} - \frac{2p}{b+d} = 0,$$

which, solving for p, yields the monopolistic (collusive) price $p^m = \frac{a}{2}$, and entails collusive profits of

$$\pi^m = \left(\frac{a}{b+d} - \frac{p_m}{b+d} \right) p_m = \left(\frac{a}{b+d} - \frac{\frac{a}{2}}{b+d} \right) \frac{a}{2} = \frac{a^2}{4(b+d)}$$

Hence, the profits that every firm obtains when colluding in prices are half of this monopoly profit

$$\pi_i^m = \frac{a^2}{8(b+d)}$$

Bertrand outcome. In this case, every firm i independently and simultaneously selects a price p_i in order to maximize its individual profits

$$\max_{p_i} \left(\frac{a}{b+d} + \frac{d}{b^2 - d^2} p_j - \frac{b}{b^2 - d^2} p_i \right) p_i$$

Taking first-order conditions with respect to p_i, we find

$$\frac{\partial \pi_i}{\partial p_i} = \frac{a}{b+d} + \frac{d}{b^2 - d^2} p_j - \frac{2b}{b^2 - d^2} p_i = 0$$

and solving for p_i we obtain firm i's best response function $p_i(p_j) = \frac{a(b-d)+dp_j}{2b}$.
A similar argument applies for firm j's best response function $p_j(p_i)$. Plugging firm
j's best response function,

$$p_j(p_i) = \frac{a(b-d)+dp_i}{2b},$$

into firm i's best response function yields

$$p_i = \frac{a(b-d)+d\left[\frac{a(b-d)+dp_i}{2b}\right]}{2b}$$

and rearranging

$$p_i = \frac{a(b-d)(2b+d)+d^2p_i}{4b^2}$$

Solving for p_i, we obtain equilibrium prices

$$p_i = \frac{a(b-d)}{2b-d} = p_j$$

Therefore, the equilibrium output level for firm i becomes

$$
\begin{aligned}
q_i &= \frac{a}{b+d} + \frac{d}{b^2-d^2}p_j - \frac{b}{b^2-d^2}p_i \\
&= \frac{a}{b+d} + \frac{d}{b^2-d^2} \cdot \frac{a(b-d)}{2b-d} - \frac{b}{b^2-d^2} \cdot \frac{a(b-d)}{2b-d} \\
&= \frac{ba}{(2b-d)(b+d)}
\end{aligned}
$$

And Bertrand equilibrium profits are thus

$$
\begin{aligned}
\pi_i^B = p_i q_i &= \frac{a(b-d)}{2b-d} \frac{ba}{(2b-d)(b+d)} \\
&= \frac{b(b-d)a^2}{(b+d)(2b-d)^2}
\end{aligned}
$$

Best deviation by firm i. Assuming that firm j selects the collusive price $p_j^m = \frac{a}{2}$, firm
i can obtain its most profitable deviation by plugging the collusive price $p_j^m = \frac{a}{2}$ into
its best response function, as follows

$$p_i\left(p_j^m\right) = \frac{a(b-d)+d\left(\frac{a}{2}\right)}{2b} = \frac{a(2b-d)}{4b}$$

Therefore, the profits that firm i obtains from such a deviation are

$$
\pi_i^D = \left[\frac{a}{b+d} + \frac{d}{b^2-d^2} \cdot \frac{a}{2} - \frac{b}{b^2-d^2} \cdot \frac{a(2b-d)}{4b} \right] \frac{a(2b-d)}{4b}
$$

$$
= \frac{a^2(2b-d)^2}{16b(b-d)(b+d)}
$$

Incentives to collude. Hence, collusion can be sustained if

$$
\pi_i^m \left(1 + \delta + \delta^2 + \cdots \right) \geq \pi_i^D + \pi_i^B \left(\delta + \delta^2 + \cdots \right)
$$

$$
\pi_i^m \frac{1}{1-\delta} \geq \pi_i^D + \pi_i^B \frac{\delta}{1-\delta}
$$

Plugging the expression of profits π_i^m, π_i^D and π_i^B we found in previous steps yields

$$
\frac{a^2}{2(b+d)} \cdot \frac{1}{1-\delta} \geq \frac{a^2(2b-d)^2}{16b(b-d)(b+d)} + \frac{\delta}{1-\delta} \cdot \frac{b(b-d)a^2}{(b+d)(2b-d)^2}
$$

Multiply both sides of the inequality by $(1 - \delta)$, we obtain

$$
\frac{a^2}{2(b+d)} \geq \frac{a^2(2b-d)^2}{16b(b-d)(b+d)} (1-\delta) + \delta \frac{b(b-d)a^2}{(b+d)(2b-d)^2}
$$

Solving for δ, we find the minimal discount factor supporting cooperation when firms sell a heterogeneous product and compete a la Bertrand,

$$
\delta \geq \frac{(d-2b)^2(2b^2-2bd+d^2)}{a^2(8b^2-8bd+d^2)}
$$

Simultaneous-Move Games with Incomplete Information

7

Introduction

This chapter introduces incomplete information in simultaneous-move games, by allowing one player to be perfectly informed about some relevant characteristic, such as the state of market demand, or its production costs; while other players cannot observe this information. In this setting, we still identify players' best responses, but we need to condition them on the available information that every player observes when formulating its optimal strategy. Once we find the (conditional) best responses for each player, we are able to describe the Nash equilibria arising under incomplete information (the so-called Bayesian Nash equilibria, BNE) of the game; as the vector of strategies simultaneously satisfying all best responses. We next define BNE, but before describe games under an incomplete information context.

Consider a game with $N \geq 2$ players, each with discrete strategy space S_i and observing information $\theta_i \in \Theta_i$, where $\theta_i \in \mathbb{R}$. Parameter θ_i describes a characteristic that only player i can privately observe, such as its production costs, the market demand, or its willingness to pay for a product. A strategy for player i in this setting, is a function of θ_i, i.e., $s_i(\theta_i)$, entailing that the strategy profile of all other $N - 1$ players is also a function of $\theta_{-i} = (\theta_1, \theta_2, \ldots, \theta_{i-1}, \theta_{i+1}, \ldots, \theta_N)$, i.e., $s_{-i}(\theta_{-i})$. We are now ready to use this notation to define a Bayesian Nash Equilibrium (BNE).

BNE. Strategy profile $(s_1^*(\theta_1), \ldots, s_N^*(\theta_N))$ is a BNE if and only if

$$EU_i\big(s_i^*(\theta_i), s_{-i}^*(\theta_{-i})\big) \geq EU_i\big(s_i(\theta_i), s_{-i}^*(\theta_i)\big)$$

for every strategy $s_i(\theta_i) \in S_i$, every $\theta_i \in \Theta_i$, and every player i. In words, the expected utility that player i obtains from selecting strategy $s_i^*(\theta_i)$ is larger than from deviating to any other strategy $s_i(\theta_i)$; a condition that holds for all players $i \in N$ for all realizations of parameter θ_i.

The original version of the chapter was revised: The erratum to the chapter is available at: 10.1007/978-3-319-32963-5_11

© Springer International Publishing Switzerland 2016
F. Munoz-Garcia and D. Toro-Gonzalez, *Strategy and Game Theory*,
Springer Texts in Business and Economics, DOI 10.1007/978-3-319-32963-5_7

As initial motivation, we explore a simple poker game in which we study the construction of the "Bayesian normal form representation" of incomplete information games, which later on will allow us to solve for the set of BNEs in the same fashion as we found Nash equilibria in Chap. 2 for complete information games, that is, underlining best response payoffs. Afterwards, we apply this solution concept to a Cournot game in which one of the firms (e.g., a newcomer) is uninformed about the state of the demand, while its competitor (e.g., an incumbent with a long history in the industry) has accurate information about such demand. In this context, we first analyze the best response function of the privately informed firm, and afterwards characterize the best response function of the uninformed firm. We then study settings in which both agents are uninformed about each other's strengths, and must choose whether to start a fight, as if they were operating "in the dark."

Exercise 1—Simple Poker Game[A]

Here is a description of the simplest poker game. There are two players and only two cards in the deck, an Ace and a King. First, the deck is shuffled and one of the two cards in dealt to player 1. That is, nature chooses the card for player 1: being the Ace with probability 2/3 and the King with probability 1/3. Player 2 has previously received a card, which both players had a chance to observe. Hence, the only uninformed player is player 2, who does not know whether his opponent has received a King or an Ace. Player 1 observes his card and then chooses whether to bet (B) or fold (F). If he folds, the game ends, with player 1 obtaining a payoff of −1 and player 2 getting a payoff of 1 (that is, player 1 loses his ante to player 2). If player 1 bets, then player 2 must decide whether to respond betting or folding. When player 2 makes this decision, she knows that player 1 bets, but she has not observed player 1's card. The game ends after player 2's action. If player 2 folds, then the payoff vector is (1, −1), meaning player 1 gets 1 and player 2 gets −1, regardless of player 1's hand. If player 2 , instead, responds betting, then the payoff depends on player 1's card: if player 1 holds the Ace then the payoff vector is (2, −2), thus indicating that player 1 wins; if player 1 holds the King, then the payoff vector is (−2, 2), reflecting that player 1 looses. Represent this game in the extensive form and in the Bayesian normal form, and find the Bayesian Nash Equilibrium (BNE) of the game.

Answer
Figure 7.1 depicts the game tree of this incomplete information game.

Player 2 has only two available strategies $S_2 = \{Bet, Fold\}$, but player 1 has four available strategies $S_1 = \{Bb, Bf, Fb, Ff\}$. In particular, for each of his strategies, the first component represents player 1's action when the card he receives is the Ace while the second component indicates his action after receiving the King. This implies that the Bayesian normal form representation of the game is given by the following 4×2 matrix (Table 7.1).

In order to find the expected payoffs for strategy profile (*Bb, Bet*), i.e., in the top left-hand side cell of the matrix, we proceed as follows:

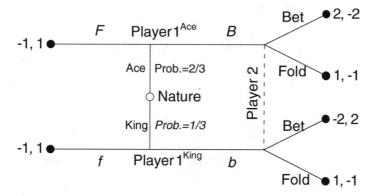

Fig. 7.1 A simple poker game

Table 7.1 Strategies in the Bayesian normal form representation of the poker game	Player 1	Player 2	
		Bet	Fold
	Bb		
	Bf		
	Fb		
	Ff		

$$EU_1 = \frac{2}{3} * 2 + \frac{1}{3} * (-2) = \frac{2}{3} \text{ and } EU_2 = \frac{2}{3} * (-2) + \frac{1}{3} * (2) = -\frac{2}{3}$$

Similarly for strategy profile (*Bf*, *Bet*),

$$EU_1 = \frac{2}{3} * (-1) + \frac{1}{3} * (2) = 0 \text{ and } EU_2 = \frac{2}{3} * (-2) + \frac{1}{3} * 1 = -1$$

For strategy profile (*Fb*, *Bet*),

$$EU_1 = \frac{2}{3}(-1) + \frac{1}{3}(-2) = -\frac{4}{3} \text{ and } EU_2 = \frac{2}{3} * 1 + \frac{1}{3} * 2 = \frac{4}{3}$$

For strategy profile (*Ff*, *Bet*), in the bottom left-hand side cell of the matrix, we obtain

$$EU_1 = \frac{2}{3}(-1) + \frac{1}{3}(-1) = -1 \text{ and } EU_2 = \frac{2}{3} * 1 + \frac{1}{3} * 1 = 1$$

In strategy profile (*Bb*, *Fold*), located in the top right-hand side cell of the matrix, we have

$$EU_1 = \frac{2}{3} * 1 + \frac{1}{3} * 1 = 1 \text{ and } EU_2 = \frac{2}{3}(-1) + \frac{1}{3}(-1) = -1$$

For strategy profile $(Bf, Fold)$,

$$EU_1 = \frac{2}{3} * 1 + \frac{1}{3}(-1) = \frac{1}{3} \text{ and } EU_2 = \frac{2}{3}(-1) + \frac{1}{3} * 1 = -\frac{1}{3}$$

Similarly for $(Fb, Fold)$,

$$EU_1 = \frac{2}{3} * (-1) + \frac{1}{3} * 1 = -\frac{1}{3} \text{ and } EU_2 = \frac{2}{3} * 1 + \frac{1}{3}(-1) = \frac{1}{3}$$

Finally, for $(Ff, Fold)$, in the bottom right-hand side cell of the matrix

$$EU_1 = \frac{2}{3} * (-1) + \frac{1}{3}(-1) = -1 \text{ and } EU_2 = \frac{2}{3} * 1 + \frac{1}{3} * 1 = 1$$

We can now insert these expected payoffs into the Bayesian normal form representation (Table 7.2).

We can now identify the best response to each player.

- For player 1, his best response when player 2 bets (in the left-hand column) is to play Bf since it yields a higher payoff, i.e., 1, than any other strategy, i.e., $BR_1(Bet) = Bf$. Similarly, when player 2 chooses to fold (in the right-hand column), player 1's best response is to play Bb, given that its associated payoff, \$1, exceeds that of all other strategies, i.e., $BR_1(Fold) = Bb$. Hence, the best responses yield an expected payoff of $\frac{2}{3}$ and 1, respectively; as depicted in the underlined payoffs of Table 7.3.
- Similarly, for player 2, when player 1 chooses Bb (in the top row), his best response is to bet, since his payoff, $\frac{-2}{3}$, is larger than that from folding, -1, i.e., $BR_2(Bb) = Bet$. If, instead, player 1 chooses Bf (in the second row), player 2's best response is to fold, $BR_2(Bf) = Fold$, since his payoff from folding, $-1/3$, is larger than from betting, -1. A similar argument applies to the case in which player 1 chooses Fb, where player 2 responds betting, $BR_2(Fb) = Bet$. Finally, when player 1 chooses Ff in the bottom row, player 2 is indifferent between responding with bet or fold since they both yield the same payoff (\$1), i.e., $BR_2(Ff) = \{Bet, Fold\}$. The payoffs that player 2 obtains from selecting these best responses are underlined in the next matrix (Table 7.3).

Table 7.2 Bayesian form representation of the poker game after inserting expected payoffs

Player 1	Player 2	
	Bet	Fold
Bb	$\frac{2}{3}, -\frac{2}{3}$	$1, -1$
Bf	$0, -1$	$\frac{1}{3}, -\frac{1}{3}$
Fb	$-\frac{4}{3}, \frac{4}{3}$	$-\frac{1}{3}, \frac{1}{3}$
Ff	$-1, 1$	$-1, 1$

Table 7.3 Underlined best response payoffs for the poker game

Player 1	Player 2	
	Bet	Fold
Bb	$\frac{2}{3}, -\frac{2}{3}$	$\underline{1}, -1$
Bf	$0, -1$	$\frac{1}{3}, -\frac{1}{3}$
Fb	$-\frac{4}{3}, \frac{4}{3}$	$-\frac{1}{3}, \frac{1}{3}$
Ff	$-1, \underline{1}$	$-1, \underline{1}$

Hence, there is a unique BNE in pure strategies where both players are selecting mutual best responses: (*Bb*, *Bet*).

Exercise 2—Incomplete Information Game, Allowing for More General Parameters[B]

Consider the Bayesian game in Fig. 7.2. First, nature selects player 1's type, either high or low with probabilities p and $1 - p$, respectively. Observing his type, player 1 chooses between x or y (when his type is high) and between x' and y' (when his type is low). Finally, player 2, neither observing player 1's type nor player 1's chosen action, responds with a or b. Note that this game can be interpreted as players 1 and 2 acting simultaneously, with player 2 being uninformed about player 1's type.

Part (a) Assume that $p = 0.75$. Find a Bayesian-Nash equilibrium.
Part (b) For each value of p, find all Bayesian-Nash equilibria.

Answer
In order to represent the game tree of Fig. 7.2 into its Bayesian normal form representation, we first need to identify the available strategies for each player. In particular, player 2 has only two strategies $S_2 = \{a, b\}$. In contrast, player 1 has four available strategies $S_1 = \{xx', xy', yx', yy'\}$ where the first component of every strategy pair denotes what player 1 chooses when his type is H and the second component reflects what he selects when his type is L. Hence, the Bayesian normal form representation of the game is given by the following 4×2 matrix (Table 7.4).

Let's find the expected utilities that each player obtains from strategy profile (xx', a), i.e., top left corner of the above matrix,

$$EU_1 = p * 3 + (1 - p) * 2 = 2 + p, \text{ and}$$
$$EU_2 = p * 1 + (1 - p) * 3 = 3 - 2 * p$$

Similarly, strategy profile (xy', a) yields expected utilities of

$$EU_1 = p * 3 + (1 - p) * 3 = 3 \text{ and}$$
$$EU_2 = p * 1 + (1 - p) * 1 = 1$$

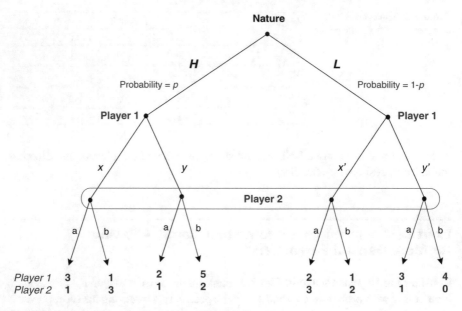

Fig. 7.2 A Bayesian game where player 2 is uninformed

Table 7.4 Bayesian normal form representation of the game	Player 1	Player 2	
		a	*b*
	xx'		
	xy'		
	yx'		
	yy'		

Table 7.5 Inserting expected payoffs in the Bayesian normal form representation of the game	Player 1	Player 2	
		a	*b*
	xx'	$2 + p, 3 - 2p$	$1, 2 + p$
	xy'	$3, 1$	$4 - 3p, 3p$
	yx'	$2, 3 - 2p$	$1 + 4p, 2$
	yy'	$3 - p, 1$	$4 + p, 2p$

Proceeding in this fashion for all remaining cells, we find the Bayesian normal form representation of the game, as illustrated in the following matrix (Table 7.5):

(a) When $p = 0.75$ (as assumed in part (a) of the exercise) the above matrix becomes (Table 7.6):

We can now identify player 1's best responses: when player 2 chooses a (in the left-hand column), $BR_1(a) = xy'$, yielding a payoff of 3; while when player

Table 7.6 Underlined best response payoffs

Player 1	Player 2	
	a	b
xx'	2.75, 1.5	1, 2.75
xy'	3, 1	1.75, 2.25
yx'	2, 1.5	4, 2
yy'	2.25, 1	4.75, 1.5

2 selects b (in the right-hand column), $BR_1(b) = yy'$, yielding a payoff of 4.75. Similarly operating for player 2, we obtain that his best responses are $BR_2(xx') = b$, with a payoff of 2.75 when player 1 selects xx' the first row; $BR_2(xy') = b$, with a payoff of 2.25 when player 1 chooses xy' in the second row; $BR_2(yx') = b$, where player 2's payoff is 2 when player 1 selects yx' in the third row; and $BR_2(yy') = b$, with a payoff of 1.5 when player 1 chooses yy' in the bottom row.

Underlining the payoff associated with the best response of each player, we find that there is a unique BNE in this incomplete information game: (yy', b).

(A) We can now approach this exercise without assuming a particular value for the probability p. Let us first analyze player 1's best responses. We reproduce the Bayesian normal form representation of the game for any probability p (Table 7.7).

Player 1's best responses: When player 2 selects the left column (strategy a), player 1 compares the payoff he obtains from xy', 3, against the payoff arising in his other strategies, i.e., $3 > 2 + p$, $3 > 2$, and $3 > 3 - p$, which hold for all values of p. Hence, player 1's best response to a is xy'. Similarly, when player 2 chooses the right-hand column (strategy b), player 1 compares the payoff from yy', $4 + p$, against that in his other available strategies, i.e., $4 + p > 1$, $4 + p > 4 - 3p$ and $4 + p > 1 + 4p$ which hold for all values of p. Hence, player 1's best response to b is yy'. Summarizing, player 1's best responses are $BR_1(a) = xy'$ and $BR_1(b) = yy'$.

Player 2's best reponses: Let us know examine player 2's to each of the four possible strategies of player 1:

Table 7.7 Bayesian normal form representation of the game, for a general probability p

Player 1	Player 2	
	a	b
xx'	$2 + p, 3 - 2p$	$1, 2 + p$
xy'	3, 1	$4 - 3p, 3p$
yx'	$2, 3 - 2p$	$1 + 4p, 2$
yy'	$3 - p, 1$	$4 + p, 2p$

When player 1 selects xx', he responds with a if $3 - 2p > 2 + p$, which simplifies to $1 > 3p$ that is, if $\frac{1}{3} > p$.

When player 1 selects xy', he responds with a if $1 > 3p$ or $\frac{1}{3} > p$ (otherwise he chooses b)

When player 1 selects yx', he responds with a if $3 - 2p > 2$, or $1 > 2p$ that is, he chooses a if $\frac{1}{2} > p$.

When player 1 selects yy', he responds with a if $1 > 2p$, or if $\frac{1}{2} > p$ (otherwise he chooses b).

Figure 7.3 summarizes player 2's best responses (either a or b), as a function of the probability p.

We can then divide our analysis into three different matrices, depending on the specific value of the probability p:

- One matrix for $p < \frac{1}{3}$
- A second matrix for $p \in \left[\frac{1}{3}, \frac{1}{2}\right]$, and
- A third matrix for $p > \frac{1}{2}$

First case: $p < \frac{1}{3}$

In this case, player 2 responds with a regardless of the action (row) selected by player 1 (see the range of Fig. 7.3 where $p < \frac{1}{3}$). This is indicated by player 2's payoffs in the matrix of Table 7.8, which are all underlined in the column where player 1 selects a. Hence, strategy a becomes a strictly dominant strategy for player

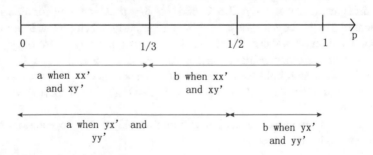

Fig. 7.3 Player 2's best responses as a function of p

Table 7.8 Bayesian normal form game with underlined best responses when $p < 1/3$

Player 1	Player 2	
	a	b
xx'	$2 + p$, $3 - 2p$	1, $2 + p$
xy'	$\underline{3}$, $\underline{1}$	$4 - 3p$, $3p$
yx'	2, $\underline{3 - 2p}$	$1 + 4p$, 2
yy'	$3 - p$, $\underline{1}$	$\underline{4 + p}$, $2p$

Table 7.9 Bayesian normal form game with underlined best responses if $1/2 \leq p \leq 1/3$

Player 1	Player 2	
	a	b
xx'	$2 + p$, $3 - 2p$	1, $\underline{2 + p}$
xy'	$\underline{3}$, 1	$4 - 3p$, $\underline{3p}$
yx'	2, $\underline{3 - 2p}$	$1 + 4p$, 2
yy'	$3 - p$, $\underline{1}$	$\underline{4 + p}$, $2p$

Table 7.10 Bayesian normal form game with underlined best responses when $p > 1/2$

Player 1	Player 2	
	a	b
xx'	$2 + p$, $3 - 2p$	1, $\underline{2 + p}$
xy'	$\underline{3}$, 1	$4 - 3p$, $\underline{3p}$
yx'	2, $3 - 2p$	$1 + 4p$, $\underline{2}$
yy'	$3 - p$, 1	$\underline{4 + p}$, $\underline{2p}$

2 in this context, while player 1's best responses are insensitive to p, namely, $BR_1(a) = xy'$ and $BR_1(b) = yy'$ (Table 7.8).

In this case, the unique BNE when $p < \frac{1}{3}$ is (xy', a)

<u>Second case</u>: $p \in \left[\frac{1}{3}, \frac{1}{2}\right]$

When $p \in \left[\frac{1}{3}, \frac{1}{2}\right]$, the intermediate range of p in Fig. 7.3 indicates that player 2 still responds with a when player 1 chooses yx' and yy', but with b when player 1 selects xx' and xy'. This is indicated in player 2's payoffs in the matrix of Table 7.9, which are underlined in the column corresponding to a when player 1 chooses yx' and yy' (last two rows), but are underlined in the right-hand column (corresponding to strategy b) otherwise (Table 7.9).

In this case there is no BNE when we restrict players to only use pure strategies.

<u>Third case</u>: $p > \frac{1}{2}$

Finally, when $p > \frac{1}{2}$, Fig. 7.3 reminds us that player 2 best responds with strategy b regardless of the action (row) selected by player 1, i.e., b becomes a strictly dominant strategy. This result is depicted in the matrix of Table 7.10, where player 2's payoffs corresponding to b are all underlined (see right-hand column) (Table 7.10).

In this case the unique BNE is (yy', b). (Note that this BNE is consistent with part (a) of the exercise, where p was assumed to be 0.75, and thus corresponds to the third case analyzed here $p > \frac{1}{2}$. Needless to say, we found the same BNE as in the third case).

Exercise 3—More Information Might Hurt[B]

Show that more information may hurt a player by constructing a two-player game with the following features: Player 1 is fully informed while player 2 is not; the game has a unique Bayesian Nash equilibrium, in which player 2's expected payoff

Table 7.11 "More information might hurt" game

<table>
<tr><td></td><td colspan="2" align="center">Player 2</td><td></td><td></td><td colspan="2" align="center">Player 2</td></tr>
<tr><td></td><td>L</td><td>R</td><td></td><td></td><td>L</td><td>R</td></tr>
<tr><td>U</td><td>0, 0</td><td>-1000,<u>50</u></td><td></td><td>U</td><td><u>100</u>, 1</td><td>1, <u>2</u></td></tr>
<tr><td>D</td><td><u>50</u>, -1</td><td><u>25</u>,<u>25</u></td><td></td><td>D</td><td>99, <u>4</u></td><td><u>2</u>, 1</td></tr>
<tr><td></td><td colspan="2" align="center">Player 2 is low type</td><td></td><td></td><td colspan="2" align="center">Player 2 is high type</td></tr>
</table>

is higher than his payoff in the unique equilibrium of any of the related games in which he knows player 1's type.

Answer
Player 2 is fully informed about the types of both players, while player 1 is only informed about his own type. The following payoff matrices describe this incomplete information game (Table 7.11).
Full Information: When both players know player 2's type, we have a unique Nash Equilibrium (NE), both when player 2's type is low and when its type is high. We next separately analyze each case.
Player 2 is low type. As depicted in the left-hand matrix, player 1 finds strategy U strictly dominated by strategy D. Similarly, for player 2, strategy L is strictly dominated by strategy R. Hence, the unique strategy pair surviving the iterative deletion of strictly dominated strategies (IDSDS) is (D, R). In addition, we know that when the set of strategies surviving IDSDS is a singleton, then such strategy profile is also the unique NE in pure strategies. Hence, the equilibrium payoffs in this case are $u_1 = u_2 = 25$.
Player 2 is high type. As depicted in the right-hand matrix, we cannot delete any strategy because of being strictly dominated. In addition, there doesn't exist any NE in pure strategies, since $BR_1(L) = U$ and $BR_1(R) = D$ for player 1, and $BR_2(U) = R$ and $BR_2(D) = L$ for player 2, i.e., there is no strategy profile that constitutes a mutual best response for both players. We, hence, need to find the MSNE for this game. Assume that player 2 randomizes between L and R with probability y and $1 - y$, respectively. Thus, player 1 is indifferent between playing U and D if:

$$EU_1(U|y) = EU_1(D|y)$$

or

$$y \cdot 100 + (1 - y) \cdot 1 = y \cdot 99 + (1 - y) \cdot 2$$

and solving for y, we find that $y = \frac{1}{2}$ which entails that player 2 chooses L with $\frac{1}{2}$ probability.

We can similarly operate for EU_2. Assuming that player 1 randomizes between L and R with probability x and $1 - x$, respectively, player 2 is made indifferent between L and R if

$$EU_2(L|x) = EU_2(R|x)$$

or

$$x \cdot 1 + (1-x) \cdot 4 = 2x + (1-x) \cdot 1$$

and solving for x, we obtain that $x = \frac{3}{4}$. This is the probability of player 1 playing L when he knows that P_2 is high type.

Hence, the unique MSNE for the case of player 2 being high type (right-hand matrix) is:

$$MSNE = \left(\frac{3}{4}U + \frac{1}{4}D, \ \frac{1}{2}L + \frac{1}{2}R\right)$$

And the associated expected payoffs for each player are:

$$EU_1 = \frac{3}{4} \cdot \underbrace{\left(\frac{1}{2} \cdot 100 + \frac{1}{2} \cdot 1\right)}_{\text{when } P_1 \text{ plays U}} + \frac{1}{4} \cdot \underbrace{\left(\frac{1}{2} \cdot 99 + \frac{1}{2} \cdot 2\right)}_{\text{when } P_2 \text{ plays D}} = \frac{3}{4} \cdot \frac{101}{2} + \frac{1}{4} \cdot \frac{101}{2} = \frac{404}{8}$$

$$= \frac{101}{2} = 50.5$$

$$EU_2 = \frac{1}{2} \cdot \underbrace{\left(\frac{3}{4} \cdot 1 + \frac{1}{4} \cdot 4\right)}_{\text{when } P_2 \text{ plays L}} + \frac{1}{2} \cdot \underbrace{\left(\frac{3}{4} \cdot 2 + \frac{1}{4} \cdot 1\right)}_{\text{when } P_2 \text{ plays R}} = \frac{1}{2} \cdot \frac{7}{4} + \frac{1}{2} \cdot \frac{7}{4} = \frac{7}{4} = 1.75$$

Therefore, $EU_1 = 50.5$ and $EU_2 = 1.75$ in the MSNE of the complete-information game in which player 2 is of a high type.

Imperfect Information Case: In this setting, we consider that player 1 is not informed about player 2's type. However, player 2 is informed about his own type, and so he has perfect information. Let π denote the probability (exogenously determined by nature) that player 2 is of low type i.e., that players interact in the left-hand matrix.

Player 1. We know that player 1 will take his decision between U and D "in the dark" since he does not observe player 2's type. Hence, player 1 will be comparing $EU_1(U|y)$ against $EU_1(D|y)$. In particular, if player 1 selects U, his expected payoff becomes:

$$EU_1(U|y) = \underbrace{\pi \cdot (-1000)}_{\substack{\text{because } P_2 \\ \text{always plays R} \\ \text{when being low} - \text{type}}} + \underbrace{(1-\pi) \cdot [y \cdot 100 + (1-y) \cdot 1]}_{\text{when } P_2 \text{ is high–type}}$$

$$= -1000\pi + (1-\pi)[100y + (1-y)]$$

Note that player 2 plays R when being a low-type because R is a strictly dominant strategy in that case. When his type is high, however, player 1 cannot anticipate whether player 2 will be playing L or R, since there is no strictly dominant strategy for player 2 in this case. We can operate similarly in order to find the expected utility that player 1 obtains from selecting D, as follows.

$$EU_1(D|y) = \underbrace{\pi \cdot 25}_{\substack{\text{because } P_2 \\ \text{always plays R} \\ \text{when being low} - \text{type}}} + \underbrace{(1 - \pi) \cdot [y \cdot 99 + (1 - y) \cdot 2]}_{\text{when } P_2 \text{ is high-type}}$$

For simplicity, assume that this probability is, in particular, $\pi = \frac{1}{25}$, and therefore $1 - \pi = \frac{24}{25}$. Hence, comparing both expected utilities, we have

$$EU_1(U|y) < EU_1(D|y)$$

$$-\frac{1000}{25} + \frac{24}{25}[99y + 1] < 1 + \frac{24}{25}[97y + 2]$$

and solving for y yields

$$\frac{-1049}{25} < \frac{-48}{25}y \rightarrow y < 21.85$$

which is always true, since $y \in [0, 1]$, thus implying that $EU_1(U) < EU_1(D)$ holds for all values of y when $\pi = \frac{1}{25}$. Hence, the uninformed player 1 chooses D.

More generally, for any value of probability π, we can easily find the difference

$$EU_1(U|y) - EU_1(D|y) = 1 + 2y - 2\pi(512 + y)$$

Hence, $EU_1(U|y) < EU_1(D|y)$ for all $\pi > \frac{-1 + 2y}{2(512 + y)} \equiv \hat{\pi}$, which holds for all $y < \frac{1}{2}$, i.e., cutoff $\hat{\pi}$ becomes negative for all $y < \frac{1}{2}$, thus implying that $\pi > \hat{\pi}$ is satisfied since $\pi \in [0, 1]$. For all $y > \frac{1}{2}$, the condition $\pi > \hat{\pi}$ is not very restrictive either since cutoff $\hat{\pi}$ increases from $\hat{\pi} = 0$ at $y = \frac{1}{2}$ to $\hat{\pi} = 0.0009$ when $y = 1$, thus implying that, for most combinations of probabilities π and y, player 1 prefers to choose D. Figure 7.4 depicts this area of (π, y)-pairs, namely that above the $\hat{\pi}$ cutoff, thus spanning most admissible values.

Player 2. Since player 2 (the informed player) can anticipate player 1's selection of D (for any value of y), player 2's best responses become:

$$BR_2(P_1 \text{ always plays } D|P_2 \text{ is low type}) = R$$

$$BR_2(P_1 \text{ always plays } D, |P_2 \text{ is high type}) = L$$

Fig. 7.4 Admissible
probability pairs

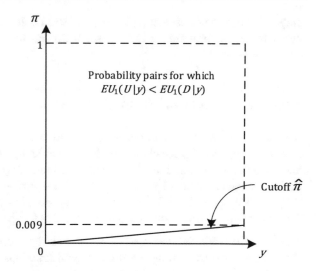

Therefore, the unique Bayesian Nash equilibrium (BNE) of the game is given by:
(D, RL), where R indicates that the informed player 2 plays R when he is a
low-type, and the second component (L) indicates what he plays when he is a
high-type. The associated payoff for the uninformed player 1 in this BNE is

$$EU_1 = \pi \cdot EU_1(D|P_2 \text{ is low type}) + (1 - \pi)EU_1(D|P_2 \text{ is high type})$$

$$= \underbrace{\frac{1}{25} \cdot 25}_{\substack{\text{when } (D,R) \text{ if } P_2 \\ \text{is low} - \text{type}}} + \underbrace{\frac{24}{25} \cdot 99}_{\substack{\text{when } (D,L) \text{ if } P_2 \\ \text{is high-type}}} = 96.04$$

Payoff comparison: Therefore, we can show that the uninformed player's payoff is
higher when he is uninformed (in the BNE) than perfectly informed (in the NE in
pure strategies), both when his opponent's type is low:

$$\underbrace{EU_1(P_2 \text{ is low type})}_{\text{informed case}} < \underbrace{EU_1}_{\text{uninformed case}} \quad \rightarrow 25 < 96.04$$

and similarly for the case in which his opponent's type is high:

$$\underbrace{EU_1(P_2 \text{ is high type})}_{\text{informed case}} < \underbrace{EU_1}_{\text{uninformed case}} \quad \rightarrow 50 < 96.04$$

As a consequence, "more information may hurt" player 1. Alternatively, he
would prefer to play the incomplete information game in which he does not know

player 2's type before selecting between U and D, than interacting with player 2 in the complete information game in which he can observe player 2's type before choosing U or D.

Exercise 4—Incomplete Information in Duopoly Markets[A]

Consider a differentiated duopoly market in which firms compete by selecting prices. Let p_1 be the price chosen by firm 1 and let p_2 be the price of firm 2. Let q_1 and q_2 denote the quantities demanded (and produced) by the two firms. Suppose that the demand for firm 1 is given by $q_1 = 22 - 2p_1 + p_2$, and the demand for firm 2 is given by $q_2 = 22 - 2p_2 + p_1$. Firm 1 produces at a constant marginal cost of 10, while firm 2's constant marginal cost is $c > 0$. Firms face no fixed costs.

Part (a) *Complete information.* Consider that firms compete in prices. Represent the normal form by writing the firms' payoff functions.
Part (b) Calculate the firms' best-response functions.
Part (c) Suppose that $c = 10$ so the firms are identical (the game is symmetric). Calculate the Nash equilibrium prices in this complete information game.
Part (d) *Incomplete information.* Now suppose that firm 1 does not know firm 2's marginal cost c. With probability ½ nature picks $c = 14$ and with probability ½ nature picks $c = 6$. Firm 2 knows its own cost (that is, it observes nature's move), but firm 1 only knows that firm 2's marginal cost is either 6 or 14 (with equal probabilities). Calculate the best-response function of player 1 and those for player 2 (one for each of its two types, $c = 6$ and $c = 14$), and find the Bayesian Nash equilibrium quantities.

Answer

Part (a) For firm 1, with constant marginal costs of production of $c = 10$, we have that its profit function is

$$\pi_1(p_1, p_2) = p_1 * q_1 - 10q_1 = (p_1 - 10)q_1 = (p_1 - 10)(22 - 2p_1 + p_2)$$
$$= 42p_1 + p_1 p_2 - 2p_1^2 - 10p_2 - 220$$

Similarly, for firm 2, with marginal costs given by $c > 0$, we have a profit function of

$$\pi_2(p_1, p_2) = p_2 * q_2 - c * q_2 = (p_2 - c)q_2 = (p_2 - c)(22 - 2p_2 + p_1)$$
$$= (22 + 2c)p_2 + p_1 p_2 - 2p_2^2 - 22c - cp_1$$

Part (b) Taking first-order conditions of firm 1's profit with respect to p_1 we obtain

$42 + p_2 - 4p_1 = 0$ and, solving for p_1, yields firm 1's best response function $p_1(p_2)$

$$= \frac{42 + p_2}{4} \leftarrow (BRF_1)$$

Thus, the best response function of firm 1 originates at $\frac{42}{4}$ and has a vertical slope of $\frac{1}{4}$.

Similarly, taking first-order conditions of $\pi_2(p_1, p_2)$ with respect to p_2 yields,

$$22 + 2c + p_1 - 4p_2 = 0$$

and solving for p_2, we obtain firm 2's best response function

$$p_2(p_1) = \frac{22 + 2c + p_1}{4} \leftarrow BRF_2$$

which originates at $\frac{22 + 2c}{4}$ and has a vertical slope of $\frac{1}{4}$. (Note that if firm 2's marginal costs were 10, then $\frac{22 + 2c}{4}$ would reduce to $\frac{42}{4}$, yielding a BRF_2 symmetric to that of firm 1.)

Part (c) In this part, the exercise assumes a marginal cost of $c = 10$. Plugging BRF_1 into BRF_2, we obtain

$$p_2 = \frac{42 + \left[\frac{42 + p_2}{4}\right]}{4}$$

And solving for p_2 yields an equilibrium price of $p_2^* = \$14$. We can finally plug $p_2^* = \$14$ into BRF_1, to find firm 1's equilibrium price

$$p_1(14) = \frac{42 + 14}{4} = \$14$$

Hence, the Nash equilibrium of this complete information game is the price pair $(p_1^*, p_2^*) = (\$14, \$14)$. Therefore, firm i's profits in the equilibrium of this complete information game are

$$\pi_i(p_i, p_j) = (14 - 10)(22 - 2 * 14 + 14) = \$32$$

Part (d) *Incomplete information game.* Starting with the informed player (Player 2), we have that:

- Firm 2's best-response function evaluated at high costs, i.e., $c = 14$, BRF_2^H, is

$$p_2^H(p_1) = \frac{22 + 2 * 14 + p_1}{4} = \frac{50 + p_1}{4}$$

- Firm 2's best-response function when having low costs, i.e., $c = 6$, BRF_2^L, is

$$p_2^L(p_1) = \frac{22 + 2 * 6 + p_1}{4} = \frac{34 + p_1}{4}$$

- The uninformed firm 1 chooses the price p_1 that maximizes its expected profits

$$\max_{p_1} \frac{1}{2}\left[42p_1 + p_1 p_2^H - 2p_1^2 - 10p_2^H\right] + \frac{1}{2}\left[42p_1 + p_1 p_2^L - 2p_1^2 - 10p_2^L\right]$$

where the first term represents the case in which firm 2 has high costs, and thus sets a price p_2^H; while the second term denotes the case in which firm 2's costs are low, setting a price p_2^L. Importantly, firm 1 cannot condition its price, p_1, on the production costs of firm 2 since, as opposed to firm 2, firm 1 does not observe this information.

Taking first order conditions with respect to p_1 we obtain

$$\frac{1}{2}\left[42 + p_2^H - 4p_1\right] + \frac{1}{2}\left[42 + p_2^L - 4p_1\right]$$

or

$$42 + \frac{1}{2}p_2^H + \frac{1}{2}p_2^L - 4p_1 = 0$$

And solving with respect to p_1 we obtain firm 1's $BRF_1(p_2^H, p_2^L)$, which depends on p_2^H and p_2^L, as follows.

$$p_1(p_2^H, p_2^L) = \frac{42 + \frac{1}{2}p_2^H + \frac{1}{2}p_2^L}{4}$$

Plugging $p_2^H(p_1)$ and $p_2^L(p_1)$ into $BRF_1(p_2^H, p_1^L)$, we obtain

$$p_1(p_2^H, p_2^L) = \frac{42 + \frac{1}{2}\left[\frac{50 + p_1}{4}\right] + \frac{1}{2}\left[\frac{34 + p_1}{4}\right]}{4}$$

And solving for p_1, we find firm 1's equilibrium price, $p_1^* = \$14$.

Finally, plugging this equilibrium price $p_1^* = \$14$ into firm 2's BRF_2^H and BRF_2^L yields equilibrium prices of

$$p_2^H(14) = \frac{50 + 14}{4} = \$16$$

when firm 2's costs are high, and

$$p_2^L(14) = \frac{34+14}{4} = \$12$$

when its costs are low. Thus, BNE of this Bertrand game under incomplete information is

$$(p_1, p_2^H, p_2^L) = (14, 16, 12)$$

In this setting, the informed player (firm 1) makes an expected equilibrium profits of

$$\pi_1\left(p_1, p_2^H, p_2^L\right) = \frac{1}{2}\left[42 * 14 + 14 * 16 - 2 * 14^2 - 10 * 16\right]$$
$$+ \frac{1}{2}\left[42 * 14 + 14 * 12 - 2 * 14^2 - 10 * 12\right]$$
$$= \$252$$

while the uninformed player (firm 2) obtains an equilibrium profit of

$$\pi_2\left(p_1, p_2^H, p_2^L\right) = (16 - 14)(22 - 2 * 16 + 16) = \$12$$

when its costs are high, i.e., $c = 14$; and

$$\pi_2\left(p_1, p_2^H, p_2^L\right) = (12 - 6)(22 - 2 * 12 + 12) = \$60$$

when its costs are low, i.e., $c = 6$.

Exercise 5—Starting a Fight Under Incomplete Information[C]

Consider two students looking for trouble on a Saturday night (hopefully, they are not game theory students). After arguing about some silly topic, they are in the verge of a fight. Each individual privately observes its own ability as a fighter (its type), which is summarized as the probability of winning the fight if both attack each other: either high, p^H, or low, p^L, where $0 < p^L < p^H < 1$, with corresponding probabilities q and $1 - q$, respectively. If a student chooses to attack and its rival does not, then he wins with probability αp^K, where $K = \{H, L\}$, and α takes a value such that $p^K < \alpha p^K < 1$. If both students attack each other, then the probability that a student with type-k wins is p^K. If a type-k fighter does not attack but the other fighter attacks, then the probability of victory for the student who did not attack decreases to only βp^K, where β takes a value such that $0 < \beta p^K < p^K$. Finally, if neither student attacks, there is no fight. A student is then more likely to win the fight the higher is his type. A student's payoff when there is no fight is 0, the benefit from winning a fight is B (e.g., being the guy in the gang), and from losing a fight is L (shame, red face). Assume that $B > 0 > L$.

a. Under which conditions there is a symmetric Bayesian-Nash equilibrium in this simultaneous-move game where every student attacks regardless of his type.
b. Find the conditions to support a symmetric BNE in which every student attacks only if his type is high.

Answer

Part (a) If one student expects the other to attack for sure (that is, both when his type is high and low), then it is optimal to attack regardless of his type. Doing so results in an expected payoff of $pB + (1 - p)L$, where p is the probability of wining the fight when both students attack, and not doing so results in an expected payoff of $\beta pB + (1 - \beta p)L$, where βp is the probability of wining the fight when the student does not attack but his rival does. Comparing these expected payoffs, we obtain that every student prefers to attack if:

$$pB + (1 - p)L > \beta pB + (1 - \beta p)L$$

rearranging yields:

$$B(p - \beta p) > L(p - \beta p)$$

or, further simplifying, $B > L$, which holds by definition since $B > 0 > L$. Therefore, attacking regardless of one's type is a symmetric Bayesian-Nash equilibrium (BNE).

Part (b) Consider a symmetric strategy profile in which a student attacks only if his type is high.

High-type student. For this strategy profile to be an equilibrium, we need that the expected utility from fighting when being a strong fighter must exceed that from not fighting. In particular,

$$q\left[p^H B + \left(1 - p^H\right)L\right] + (1 - q)\left[\alpha p^H B + \left(1 - \alpha p^H\right)L\right]$$
$$\geq q[\beta p^H B + (1 - \beta p^H)L] + (1 - q)0$$

Let's start analyzing the *left-hand side* of the inequality:

- The first term represents the expected payoff that a type-p^H student obtains when facing a student who is also type p^H (which occurs with probability q). In such a case, this equilibrium prescribes that both students attack (since they are both p^H), and the student we analyze wins the fight with probability p^H (and alternatively losses with probability $1 - p^H$).

- The second term represents the expected payoff that a type p^H student obtains when facing a student who is, instead, type p^L (which occurs with probability $1 - q$). In this case, his rival doesn't attack in equilibrium, increasing the probability of wining the fight for the student we analyze to αp^H, where $\alpha p^H > p^H$.

Let us now examine the *right-hand side* of the inequality, which illustrates the expected payoff that student p^H obtains when he deviates from his equilibrium strategy, i.e., he does not attack despite having a type p^H.

- The first term represents the expected payoff that a type-p^H student obtains when facing a student who is also p^H (which occurs with probability q). In such a case, the student we analyze doesn't attack while his rival (also of type H) attacks, lowering the chances that the former wins the fight to βp^H, where $\beta p^H < p^H$.
- If, in contrast, the type of his rival is p^L, then no individual attacks and their payoffs are both zero.

Rearranging the above inequality, we obtain:

$$q\left[p^H B + \left(1 - p^H\right)L\right] + (1 - q)[\alpha p^H B + (1 - \alpha p^H)L]$$
$$\geq q[\beta p^H B + \left(1 - \beta p^H\right)L] + (1 - q)0$$

Factoring out B and L, we have

$$B[qp^H + \alpha p^H - q\alpha p^H - q\beta p^H] \geq L[qp^H + \alpha p^H - q\alpha p^H - q\beta p^H - (1 - q)]$$

and solving for B, yields

$$B \geq L\left(1 - \frac{1 - q}{p^H[q(1 - \beta) + (1 - q)\alpha]}\right)$$

Since $(1 - q) > 0$ and $p^H[q(1 - \beta) + (1 - q)\alpha] > 0$, the term in parenthesis satisfies $\left(1 - \frac{(1-q)}{p^H[q(1-\beta) + (1-q)\alpha]}\right) < 1$, and this condition holds by the initial condition $B > L$. For instance, if $p^H = 0.8$, $\alpha = 1.3$, and $\beta = 0.8$, the above inequality becomes

$$B \geq L\left(\frac{1 + 3q}{22q - 26}\right).$$

If, for instance, losing the fight yields a payoff of $L = -2$, then Fig. 7.5 depicts the (B, q)-pairs for which the high-type student starts a fight, i.e., shaded region. Intuitively, the benefit from winning, B, must be relatively high. This is likely when the probability of the other student being strong, q, is relatively low. For instance, when $q = 0$, B only needs be larger than 1/13 in this numerical example for the student to start a fight. However, when his opponent is likely strong, i.e., q approaches 1, the benefit from winning the fight must be much higher (large value of B) for the high-type student to start a fight.

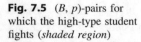

Fig. 7.5 (B, p)-pairs for which the high-type student fights (*shaded region*)

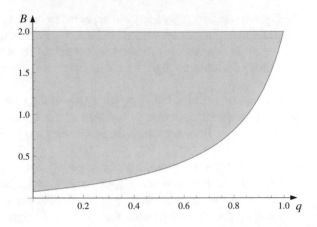

Low-type student. Let us now examine the student of type p^L. Recall that, according to the equilibrium we analyze, this student prefers to *not attack*. That is, his expected utility must be larger from not attacking than attacking, as follows.

$$q[\beta p^L B + (1 - \beta p^L)L] + (1 - q)0$$
$$\geq q[p^L B + (1 - p^L)L] + (1 - q)[\alpha p^L B + (1 - \alpha p^L)L].$$

An opposite intuition as above can be constructed for this inequality, i.e., a student prefers not to attack when it type is p^L. In this case, rearranging we obtain:

$$q[\beta p^L B + (1 - \beta p^L)L] + (1 - q)0$$
$$\geq q[p^L B + (1 - p^L)L] + (1 - q)[\alpha p^L B + (1 - \alpha p^L)L]$$

Factoring out B and L, and solving for B, yields

$$B \geq L\left(1 - \frac{1 - q}{p^L[q\beta - \alpha + q\alpha - q]}\right)$$

Since $(1 - q) > 0$ and $p^L[q\beta - \alpha + q\alpha - q] > 0$, the term in parenthesis satisfies

$$\left(1 - \frac{1 - q}{p^L[q\beta - \alpha + q\alpha - q]}\right) < 1$$

As a consequence, the above inequality is also implied by the initial condition $B > L$. Following a similar numerical example as above, where $p^L = 0.4, \alpha = 1.3, \beta = 0.8$, we obtain $B \geq \frac{2L(18q - 19)}{11q - 13}$. Using the same numerical example as for the high-type player, this yields a cutoff $B \geq \frac{76 - 72q}{11q - 13}$, which is negative for all values of q, and thus satisfied for any positive benefit from winning the fight, $B > 0$.

Auctions

8

Introduction

In this chapter we examine different auction formats, such as first-, second-, third- and all-pay auctions. Auctions are a perfect setting in which to apply the Bayesian Nash Equilibrium (BNE) solution concept learned in Chap. 7, since competing bidders are informed about their private valuation for the object for sale, but are uninformed about each other's valuations. Since, in addition, bidders are asked to simultaneously and independently submit their bids under an incomplete information environment; we can use BNE to identify equilibrium behavior, namely, equilibrium bidding strategies. In order to provide numerical examples, the following exercises often depict equilibrium bids for the case in which valuations are distributed according to a uniform distribution.

We start analyzing of the first-price auction (whereby the bidder submitting the highest bid wins the auction and must pay his bid), and show how to find equilibrium bids using two approaches (first order conditions, and the envelope theorem). We also investigate the effect of a larger number of competing bidders on players' equilibrium bids, showing that players become more aggressive as more bidders compete for the object. Exercise 8.2 then examines the second-price auction, where the winner of the object is still the bidder submitting the highest bid. However, the winner does not pay his own bid, but the second highest bid. Exercises 8.3 and 8.4 move to equilibrium bids in the so-called "all pay auction," whereby all bidders, irrespective of whether they win the object for sale, must pay the bid they submitted.

We afterwards compare equilibrium bidding functions in the first-, second- and all-pay auction. In Exercise 8.5, we explore third-price auctions, where the winner is the bidder who submits the highest bid (as in all previous auction formats), but he/she only pays the third highest bid. Finally, we introduce the possibility that

The original version of the chapter was revised: The erratum to the chapter is available at: 10.1007/978-3-319-32963-5_11

© Springer International Publishing Switzerland 2016
F. Munoz-Garcia and D. Toro-Gonzalez, *Strategy and Game Theory*,
Springer Texts in Business and Economics, DOI 10.1007/978-3-319-32963-5_8

bidders are risk averse, and examine how equilibrium bidding functions in the first-
and second-price auctions are affected by players' risk aversion.

Exercise 1—First Price Auction with N Bidders[B]

Consider a first-price sealed-bid auction in which bidders simultaneously and
independently submit bids and the object goes to the highest bidder at a price equal
to his/her bid. Suppose that there are $N \geq 2$ bidders and that their values for the
object are independently drawn from a uniform distribution over [0,1]. Player i's
payoff is $v - b_i$ when she wins the object by bidding b_i and her value is v; while her
payoff is 0 if she does not win the object.

Part (a) Formulate this auction as a Bayesian game.
Part (b) Let $b_i(v)$ denote the bid made by player i when his valuation for the object
is v. Show that there is a Bayesian Nash Equilibrium in which bidders submit a bid
which is a linear function of their valuation, i.e., $b_i(v) = \alpha + \beta v$ for all i and v,
where parameters α and β satisfy $\alpha, \beta > 0$. Determine the values of α and β.

Solution
First, note that the expected payoff of every bidder i with valuation v_i is given by:

$$
u_i(b_i, b_{-i}, v_i) = \begin{cases} v_i - b_i & \text{if } b_i > \max_{j \neq i} b_j \\ \frac{v_i - b_i}{2} & \text{if } b_i = \max_{j \neq i} b_j \\ 0 & \text{if } b_i < \max_{j \neq i} b_j \end{cases}
$$

where, in the top line, bidder i is the winner of the auction, since his bid, b_i, exceeds
that of the highest competing bidder, $\max_{j \neq i} b_j$. In this case, his net payoff after
paying b_i for the object is $v_i - b_i$. The bottom line, bidder i loses the auction, since
his bid is lower than that of the highest competing bidder. In this case, his payoff is
zero. Finally, in the middle line, there is a tie, since bidder i's bid coincides with
that of the highest competing bidder. In this case, the object is assigned with equal
probability among the two winning bidders.

Part (a) To formulate this first price auction (FPA) as a Bayesian game we need to
identify the following elements:

- Set of N players (bidders).
- Set of actions available to bidder i, A_i, which in this case is the space of
  admissible bids $b_i \in [0, \infty)$, i.e., a positive real number.
- Set of types of every bidder i, T_i, which in this case is just his set of valuations
  for the object on sale, i.e., $v_i \in [0, 1]$.
- A payoff function u_i, as described above in $u_i(b_i, b_{-i}, v_i)$.

- The probability that bidder i wins the auction is, intuitively, the probability that his bid exceeds that of the highest competing bidder, i.e., $p_i = prob_i(win) = prob_i(b_i \geq \max_{j \neq i} b_j)$.

  Hence, the Bayesian game is compactly defined as: $G = <N, (A_i)_{i \in N}, T_i, (u_i)_{i \in N}, p_i >$

Part (b) Let $b_i(v_i)$ denote the bid submitted by bidder i with type (valuation) given by v_i. First, note that bidder i will not bid above his own valuation, given that it constitutes a strictly dominated strategy: if he wins he would obtain a negative payoff (since he has to pay for the object a price above his own valuation), while if he loses he would obtain the same payoff submitting a lower bid. Therefore, there must be some "bid shading" in equilibrium, i.e., bidder i's bid lies below his own valuation, $b_i < v$, or alternatively, $b_i(v) = \alpha + \beta v < v$. Hence, we need to show that in a FPA there exists a BNE in which the optimal bidding function is linear and given by $b_i(v) = \alpha + \beta v$ for every bidder i, and for any valuation v.

Proof First, we will find the optimal bidding function using the Envelope Theorem. Afterwards, we will show that the same bidding function can be found by differentiating the bidder's expected utility function with respect to his bid i.e., the so-called first-order approach (or direct approach). For generality, the proof does not make use of the uniform distribution $U \sim [0, 1]$, i.e., $F(v) = v$, until the end. (This would allow you to use other distribution functions different from the uniform distribution.)

Envelope Theorem Approach

Let's take a bidder i whose valuation is given by v_i and who is considering to bid according to another valuation $z_i \neq v_i$ in order to obtain a higher expected payoff. His expected payoff of participating in the auction will then be given by:

$$EU_i(v, z) = prob_i(win)[v_i - b_i(z_i)]$$

where the probability of winning is given by:

$$prob_i(win) = prob_i\left(\max_{j \neq i} b_j(v_j) < b_i(z_i)\right) = prob_i\left(\max_{j \neq i} v_j < z_i\right)$$

given that we assume that the bidding function, $b_i(\cdot)$, is monotonically increasing, and all bidders use the same bidding function. [Note that this does not imply that bidders submit the same bid, but just that they use the same function in order to determine their optimal bid.] In addition, the probability that valuation z_i lies above of the other bidder's valuation, v_j, is $F(z_i) = prob(v_j < z_i)$. Therefore, the probability that z_i exceeds all other $N - 1$ competing bidders' valuations is $F(z_i) \cdot F(z_i) \ldots F(z_i) = F(z_i)^{N-1}$, thus implying that the expected utility from participating in the auction can be rewritten as

$$EU_i(v_i, z_i) = F(z_i)^{N-1}[v_i - b_i(z_i)].$$

Bidder i then chooses to bid according to a valuation z_i that maximizes his expected utility

$$\frac{dEU_i(v, z)}{dz_i} = (N - 1)F(z_i)^{N-2}f(z_i)\left[v_i - \hat{b}(z_i)\right] - F(z_i)^{N-1}\hat{b}'(z_i) = 0$$

For $\hat{b}_i(v)$ to be an optimal bidding function, it should be optimal for the bidder *not* to pretend to have a valuation z_i different from his real one, v_i. Hence $z_i = v_i$ and $b(z_i) = \hat{b}(v)$, implying that the above first-order condition becomes

$$(N - 1)F(v)^{N-2}f(z_i)\left[v_i - \hat{b}(v)\right] - F(v)^{N-1}\hat{b}'(v) = 0$$

And rearranging

$$(N - 1)F(v)^{N-2}f(v)\hat{b}(v) + F(v)^{N-1}\hat{b}'(v) = (N - 1)F(v)^{N-2}f(v)v$$

Note that the left-hand side of the above expression can be alternative represented as $\frac{dF(v)^{N-1}\hat{b}(v)}{dv}$. Hence we can rewrite the above expression as

$$\frac{dF(v)^{N-1}\hat{b}(v)}{dv} = (N - 1)F(v)^{N-2}f(v)v$$

Applying integrals to both sides of the equality, we obtain:

$$F(v)^{N-1}\hat{b}(v) = \int_0^v (N - 1)F(x)^{N-2}f(x)x\,dx$$

Hence, solving for the optimal bidding function $\hat{b}(v)$, we find

$$\hat{b}(v_i) = \frac{1}{F(v)^{N-1}} \int_0^v (N - 1)F(x)^{N-2}f(x)x\,dx$$

Finally, since $dF(x)^{N-1} = (N - 1)F(x)^{N-2}f(x)$, we can more completely express the optimal bidding function in the FPA as:

$$\hat{b}(v) = \frac{1}{F(v)^{N-1}} \int_0^v x\,dF(x)^{N-1}$$

Direct Approach

We can confirm our previous result by finding the optimal bidding function without using the Envelope Theorem. Similarly as in our above approach, bidder i wins the object when his bid, b_i, exceeds that of the highest competing bidder that is

$$prob(win) = prob(\beta(Y_1) \leq b_i)$$

where Y_1 is the highest valuation among the $N - 1$ remaining bidders (the highest order statistic), and $\beta(\cdot)$ is the symmetric bidding function that all bidders use. We can, hence, invert this bidding function, to express bidder i's probability of winning in terms of valuations (rather than in terms of bids),

$$prob(\beta(Y_1) \leq b_i) = prob(Y_1 \leq \beta^{-1}(b_i)) = G(\beta^{-1}(b_i)) \qquad (8.1)$$

where in the first equality we apply the inverse of the bidding function $\beta^{-1}(\cdot)$ on both sides of the inequality inside the parenthesis. Intuitively, we move from expressing the probability of winning in terms of bids to expressing it in terms of valuations, where Y_1 is the valuation of the highest competing bidder and $\beta^{-1}(b_i)$ is bidder i's valuation. Figure 8.1 depicts the bidding function of bidder i: when he has a valuation v_i for the object on sale, his bid becomes $\beta(v_i)$. If, observing such a bid, we seek to infer his initial valuation, we would move from the vertical to the horizontal axes on Fig. 8.1 by inverting the bidding function, i.e., $\beta^{-1}(v_i) = v_i$.

The second equality of expression (8.1) evaluates the probability that the valuation of all competing bidders is lower than the valuation of bidder i (given by $\beta^{-1}(b_i) = v_i$).

Hence, the expected payoff function for a bidder with valuation v_i can be rewritten as:

$$EU_i(v_i) = G(\beta^{-1}(b_i)) [v_i - b_i]$$

Fig. 8.1 Bidding function

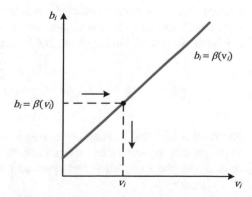

We can now take first-order conditions with respect to bidder i's bid, b_i, in order to find which is his optimal bid,

$$\frac{dEU_i(v_i)}{db_i} = \frac{g\left(\beta^{-1}(b_i)\right)}{\beta'\left(\beta^{-1}(b_i)\right)}[v_i - b_i] - G\left(\beta^{-1}(b_i)\right) = 0$$

and rearranging,

$$g(\beta^{-1}(b_i))[v_i - b_i] - \beta'(\beta^{-1}(b_i))G\left(\beta^{-1}(b_i)\right) = 0$$

Given that in equilibrium we have that $\beta(v) = b_i$ we can invert this function to obtain $v = \beta^{-1}(b_i)$; as depicted in Fig. 8.1. As a consequence, we can rewrite the above expression as:

$$g(v_i)(v_i - \beta(v_i)) - \beta(v_i)G(v_i) = 0,$$

Or more compactly as,

$$g(v_i)\beta(v_i) + \beta(v_i)G(v_i) = g(v_i)v_i$$

Similarly as in our proof using the Envelope Theorem, note that the left-hand side can be alternatively expressed as $\frac{d[\beta(v_i)G(v_i)]}{dv_i}$, implying that the equality can be rewritten as:

$$\frac{d[\beta(v_i)G(v_i)]}{dv_i} = g(v_i)v_i$$

Integrating on both sides, yields

$$\beta(v_i)G(v_i) = \int_0^v g(y)y\,dy + C,$$

where C is the integration constant and it is zero given that $\beta(0) = 0$, i.e., a bidder with a zero valuation for the object can be assumed to submit a zero bid. Hence, solving for the optimal bid $\beta(v_i)$ we find that the optimal bidding function in a FPA is:

$$\beta(v_i) = \frac{1}{G(v_i)}\int_0^v g(y)y\,dy$$

Equivalent Bidding Functions. Let us now check that the two optimal bidding functions that we found (the first one with the Envelope Theorem approach, and the second with the so-called Direct approach) are in fact equivalent.

First note that $G(x) = F(x)^{N-1}$, and differentiating with respect to x yields $G'(x) = (N-1)F(x)^{N-2}f(x)$. Hence, substituting into the expression of the optimal bidding function that we found using the Direct approach, we obtain:

$$\beta(v_i) = \frac{1}{G(v_i)} \int_0^v g(y)y\, dy = \frac{1}{F(v)^{N-1}} \int_0^v (N-1)F(x)^{N-2}f(x)x\, dx$$

And noting that $(N-1)F(x)^{N-2}f(x) = \frac{dF(x)^{N-1}}{dx}$, we can express this bidding function as $\beta(v_i) = \frac{1}{F(v)^{N-1}} \int_0^v x\, dF(x)^{N-1}$, which exactly coincides with the optimal bidding function we found using the Envelope Theorem. Hence, $\beta(v_i) = \hat{b}(v_i)$, and both methods are equivalent.

Uniformly Distributed Valuation. Given that in this case we know that the valuations are drawn from a uniform cumulative distribution $F(x) = x$ with density $f(x) = 1$ and support $[0,1]$, we obtain that:

$$F(v)^{N-1} = v^{N-1}, \text{ and } dF(v)^{N-1} = (N-1)x^{N-1}dx$$

Therefore, the optimal bidding function $\hat{b}(v_i)$ in this case becomes

$$\hat{b}(v_i) = \frac{1}{v^{N-1}} \int_0^v x(N-1)x^{N-2}dx = \frac{1}{v^{N-1}} \int_0^v (N-1)x^{N-1}dx$$

rearranging,

$$= \frac{(N-1)}{v^{N-1}} \int_0^v x^{N-1}dx = \frac{(N-1)}{v^{N-1}} \left[\frac{x^N}{N}\right]_0^v$$

$$= \frac{(N-1)}{v^{N-1}} \frac{v^N}{N} = \frac{N-1}{N}v$$

Comparative Statics: More Bidders. Hence, the optimal bidding function in a FPA with N bidders with uniformly distributed valuations is

$$\hat{b}(v_i) = \frac{N-1}{N}v_i$$

Therefore, when we only have $N = 2$ bidders, their symmetric bidding function is $\frac{1}{2}v$. Interestingly, as N increases, bidders also increase their bid, i.e., from $\frac{1}{2}v$ when $N = 2$, to $\frac{2}{3}v$ when $N = 3$, and to $\frac{99}{100}v$ when $N = 100$; as depicted in Fig. 8.2. Intuitively, their bids get closer to their actual valuation for the object i.e., the equilibrium bidding functions approach the 45°-line where $b_i = v_i$. Hence, bidders reduce their "bid shading" as more bidders compete for the same object.

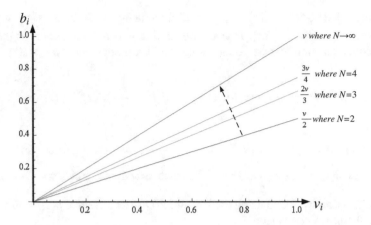

Fig. 8.2 Equilibrium bidding function in FPA with more bidders

Hence,
the optimal bidding function corresponds to the linear function
$b(v_i) = \alpha + \beta v$, where:

- $\alpha = 0$ (vertical intercept is zero) because we have assumed that the optimal bid
  for a bidder with a valuation of zero is zero, $b(0) = 0$.

- $\beta = \frac{1}{F(v)^{N-1}} \int_0^v x \, dF(x)^{N-1}$, which is the slope of the bidding function, e.g., under
  uniformly distributed valuations $\beta = \frac{N-1}{N}$. It represents to what extent the bidder
  shades his own valuation. In the case with two bidders, every player just bids
  half of his valuation, whereas with more competing bidders, bidder i would
  submit bids closer to his own valuation; graphically pivoting the bidding
  function towards the 45°-line where no bid shading occurs.

Exercise 2—Second Price Auction[A]

Consider a setting where every bidder privately observes his/her own valuation for
the object. In addition, assume that N bidders interact in an auction in which they
simultaneously and independently submit their own bids, the highest bidder wins
the object, but he/she pays the second-highest bid (this is the so-called second-price
auction, SPA). All other bidders pay zero.

a. Argue that it is a weakly dominant strategy for a bidder to bid her value in the
   SPA.
b. Are optimal bids affected if the joint distribution of bidders' values exhibits
   correlation, i.e., if bidder i observes that he has a high valuation for the object he

can infer that other bidders must also have a high valuation (in the case of positively correlated values) or that, instead, they must have a low valuation for the object (in the case of negatively correlated values)?

c. Are optimal bids affected by the number of bidders participating in the auction? Are optimal bids affected by bidders' risk aversion?

Part (a) Let $r_i = \max_{j \neq i} b_j$ be the maximum bid submitted by the other players (different from bidder i). Suppose first that player i is considering bidding *above* his own valuation, i.e., $b_i > v_i$. We have the following possibilities:

- If $r_i \geq b_i > v_i$, bidder i does not win the object (or he does so in a tie with other players). However, his expected payoff is zero (if he loses the auction) or negative (if he wins in a tie), thus implying that he obtains the same (or lower) payoff than when bids his own valuation, v_i.
- If $b_i > r_i > v_i$, player i wins the object but gets a payoff of $v_i - r_i < 0$; he would be strictly better off by bidding v_i and not getting the object (i.e. obtaining a payoff of zero).
- If $b_i > v_i > r_i$, player i wins the object and gets a payoff of $v_i - r_i \geq 0$. However, his payoff would be exactly the same if he were to bid his own valuation v_i.

We can apply a similar argument to the case in which bidder i considers bidding *below* his valuation, $b_i < v_i$.

- When $r_i > v_i > b_i$, bidder i loses the object since the highest competing bid, r_i, is higher than his bid, b_i. In this setting, bidder i's expected payoff would be unchanged if he were to submit a bid equal to his own valuation, $b_i = v_i$, instead of $b_i < v_i$.
- If, however, the highest competing bid, r_i, lies between bidder i's valuation and his bid, i.e., $v_i > r_i > b_i$, the bidder would be forgoing a positive payoff by underbidding. He currently loses the object (and gets a zero payoff) whereas if he bids $b_i = v_i$, he would win the object and obtain a positive payoff of $v_i - r_i > 0$. Therefore, bidder i does not have incentives to submit a bid below his valuation for the object.

This argument, hence, establishes that bidding one's valuation is a *weakly dominant* strategy, i.e., submitting bids that lie above or below his valuation will provide him with exactly the same utility level as he obtains when submitting his valuation, or a strictly lower utility; thus implying that the bidder does not have strict incentives to deviate from $b_i = v_i$.

Part (b) By the definition of weak dominance, this means that bidding one's valuation is weakly optimal no matter what the bidding strategies of the other players are. It is, therefore, irrelevant how the other players' strategies are related to their own valuations, or how their valuations are related to bidder i's own valuation. In other words, bidding one's valuation is weakly optimal irrespectively of the presence of correlation (positive or negative) in the joint distribution of player's values.

Part (c) The above equilibrium bidding strategy in the SPA is, importantly, unaffected by the number of bidders who participate in the auction, N, or their risk-aversion preferences. In terms of Fig. 8.2 in Exercise 8.1, this result implies that the equilibrium bidding function in the SPA, $b_i(v_i) = v_i$, coincides with the 45°-line where players do not shade their valuation for the object. Intuitively, since bidders anticipate that they will not have to pay the bid they submitted, but that of the second-highest bidder, every bidder becomes more aggressive in his bidding behavior. In particular, our above discussion considered the presence of N bidders, and an increase in their number does not emphasize or ameliorate the incentives that every bidder has to submit a bid that coincides with his own valuation for the object, $b_i(v_i) = v_i$. Furthermore, the above results remain valid when bidders evaluate their net payoff according to a concave utility function, such as $u(w) = w^\alpha$, where $\alpha \in (1, 0)$, thus, exhibiting risk aversion. Specifically, for a given value of the highest competing bid, r_i, bidder i's expected payoff from submitting a bid $b_i(v_i) = v_i$ would still be weakly larger than from deviating to a bidding strategy strictly above his own valuation for the object, $b_i(v_i) > v_i$, or strictly below it, $b_i(v_i) < v_i$.

Exercise 3—All-Pay Auction[B]

Modify the first-price, sealed-bid auction in Exercise 8.1 so that the loser also pays his bid (but does not win the object). This modified auction is the so-called the all-pay auction (APA), since all bidders must pay their bids, regardless whether they win the object or not.

Part (a) Show that, in the case of two bidders, there is a Bayesian Nash Equilibrium in which every bidder i's optimal bidding function is given by $b_i(v) = \gamma + \delta v + \phi v^2$ for all i and v where $\gamma, \delta, \phi \geq 0$, i.e., every bidder submits a bid which is a convex function of his private valuation v for the object.

Part (b) Analyze how the optimal bidding function varies in the number of bidders, N.

Part (c) How do players' bids in the APA compare to those in the first price auction? What is the intuition behind this difference in bids?

Solution

In this case, the payoff function for bidder i with valuation v_i will be:

$$u_i(b_i, b_{-i}, v_i) = \begin{cases} v_i - b_i & \text{if } b_i > \max_{j \neq i} b_j \\ \frac{v_i - b_i}{2} & \text{if } b_i = \max_{j \neq i} b_j \\ -b_i & \text{if } b_i < \max_{j \neq i} b_j \end{cases}$$

This payoff function coincides with that of the FPA, except for the case in which bidder i's bid is lower than that of the highest competing bidder, i.e., $b_i < \max_{j \neq i} b_j$, and thus loses the auction (see bottom row), since in the APA bidder i must still pay his bid, implying that his payoff in this event is $-b_i$; as opposed to a payoff zero in the FPA.

We need to show that there exists a BNE in which the optimal bidding function in the APA is given by the convex function $b_i(v) = \gamma + \delta v + \phi v^2$, for every bidder i with valuation v. (Similarly as in Exercise 8.1, we only consider that bidders' valuations are distributed according to a uniform distribution function $U \sim [0, 1]$ after finding the optimal bidding function for any cumulative distribution function $F(x)$.)

Proof The expected payoff of a bidder i with a real valuation of v_i but bidding according to a different valuation of $z_i \neq v_i$ is:

$$EU_i(v, z) = prob_i(win)(v_i - b_i(z_i)) + (1 - prob_i(win))(-b_i(z_i))$$

where, if winning, bidder i obtains a net payoff $v_i - b_i(z_i)$; but, if losing, which occurs with probability $1 - prob_i(win)$, he must pay the bid he submitted. Similarly to the FPA, his probability of wining is also given by

$$prob_i(win) = prob_i\left(\max_{j \neq i} b_j(v_j) < b_i(z_i)\right) = prob_i\left(\max_{j \neq i} v_j < z_i\right) = F(z_i)^{N-1},$$

since we assume a symmetric and monotonic bidding function. Hence, the expected utility from participating in a APA can be rewritten as

$$EU_i(v, z) = F(z_i)^{N-1}(v_i - b_i(z_i)) + \left[1 - F(z_i)^{N-1}\right](-b_i(z_i))$$

or, rearranging, as:

$$EU_i(v, z) = F(z_i)^{N-1}v_i - b_i(z_i)$$

Intuitively, if winning, bidder i enjoys his valuation for the object, v_i, but he must pay the bid he submitted both when winning and losing the auction. For completeness, we next show that bidding function $b_i(v) = \gamma + \delta v + \phi v^2$ is optimal in the APA using first the Envelope Theorem approach and afterwards using the so-called Direct approach.

Envelope Theorem Approach

Since the bidder is supposed to be utility maximizing, we can find the following first-order condition using the Envelope Theorem:

$$\frac{dEU_i(v, z)}{dz_i} = (N - 1)F(z_i)^{N-2}f(z_i)v_i - \frac{db_i(z_i)}{dz_i} = 0$$

Hence, for $b_i(v)$ to be an optimal bidding function, it should not be optimal for the bidder to pretend to have a valuation z_i different from his real one, v_i. Hence $z_i = v_i$, which entails that the above first-order condition becomes

$$(N-1)F(v_i)^{N-2}f(v_i)v_i = \frac{d\hat{b}_i(v_i)}{dv_i}$$

Given that this equality holds for every valuation, v, we can apply integrals on both sides of the equality to obtain,

$$\hat{b}(v_i) = \int_0^v (N-1)F(x)^{N-2}f(x)xdx + C,$$

where C is the constant of integration, which equals 0 given that $\hat{b}(0) = 0$. Hence, using the property that $dF(x)^{N-1} = (N-1)F(x)^{N-2}f(x)$, we obtain that the optimal bidding function for this APA is:

$$\hat{b}(v_i) = \int_0^v xdF(x)^{N-1}$$

Direct Approach

We can alternatively find the above bidding function by using a methodology that does not require the Envelope Theorem. We know that bidder i wins the auction if his bid, b_i, exceeds that of the highest competing bidder, $\beta(Y_i)$, where Y_1 represents the highest valuation among the $N-1$ remaining bidders (the first order statistic). Hence, the probability of winning is given by:

$$prob(win) = prob(\beta(Y_1) \le b_i) = prob\left(Y_1 \le \beta^{-1}(b_i)\right) = G(\beta^{-1}(b_i))^{N-1}$$

(for a discussion of this expression, see Exercise 8.1). Thus, the expected payoff function for a bidder with valuation v_i is given by:

$$EU_i(v_i) = G(\beta^{-1}(b_i))(v_i - b_i) + \left[1 - G(\beta^{-1}(b_i))\right](-b_i)$$

or, rearranging,

$$EU_i(v_i) = G(\beta^{-1}(b_i))v_i - b_i$$

We can now take first-order conditions with respect to the optimal bid of this player, b_i, obtaining

$$\frac{dEU_i(v_i)}{db_i} = \frac{g\left(\beta^{-1}(b_i)\right)}{\beta'\left(\beta^{-1}(b_i)\right)} v_i - 1 = 0$$

Multiplying both sides by $\beta'\left(\beta^{-1}(b_i)\right)$, and rearranging,

$$g\left(\beta^{-1}(b_i)\right)v_i - \beta'\left(\beta^{-1}(b_i)\right) = 0$$

Given that in equilibrium we have that $\beta(v_i) = b_i$, we can invert this bidding function to obtain $v_i = \beta^{-1}(b_i)$, Therefore, we can rewrite the above expression as:

$$g(v_i)v_i - \beta'(v_i) = 0 \text{ , or as } \beta'(v_i) = g(v_i)v_i$$

Given that this equality holds for any value of v, we can integrate on both sides of the equality, to obtain $\beta(v) = \int_0^v g(y)y dy + C$, where C is the integration constant which is zero given that $\beta(0) = 0$. Thus, the optimal bidding function in an APA is:

$$\beta(v) = \int_0^v g(y)y \, dy$$

Equivalent Bidding Functions. Let us now check that the two optimal bidding functions that we found (the first using the Envelope Theorem approach, and the second with the Direct approach) are in fact equivalent. First, note that $G(x) = F(x)^{N-1}$ and, hence, its derivative is $g(x) = (N-1)F(x)^{N-2}f(x)$. Substituting $g(x)$ into the expression of the optimal bidding function that we found using the Direct approach, yields

$$\beta(v_i) = \int_0^v x(N-1)F(x)^{N-2}f(x)dx = \int_0^v xdF(x)^{N-1}$$

since $(N-1)F(x)^{N-2}f(x) = dF(x)^{N-1}$. Hence, $\beta(v_i) = \hat{b}(v_i)$ and both methods produce equivalent bidding functions.

Uniform distribution. Given that in this case we know that the valuations are drawn from a uniform distribution $F(x) = x$ with density $f(x) = 1$ and support $[0,1]$, $F(x)^{N-1} = x^{N-1}$ and its derivate is $dF(x)^{N-1} = (N-1)x^{N-2}dx$, implying that the optimal bidding function becomes

$$\hat{b}(v_i) = \int_0^v x(N-1)x^{N-2}dx = (N-1)\int_0^v x^{N-1}dx = (N-1)\left[\frac{x^N}{N}\right]_0^v = \frac{N-1}{N}v^N$$

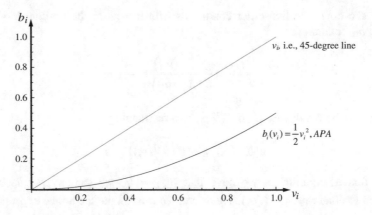

Fig. 8.3 Optimal bidding function in the APA

Therefore, in the case of $N = 2$ bidders, the optimal bidding function for the APA will be $\hat{b}(v_i) = \frac{1}{2}v_i^2$, as depicted in Fig. 8.3.

Hence, this (convex) optimal bidding function corresponds to $b_i(v_i) = \gamma + \delta v + \phi v^2$ where $\gamma = 0$ (vertical intercept is zero) because we have assumed that the optimal bid for a bidder with a valuation of zero is zero, i.e., $\hat{b}_i(0) = 0$; $\delta = 0$ and $\phi = 1/2$, which implies that the optimal bidding function is non linear.

(b) When increasing the number of bidders, N, we obtain a more convex function. Figure 8.4 depicts the optimal bidding function in the APA with two bidders, $\frac{v^2}{2}$, three bidders, $\frac{2}{3}v^3$, and ten bidders, $\frac{9}{10}v^{10}$. Intuitively, as more bidders compete for the object, bid shading becomes substantial when bidder i has a relatively low valuation, but induces him to bid more aggressively when his valuation is likely the highest among all other players, i.e., when $v_i \to 1$.

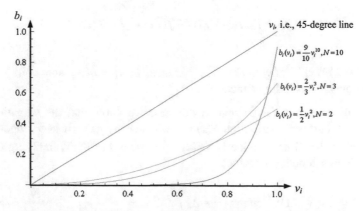

Fig. 8.4 Comparative statics of the optimal bidding function in the APA

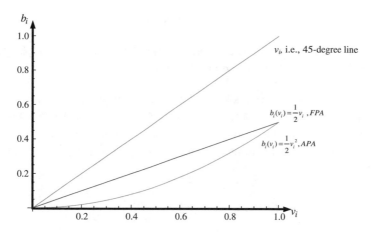

Fig. 8.5 Optimal bidding functions in FPA versus APA

(c) Figure 8.5 compares the optimal bidding function (for the case of uniformly distributed valuations and two bidders) in the FPA against that in the APA.

As we can see in Figure 8.5, every bidder in the APA bids less aggressively than in the FPA, since he has to pay the bid he submits, i.e., the optimal bidding function for the APA lies below that of the FPA. More formally, the optimal bidding function in the FPA, $b^{FPA}(v) = \frac{N-1}{N} v$, lies above that in the APA, $b^{FPA}(v) = \frac{N-1}{N} v^N$, since the difference

$$b^{FPA}(v) - b^{FPA}(v) = \frac{(N-1)(v - v^N)}{N}$$

is positive, since $N \geq 2$ and $v > v^N$, graphically, the 45°-line lies above any convex function such as v^N.

Exercise 4—All-Pay Auctions (Easier Version)[A]

Consider the following all-pay auction with two bidders privately observing their valuation for the object. Valuations are uniformly distributed, i.e., $v_i \sim U[0,1]$. The player submitting the highest bid wins the object, but all players must pay the bid they submitted. Find the optimal bidding strategy, taking into account that it is of the form $b_i(v_i) = m \cdot v_i^2$, where m denotes a positive constant.

Solution

- Bidder i's expected utility from submitting a bid of x dollars in an all-pay auction is

$$EU_i(x|v_i) = prob(win) \cdot v_i - x$$

where, if winning, bidder i gets the object (which he values at v_i), but he pays his bid, x, both when winning and losing the auction

- Let us now specify the probability of winning, $prob(win)$. If bidder i submits a bid \$$x$ using a bidding function $x = m \cdot v_i^2$, we can recover the valuation v_i that generated such a bid, i.e., solving for v_i in $x = m \cdot v_i^2$ we obtain $\sqrt{\frac{x}{m}} = v_i$. Hence, since valuations are distributed according to $v_i \sim U[0,1]$, the probability of winning is given by

$$prob(v_j < v_i) = prob\left(v_j < \sqrt{\frac{x}{m}}\right) = \sqrt{\frac{x}{m}}.$$

- Therefore, the above expected utility becomes

$$EU_i(x|v_i) = \sqrt{\frac{x}{m}} \cdot v_i - x$$

Taking first-order conditions with respect to the bid, x, we obtain

$$-1 + \frac{v_i \sqrt{\frac{x}{m}}}{2x} = 0$$

and solving for x, we find bidder i's optimal bidding function in the all-pay auction

$$b_i(v_i) = \frac{1}{4m} \cdot v_i^2$$

- If, for instance, $m = 0.5$, this bidding function becomes $b_i(v_i) = \frac{1}{2} \cdot v_i^2$. Figure 8.6 depicts this bidding function, comparing it with bids that coincide with players' valuation for the object, i.e., $b_i = v_i$ in the 45°-line. (For a more general identification of optimal bidding functions in APAs, see Exercise #3.)

Exercise 5—Third-Price Auction[A]

Consider a third-price auction, where the winner is the bidder who submits the highest bid, but he/she only pays the third highest bid. Assume that you compete against two other bidders, whose valuations you are unable to observe, and that your valuation for the object is \$10. Show that bidding above your valuation (with a bid of, for instance, \$15) can be a best response to the other bidders' bid, while submitting a bid that coincides with your valuation (\$10) might not be a best response to your opponents' bids.

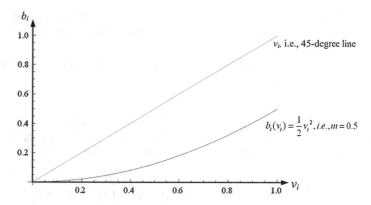

Fig. 8.6 Optimal bidding function in the APA

Solution

- If you believe that the other players' bids are \$5 and \$12, then submitting a bid that coincides with your valuation, \$10, will only lead you to lose the auction, yielding a payoff of zero. If, instead, you submit a bid of \$15, your bid becomes the highest of the three, and you win the auction. Since, upon winning, you pay the third highest bid, which in this case is only \$5, your net payoff is $v_i - p = 10 - 5 = 5$.
- Therefore, submitting a bid above your own valuation for the object can be a best response in the third-price auction. Intuitively, this occurs if one of your competitors submits a bid above your own valuation, while the other submits a bid below your valuation.

Exercise 6—FPA with Risk-Averse Bidders[B]

In previous exercises, we assumed that both the seller and all bidders are risk neutral. Let us next analyze how our equilibrium results would be affected if bidders are risk averse, i.e., their utility function is concave in income, w, and given by $u(w) = w^\alpha$, where $0 < \alpha \leq 1$ denotes bidder i's risk-aversion parameter. In particular, when $\alpha = 1$ he is risk neutral, while when α decreases, he becomes risk averse.[1]

---

[1]An example you have probably encountered in intermediate microeconomics courses includes the concave utility function $u(x) = \sqrt{x}$ since $\sqrt{x} = x^{1/2}$. As a practice, note that the Arrow-Pratt coefficient of absolute risk aversion $r_A(x) = -\frac{u''(x)}{u'(x)}$ for this utility function yields $\frac{1-\alpha}{x}$, confirming that, when $\alpha = 1$, the coefficient of risk aversion becomes zero, but when $0 < \alpha < 1$, the coefficient is positive. That is, as α approaches zero, the function becomes more concave.

Part (a) Find the optimal bidding function, $b(v_i)$, of every bidder i in a FPA where $N = 2$ bidders compete for the object and where valuations are uniformly distributed $U \sim [0, 1]$.

Part (b) Explain how this bidding function is affected when bidders become more risk averse.

Solution

First, note that the probability of winning the object is unaffected, since, for a symmetric bidding function $b_i(v_i) = a \cdot v_i$ for bidder i, where $a \in (0, 1)$, the probability that bidder i wins the auction against another bidder j is

$$prob(b_i > b_j) = prob(b_i > a \cdot v_j) = prob\left(\frac{b_i}{a} > v_j\right) = \frac{b_i}{a}$$

where the first equality is due to the fact that $b_j = a \cdot v_j$, the second equality rearranges the terms in the parenthesis, and the last equality makes use of the uniform distribution of bidders' valuations. Therefore, bidder i's expected utility from participating in this auction by submitting a bid b_i when his valuation is v_i is given by

$$EU_i(b_i|v_i) = \frac{b_i}{a} \times (v_i - b_i)^\alpha$$

where, relative to the case of risk-neutral bidders analyzed in Exercise 8.1, the only difference arises in the evaluation of the net payoff from winning, $v_i - b_i$, which it is now evaluated with the concave function $(v_i - b_i)^\alpha$. Taking first-order conditions with respect to his bid, b_i, yields

$$\frac{1}{a}(v_i - b_i)^\alpha - \frac{b_i}{a}\alpha(v_i - b_i)^{\alpha-1} = 0,$$

and solving for b_i, we find the optimal bidding function,

$$b_i(v_i) = \frac{v_i}{1 + \alpha}.$$

Importantly, this optimal bidding function embodies as a special case the context in which bidders are risk neutral (as in Exercise 8.1). Specifically, when $\alpha = 1$, bidder i's optimal bidding function becomes $b_i(v_i) = \frac{v_i}{2}$. However, when his risk aversion increases, i.e., α decreases, bidder i's optimal bidding function increases. Specifically,

$$\frac{\partial x(v_i)}{\partial \alpha} = -\frac{v_i}{(1 - a)^2}$$

which is negative for all parameter values. In the extreme case in which α decreases to $\alpha \to 0$, the optimal bidding function becomes $x(v_i) = v_i$, and players do not

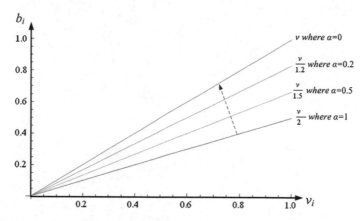

Fig. 8.7 Risk aversion and optimal bidding strategies in the FPA

practice bid shading. Figure 8.7 illustrates the increasing pattern in players' bidding function, starting from $\frac{v_i}{2}$ when bidders are risk neutral, $\alpha = 1$, and approaching the 45-degree line (no bid shading) as players become more risk averse.

Intuitively, a risk-averse bidder submits more aggressive bids than a risk-neutral bidder in order to minimize the probability of losing the auction. In particular, consider that bidder i reduces his bid from b_i to $b_i - \varepsilon$. In this context, if he wins the auction, he obtains an additional profit of ε, since he has to pay a lower price for the object he acquires. However, by lowering his bid, he increases the probability of losing the auction. Importantly, for a risk-averse bidder, the positive effect of slightly lowering his bid, arising from getting the object at a cheaper price, is offset by the negative effect of increasing the probability that he loses the auction. In other words, since the possible loss from losing the auction dominates the benefit from acquiring the object at a cheaper price, the risk-averse bidder does not have incentives to reduce his bid, but rather to increase it, relative to risk-neutral bidders.

Perfect Bayesian Equilibrium and Signaling Games

9

Introduction

This chapter examines again contexts of incomplete information but in sequential move games. Unlike simultaneous-move settings, sequential moves allow for players' actions to convey or conceal the information they privately observe to players acting in subsequent of the game and who did not have access to such information (uninformed players). That is, we explore the possibility that players' actions may signal certain information to other players acting latter on in the game.

In order to analyze this possibility, we first examine how uninformed players form beliefs upon observing certain actions, by the use of the so-called Bayes' rule of conditional probability. Once uninformed players update their beliefs, they optimally respond to other players' actions. If all players' actions are in equilibrium, then such strategy profile constitutes a Perfect Bayesian Equilibrium (PBE) of the sequential-move game under incomplete information; as we next define.

PBE. A strategy profile for N players $(s_1^*(\theta_1), \ldots, s_N^*(\theta_N))$ and a system of beliefs $(\mu_1, \mu_2, \ldots, \mu_N)$ are a PBE if:

(a) the strategy of every player i, $s_i^*(\theta_i)$, is optimal given the strategy profile selected by other players $s_{-i}^*(\theta_{-i})$ and given player i's beliefs, μ_i; and

(b) the beliefs of every player i, μ_i, are consistent with Bayes' rule (i.e., they follow Bayesian updating) wherever possible.

In particular, if the first-mover's actions differ depending on the (private) information he observes, then his actions "speak louder than words," thus giving rise to a separating PBE. If, instead, the first-mover behaves in the same manner regardless of the information he observes, we say that such strategy profile forms a pooling PBE.

The original version of the chapter was revised: The erratum to the chapter is available at: 10.1007/978-3-319-32963-5_11

© Springer International Publishing Switzerland 2016
F. Munoz-Garcia and D. Toro-Gonzalez, *Strategy and Game Theory*,
Springer Texts in Business and Economics, DOI 10.1007/978-3-319-32963-5_9

We first study games in which both first and second movers have only two available actions at their disposal, thus allowing for a straightforward specification of the second mover's beliefs. To facilitate our presentation, we accompany each of the possible strategy profiles (two separating and two pooling) with their corresponding graphical representation, which helps focus the reader's attention on specific nodes of the game tree. We then examine a simplified version of Spence's (1974) job market signaling game where workers can only choose two education levels and firms can only respond with two possible actions: hiring the worker as a manager or as a cashier.

Afterwards, we study more elaborate signaling games in which the uninformed agent can respond with more than two actions, and then we move to more general settings in which both the informed and the uninformed player can choose among a continuum of strategies. In particular, we consider an industry where the incumbent firm observes its production costs, while the potential entrant does not, but uses the incumbent's pricing decision as a signal to infer this incumbent's competitiveness.

We finally examine signaling games where the privately informed player has three possible types, and three possible messages, and investigate under which conditions a fully separating strategy profile can be sustained as a PBE of the game (where every type of informed player chooses a different message), and under which circumstances only a partially separating profile can be supported (where two types choose the same message and the third type selects a different message).

We end the chapter with an exercise that studies equilibrium refinements in incomplete information games, applying the Cho and Kreps' (1987) Intuitive Criterion to a standard signaling game where the informed player has only two types and two available messages. In this context, we make extensive use of figures illustrating equilibrium and off-the-equilibrium behavior, in order to facilitate the presentation of off-the-equilibrium beliefs, and its subsequent restriction as specified by the Intuitive Criterion.

Exercise 1—Finding Separating and Pooling Equilibria[A]

Consider the following game of incomplete information: nature first determines player 1's type, either high or low, with equal probabilities. Then player 1, observing his own type, decides whether to choose left (L) or Right (R). If he chooses left, the game ends and players' payoffs become $(u_1, u_2) = (2, 0)$, regardless of player 1's type. Figure 9.1 depicts these payoffs in the left-hand side of the tree. If, instead, player 1 chooses Right, player 2 is called on to move. In particular, without observing whether player 1's type is high or low, player 2 must respond by either selecting Up (U) or Down (D). As indicated by the payoffs in the right-hand side of the tree, when player 1's type is high, player 2 prefers to play Up. The opposite preference ranking applies when player 1's type is low.

Part (a) Does this game have a Separating Perfect Bayesian Equilibrium (PBE)?
Part (b) Does this game have a Pooling PBE?

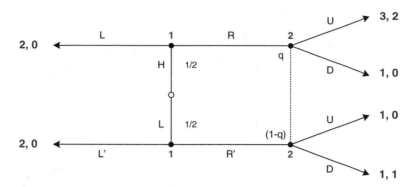

Fig. 9.1 Game tree where player 2 is uninformed

Answer

Part (a) *Separating PBE*

First type of separating equilibrium: RL′

Figure 9.2 shades the branches corresponding to the separating strategy profile RL′, whereby player 1 chooses R when his type is high (in the upper part of the game tree), but selects L′ when his type is Low (see lower part of the game tree).

First step (use Bayes' rule to specify the second mover's beliefs):

After observing that player 1 chose right, the second-mover's beliefs of dealing with a high-type player 1 (upper node of his information set) are

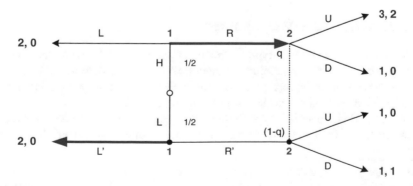

Fig. 9.2 Separating strategy profile RL′

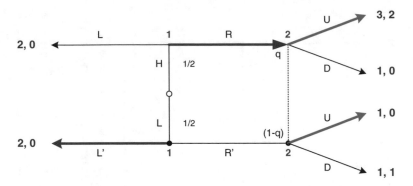

Fig. 9.3 Optimal response by player 2 (U) in strategy profile RL'

$$q = p(H|R) = \frac{p(H) * p(R|H)}{p(R)} = \frac{\frac{1}{2} \times 1}{\frac{1}{2} \times 1 + \frac{1}{2} \times 0} = 1$$

Intuitively, this implies that the second mover, after observing that player 1 chose Right, assigns full probability to such a message originating from an H-type of first mover. In other words, only a H-type would be selecting Right in this separating strategy profile.

Second step (focus on the second mover):

After observing that player 1 chose Right, and given the second-mover's beliefs specified above, the second mover is essentially convinced on being on the upper node of the information set. At this point, he obtains a higher payoff responding with U, which provides him a payoff of 2, than with D, which only yields a payoff of 0. Hence, the second mover responds with U. To keep track of player 2's optimal response in this strategy profile, Fig. 9.3 shades the branches corresponding to U.[1]

Third step (we now analyze the first mover):

When the first mover is H-type (in the upper part of Fig. 9.3), he prefers to select R (as prescribed in this separating strategy profile) than deviate towards L, given that he anticipates that player 2 will respond with U. Indeed, player 1's payoff from choosing R, 3, is larger than that from deviating to L', 2.

When the first mover is L-type (in the lower part of Fig. 9.3), he prefers to select L' (as prescribed in this separating strategy profile) than deviate towards R', since he

---

[1]Note that player 2 responds with U after observing that player 1 chooses Right, and that such response cannot be made conditional on player 1's type (since player 2 does not observe his opponent's type). In other words, player 2 responds with U both when player 1's type is high and when it is low (for this reason, both arrows labeled with U are shaded in Fig. 9.3 in the upper and lower part of the game tree).

can also anticipate that player 2 will respond with U. Indeed, player 1's payoff from choosing L', 2, is larger than that from his deviation towards R', 1.

Then, this separating strategy profile can be sustained as a PBE where (RL', U) and beliefs are $q = 1$.

Second type of separating equilibrium, LR'

Figure 9.4 depicts this separating strategy profile, in which the high-type of player 1 now chooses L while the low-type selects R', as indicated by the thick shaded arrows of the figure.

First step (use Bayes' rule to determine the second mover's beliefs):

After observing that the first mover chose Right, the second-mover's beliefs of dealing with a high-type of player 1 are

$$q = \frac{p(H) * p(R|H)}{p(R)} = \frac{\frac{1}{2} \times 0}{\frac{1}{2} \times 1 + \frac{1}{2} \times 0} = 0$$

Intuitively, this implies that the second mover, after observing that the first mover chose Right, assigns full probability to such a message originating from a low-type. Alternatively, a message of "Right" can never originate from an H-type in this separating strategy profile, i.e., $q = 0$.

Second step (examine the second mover):

After observing that the first mover chose Right, and given that the second-mover's beliefs specify that he is convinced of being in the lower node of his information set, he responds with D, which provides him a payoff of 1, rather than with U, which provides him a payoff of 0 (conditional on being in the lower node); see the lower right-hand corner of Fig. 9.4 for a visual reference. To keep

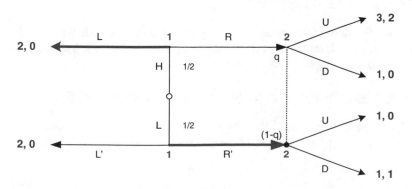

Fig. 9.4 Separating strategy profile LR'

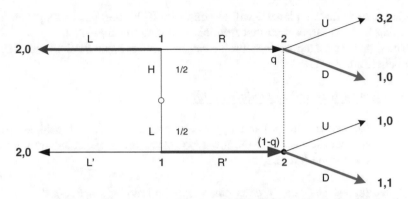

Fig. 9.5 Optimal response of player 2 (D) in strategy profile LR'

track of the optimal response of D from player 2, Fig. 9.5 shades the D branches in blue color.

Third step (analyze the first mover):

When the first mover is H-type (upper part of the game tree in Fig. 9.5), he prefers to select L (as prescribed in this separating strategy profile) since L yields him a payoff of 2 while deviating towards R would only provide a payoff of 1, as he anticipates player 2 will respond with D.

When the first mover is L-type (lower part of the game tree), however, he prefers to deviate towards L' than selecting R' (the strategy prescribed for him in this separating strategy profile) since his payoff from R' is only 1 whereas that of deviating towards L' is 2. Then, this separating strategy profile *cannot* be sustained as a PBE, since one of the players (namely, the low type) has incentives to deviate.

Part (b) *Does this game have a pooling PBE?*

First type of pooling equilibrium: LL'

Figure 9.6 depicts this pooling strategy profile in which both types of player 1 choose Left, i.e., L for the high type in the upper part of the game tree and L' for the low type in the lower part of the tree.

First step (use Bayes' rule to determine the second mover's beliefs):

Upon observing that the first mover chose Right (which occurs off-the-equilibrium, as indicated in the figure), the second-mover's beliefs are

$$q = \frac{0.5 * 0}{0.5 * 0 + 0.5 * 0} = \frac{0}{0}$$

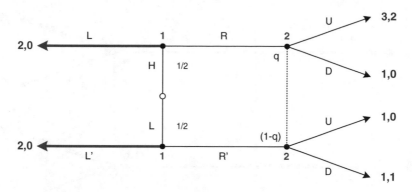

Fig. 9.6 Pooling strategy profile LL′

Thus, beliefs cannot be updated using Bayes' rule, and must be left undefined in the entire range of probabilities, $q \in [0, 1]$.

Second step (examine the second mover):

Let us analyze the second-mover's optimal response. Note that the second mover is only called on to move if the first mover chooses Right, which occurs off-the-equilibrium path. In such an event, the second mover must compare his expected utility from responding with U versus that of selecting D, as follows

$$EU_2(U|R) = 2 \times q + 0 \times (1-q) = 2q$$

$$EU_2(D|R) = 0 \times q + 1 \times (1-q) = 1 - q$$

Hence, $EU_2(U|R) > EU_2(D|R)$ holds if $2q > 1 - q$, or simply $q > \frac{1}{3}$.

Third step (analyze the first mover):

Let us next divide our following analysis of the first mover (since we have already examined the second mover) into two cases depending on the optimal response of the second mover: Case 1: $q > \frac{1}{3}$, which implies that the second mover responds selecting U; and Case 2: $q < \frac{1}{3}$, leading the second mover to respond with D.

CASE 1: $q > \frac{1}{3}$.

Figure 9.7 depicts the optimal response of the second mover (U) in this case where $q > \frac{1}{3}$, where we shaded the arrows corresponding ot U.

When the first mover is H-type (in the upper part of the tree), he prefers to deviate towards R rather than selecting L (as prescribed in this strategy profile) since his payoff from deviating to R, 3, is larger than that from L, 2, given that he anticipates that the second mover will respond with U since $q > \frac{1}{3}$. This is already sufficient to conclude that the pooling strategy profile LL′ cannot be sustained as a

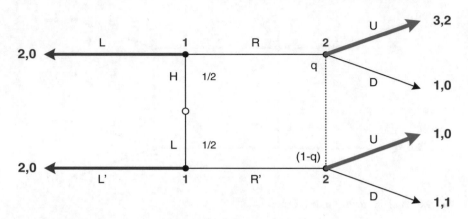

Fig. 9.7 Optimal response of player 2 (U) in strategy profile LL′

PBE of the game. When off-the-equilibrium beliefs are relatively high, i.e., $q > \frac{1}{3}$. (As a practice, you can check that the L-type of first mover does not have incentives to deviate, since his payoff from L′, 2, is larger than that from R′, 1. However, since we have found a type of player with incentives to deviate, we conclude that the pooling strategy profile LL′ cannot be supported as a PBE).

CASE 2: $q < \frac{1}{3}$.

In this setting the second mover responds with D, as depicted in Fig. 9.8, where the arrows corresponding to D are thick and shaded in blue color.

When the first mover is H-type, he prefers to select L (as prescribed in this strategy profile), which provides him with a payoff of 2, rather than deviating towards R, which only yields a payoff of 1, as indicated in the upper part of the figure (given that player 1 anticipates that player 2 will respond with D).

Similarly, when the first mover is L-type, he prefers to select L′ (as prescribed in this strategy profile), which yields a payoff of 2, rather than deviating towards R′, which only yields a payoff of 1, as indicated in the lower part of the figure.

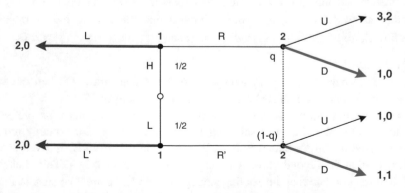

Fig. 9.8 Optimal response of player 2(D) in strategy profile LL′

Hence, the pooling strategy profile LL' can be sustained as a PBE of the game when off-the-equilibrium beliefs are relatively low, i.e., $q < \frac{1}{3}$.

Second type of pooling equilibrium: RR'

Figure 9.9 illustrates that pooling strategy profile in which both types of player 1 choose to move Right; as indicated by the thick arrows.

First step (use Bayes' rule to define he second mover's beliefs) :

After observing that the first mover chose Right (which occurs in-equilibrium, as indicated in the shaded branches of the figure), the second-mover's beliefs are

$$q = \frac{\frac{1}{2}}{\frac{1}{2} + \frac{1}{2}} = \frac{1}{2}$$

Therefore, the second-mover's updated beliefs after observing Right coincide with the prior probability distribution over types. Intuitively, this indicates that, as both types of player 1 choose to play Right, then the fact of observing Right does not provide player 2 more precise information about player 1's type than the probability with which each type occurs in nature, i.e., $\frac{1}{2}$. In this case, we formally say that the prior probability of types (due to nature) coincides with the posterior probability of types (after applying Bayes' rule) or, more compactly, that priors and posteriors coincide.

Second step (examine the second mover):

Let us analyze the second-mover's optimal response after observing that the first mover chose Right. The second mover must compare his expected utility from responding with U versus that of selecting D, as follows

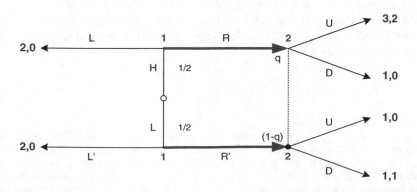

Fig. 9.9 Pooling strategy profile RR'

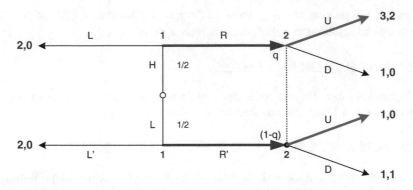

Fig. 9.10 Optimal response for player 2 (U) in strategy profile RR′

$$EU_2(U|R) = 2 \times \frac{1}{2} + 0 \times \left(1 - \frac{1}{2}\right) = 1$$

$$EU_2(D|R) = 0 \times \frac{1}{2} + 1 \times \left(1 - \frac{1}{2}\right) = \frac{1}{2}$$

Hence $EU_2(U|R) > EU_2(D|R)$ since $1 > \frac{1}{2}$, and the second mover responds with U.

Third step (analyze the first mover):

Given that the first mover can anticipate that the second mover will respond with U (as analyzed in our previous step), we can shade this branch for player 2, as depicted in Fig. 9.10.

When the first mover is H-type, he prefers to select R (as prescribed in this pooling strategy profile), which provides him a payoff of 3, rather than deviating towards L, which only yields a payoff of 2. However, when the first mover is L-type, he prefers to deviate towards L′, which yields a payoff of 2, rather than selecting R′ (as prescribed in this strategy profile), which only provides a payoff of 1.[2]

Hence, the pooling strategy profile in which both types of player 1 choose Right *cannot* be sustained as a PBE of the game since the L-type of first mover has incentives to deviate (from R′ to L′).

---

[2]Importantly, the low-type of player 1 finds R′ to be strictly dominated by L′, since R′ yields a lower payoff than L′ *regardless* of player 2's response. This is an important result, since it allows us to discard all strategy profiles in which R′ is involved, thus leaving us with only two potential candidates for PBE: the separating strategy profile RL′ and the pooling strategy profile LL′. Nonetheless, and as a practice, this exercise examined each of the four possible strategy profiles in this game.

Exercise 2—Job-Market Signaling Game[B]

Let us consider the sequential game with incomplete information depicted in Fig. 9.11. A worker privately observes whether he has a High productivity or a Low productivity, and then decides whether to acquire some education, such as a college degree, which he will subsequently be able to use as a signal about his innate productivity to potential employers. For simplicity, assume that education is not productivity enhancing. A firm that considers hiring the candidate, however, does not observe the real productivity of the worker, but only knows whether the worker decided to acquire a college education or not (the firm can observe whether the candidate has a valid degree). In this setting, the firm must respond hiring the worker as a manager (M) or as a cashier (C).[3]

Part (a) *Consider the separating strategy profile [NE']*

Figure 9.12 depicts the separating strategy profile in which only the low productivity worker chooses to acquire education. (We know it's a rather crazy strategy profile, but we have to check for all possible profiles!)

1. Responder's beliefs:
   Firm's updated beliefs about the worker's type are $p = 1$ after observing No education, since such a message must only originate from a high-productivity worker in this strategy profile; and $q = 0$ after observing Education, given that such a message is only sent by low-productivity workers in this strategy profile. Graphically, $p = 1$ implies that the firm is convinced to be in the upper node

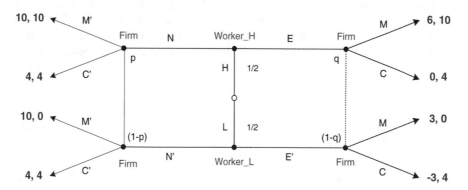

Fig. 9.11 The job-market signaling game

---

[3]This game presents a simplified version of Spence's (1974) labor market signaling game, where the worker can be of only two types (either high or low productivity), and the firm manager can only respond with two possible actions (either hiring the candidates as a manager, thus offering him a high salary, or as a cashier, at a lower salary). A richer version of this model, with a continuum of education levels and a continuum of salaries, is presented in the next chapter.

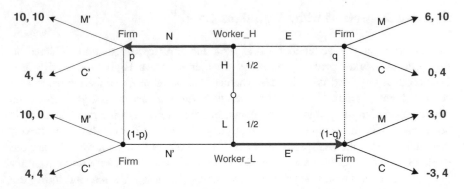

Fig. 9.12 Separating strategy profile NE'

after observing no education (in the left-hand side of the tree); whereas $q = 0$ entails that the firm believes to be in the lower node upon observing an educated worker (in the right-hand side of the tree).

2. Firms' optimal response given their updated beliefs:
   After observing "No Education" the firm responds with M' since, conditional on being in the upper left-hand node of Fig. 9.13, the firm's profit from M', 10, exceeds that from C', 4; while after observing "Education" the firm chooses C, receiving 4, instead of M, which yields a zero payoff. Figure 9.13 summarizes the optimal responses of the firm, M' and C.

3. Given steps 1 and 2, the worker's optimal actions are:
   When the worker is a high-productivity type, he does not deviate from "No Education" (behaving as prescribed) since his payoff from no education and be recognized as a high productivity worker by the firm (thus being hired as a manager), 10, exceeds that from acquiring education and be identified as a low-productivity worker (and thus be hired as a cashier), 0; as indicated in the upper part of Fig. 9.13.

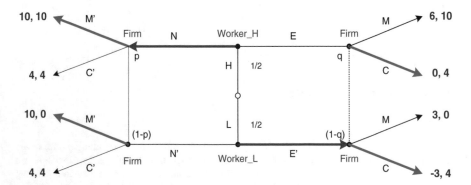

Fig. 9.13 Optimal responses of the firm in strategy profile NE'

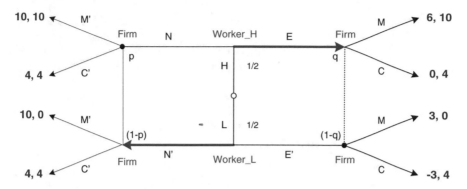

Fig. 9.14 Separating strategy profile EN′

When the worker is of low type, he deviates from "Education", which would only provide him a payoff of −3, to "No Education", which gives him a larger payoff of 10; as indicated in the lower part of Fig. 9.13. Intuitively, by acquiring education, the low-productivity worker is only identifying himself as a low-productivity candidate, which induces the firm to hire him as a cashier. By, instead, deviating towards No Education this worker "mimics" the high-productivity worker.

Therefore, this separating strategy profile *cannot* be supported as a PBE, since we found that at least one type of worker had incentives to deviate.

Part (b) *Consider now the opposite separating strategy profile [EN′]*

Figure 9.14 depicts this (more natural!) separating strategy profile, whereby only the high productivity worker acquires education, as indicated in the thick arrows E and N′.

1. Responder's beliefs:
   Firm's beliefs about the worker's type can be updated using Bayes' rule, as follows:

$$q = p(H|E) = \frac{p(H) \times p(E|H)}{p(E)} = \frac{\frac{1}{2} \times 1}{\frac{1}{2} \times 1 + (1 - \frac{1}{2}) \times 0} = 1$$

   Intuitively, upon observing that the worker acquires education, the firm infers that the worker must be of high productivity, i.e., $q = 1$, since only this type of worker acquires education in this separating strategy profile; while No education conveys the opposite information, i.e., $p = 0$, thus implying that the worker is not of high productivity but instead of low productivity. Graphically, the firm restricts its attention to the upper right-hand corner, i.e., $q = 1$, and to the lower left-hand corner, i.e., $p = 0$.

2. Firms' optimal response given the firm's updated beliefs:
   After observing "Education" the firm responds with M. Graphically, the firm is convinced to be located in the upper right-hand corner of the game tree since

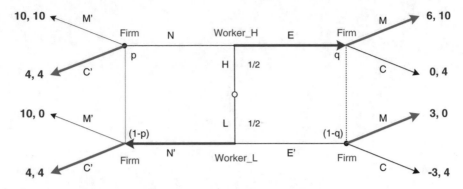

Fig. 9.15 Optimal responses of the firm in strategy profile EN′

$q = 1$. In this corner, the best response of the firm is M, which provides a payoff of 10, rather than C, which only yields a payoff of 4.

After observing "No Education" the firm responds with C′. In this case the firm is convinced to be located in the lower left-hand corner of the game tree given that $p = 0$. In such a corner, the firm's best response is C′, providing a payoff of 4, rather than M′, which yields a zero payoff. Figure 9.15 illustrates these optimal responses for the firm.

3. Given the previous steps 1 and 2, let us now find the worker's optimal actions: When he is a high-productivity worker, he acquires education (as prescribed in this strategy profile) since his payoff from E (6) is higher than that from deviating to N (4); as indicated in the upper part of Fig. 9.15.

When he is a low-productivity worker, he does not deviate from "No Education", i.e., N′, since his payoff from No Education, 4, is larger than that from deviating to Education, 3; as indicated in the lower part of the game tree. Intuitively, even if acquiring no education reveals his low productivity to the firm, and ultimately leads him to be hired as a cashier, his payoff is larger than what he would receive when acquiring education (even if such education helped him be hired as a manager). In short, the low-productivity worker finds too costly to acquire education.

Then, the separating strategy profile [N′E, C′M] can be supported as a PBE of this signaling game, where firm's beliefs are $q = 1$ and $p = 0$.

Part (c) *Pooling strategy profile [EE′] in which both types of workers acquire education.*

Figure 9.16 describes the pooling strategy profile in which all types of workers acquire education by graphically shading branches E and E′ of the tree.

1. Responder's beliefs:
   Upon observing the equilibrium message of Education, the firm cannot further update its beliefs about the worker's type. That is, its beliefs are

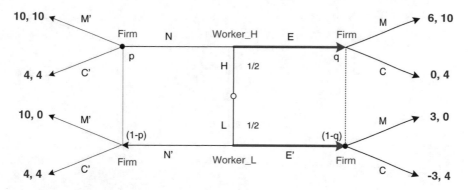

Fig. 9.16 Pooling strategy profile EE′

$$q = p(H|E) = \frac{p(H) * p(E|H)}{p(E)} = \frac{\frac{1}{2} * 1}{\frac{1}{2} * 1 + \frac{1}{2} * 1} = \frac{1}{2}$$

which coincide with the prior probability of the worker being of high productivity, i.e., $q = \frac{1}{2}$. Intuitively, since both types of workers are acquiring education, observing a worker with education does not help the firm to further restrict its beliefs. After observing the off-the-equilibrium message of No Education, the firm's beliefs are

$$p = p(H|NE) = \frac{p(H) * p(NE|H)}{p(NE)} = \frac{\frac{1}{2} * 0}{\frac{1}{2} * 0 + \frac{1}{2} * 0} = \frac{0}{0}$$

and must then be left unrestricted, i.e., $p \in [0, 1]$.

2. Firms' optimal response given its updated beliefs:
Given the previous beliefs, after observing "Education" (in equilibrium): if the firm responds hiring the worker as a manager (M), it obtains an expected payoff of

$$EU_F(M) = \frac{1}{2} \times 10 + \frac{1}{2} \times 0 = 5$$

if, instead, the firm hires him as a cashier (C), its expected payoff is only

$$EU_F(C) = \frac{1}{2} \times 4 + \frac{1}{2} \times 4 = 4$$

Thus inducing the firm to hire the worker as a Manager (M).
After observing "No Education" (off-the-equilibrium): the firm obtains a expected payoffs of

$$EU_F(M') = p \times 10 + (1 - p) \times 0 = 10p, \text{ and}$$

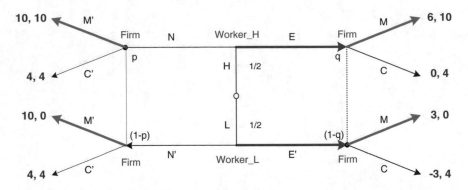

Fig. 9.17 Optimal responses to strategy profile EE′—Case 1

when it hires the worker as a manager (M′), and

$$EU_F(C') = p \times 4 + (1 - p) \times 4 = 4$$

when it hires him as a cashier (C′). (Note that, since firm's beliefs upon observing the off-the-equilibrium message of No Education, p, had to be left unrestricted, we must express the above expected utilities as a function of p.) Hence, the firm prefers to hire him as a manager (M′) after observing no education if and only if $10p > 4$, or $p > 2/5$. Otherwise, the firm hires the worker as a cashier (C′).

3. Given the previous steps 1 and 2, the worker's optimal actions must be divided into two cases (one where the firm's off-the-equilibrium beliefs satisfy $p > 2/5$, thus implying that the firm responds hiring the worker as a manager when he does not acquire education, and another case in which $p \le 2/5$ entailing that the firm hires the worker as a cashier upon observing that he did not acquire education):

   Case 1: When $p > 2/5$ the firm responds hiring him as a manager when he does not acquire education (M′), as a depicted in Fig. 9.17 (see left-hand side of the figure).

   In this setting, if the worker is a high-productivity type, he deviates from "Education," where he only obtains a payoff of 6, to "No Education," where his payoff increases to 10; as indicated in the upper part of the game tree. As a consequence, the pooling strategy profile in which both workers acquire education (EE′) cannot be sustained as a PBE when off-the-equilibrium beliefs satisfy $p > 2/5$.

   Case 2: When $p \le 2/5$ the firm responds hiring the worker as a cashier (C′) after observing the off-the-equilibrium message of No education, as a depicted in the shaded branches on the left-hand side of Fig. 9.18.

   In this context, if the worker is a low-productivity type, he deviates from "Education," where his payoff is only 3, to "No Education," where his payoff increases to 4; as indicated in the lower part of the game tree. Therefore, the pooling strategy profile in which both types of workers acquire education (EE′)

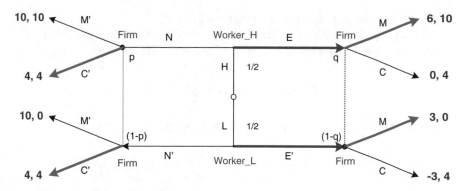

Fig. 9.18 Optimal responses in strategy profile EE'—Case 2

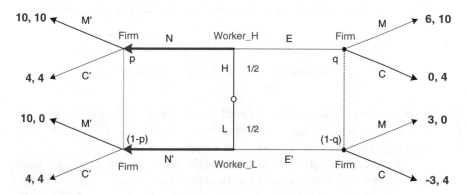

Fig. 9.19 Pooling strategy profile NN'

cannot be sustained when off-the-equilibrium beliefs satisfy $p \leq 2/5$. In summary, this strategy profile cannot be sustained as a PBE for *any* off-the-equilibrium beliefs that the firm sustains about the worker's type.

Part (d) *Let us finally consider the pooling strategy profile [NN'], in which neither type of worker acquires education.*

Figure 9.19 depicts the pooling strategy profile in which both types of worker choose No Education by shading branches N and N' (see thick arrows).

1. Responder's beliefs:
   Analogously to the previous pooling strategy profile, the firm's equilibrium beliefs (after observing No Education) coincide with the prior probability of a high type, $p = \frac{1}{2}$; while its off-the-equilibrium beliefs (after observing Education) are left unrestricted, i.e., $q \in [0, 1]$.

2. Firms' optimal response given its updated beliefs:
   Given the previous beliefs, after observing "No Education" (in equilibrium): if
   the firm hires the worker as a manager (M'), it obtains an expected payoff of

$$EU_F(M') = \frac{1}{2} \times 10 + \frac{1}{2} \times 0 = 5$$

and if it hires him as a cashier (C') its expected payoff is only

$$EU_F(C') = \frac{1}{2} \times 4 + \frac{1}{2} \times 4 = 4,$$

leading the firm to hire the worker as a manager (M') after observing the
equilibrium message of No Education.
After observing "Education" (off-the-equilibrium): the expected payoff the firm
obtains from hiring the worker as a manager (M) or cashier (C) are, respectively,

$$EU_F(M) = q \times 10 + (1 - q) \times 0 = 10q, \text{ and}$$

$$EU_F(C) = q \times 4 + (1 - q) \times 4 = 4$$

Hence, the firm responds hiring the worker as a manager (M) after observing the
off-the-equilibrium message of Education if and only if $10q > 4$, or $q > \frac{2}{5}$.
Otherwise, the firm hires the worker as a cashier.

3. Given the previous steps 1 and 2, let us find the worker's optimal actions. In this
   case, we will also need to split our analysis into two cases (one in which $q > \frac{2}{5}$
   and thus the firm hires the worker as a manager upon observing the
   off-the-equilibrium message of Education, and the case in which $q \leq \frac{2}{5}$, in which
   the firm responds hiring him as a cashier):
   Case 1: When $q > \frac{2}{5}$, the firm responds hiring the worker as a manager
   (M) when he acquires education, as depicted in Fig. 9.20 (see thick shaded
   arrows on the right-hand side of the figure).
   If the worker is a high-productivity type, he plays "No Education" (as prescribed
   in this strategy profile) since his payoff from doing so, 10, exceeds that from
   deviating to E, 6; as indicated in the shaded branches of upper part of the game
   tree. Similarly, if he is a low-productivity type, he plays "No Education" (as
   prescribed) since his payoff from this strategy, 10, is higher than from deviating
   towards E', 3; as indicated in the lower part of the game tree.
   Hence, the pooling strategy profile in which no worker acquires education,
   [NN', M'M], can be supported as a PBE when off-the-equilibrium beliefs satisfy
   $q > \frac{2}{5}$.
   Case 2: When $q \leq \frac{2}{5}$, the firm responds hiring the worker as a cashier (C) upon
   observing the off-the-equilibrium message of Education, as depicted in the thick
   shaded branches at the right-hand side of Fig. 9.21.

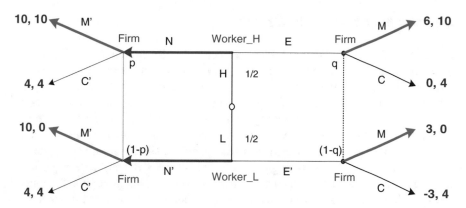

Fig. 9.20 Optimal responses in strategy profile NN′—Case 1

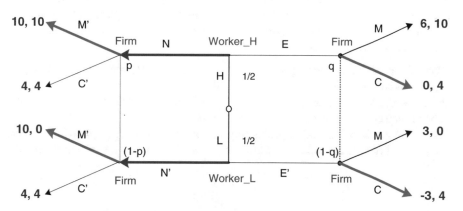

Fig. 9.21 Optimal responses in strategy profile NN′—Case 2

If the worker is a high-productivity type, he does not deviate from No Education since his payoff from N, 10, exceeds the zero payoff he would obtain by deviating to E; as indicated in the shaded branches at the upper part of the game tree. Similarly, if he is a low-productivity worker, he does not deviate from No Education since his (positive) payoff from N′ (10) is larger than the negative payoff he would obtain by deviating to E′, −3; as indicated in the lower part of the game tree.

Therefore, the pooling strategy profile in with no type of worker acquires education, [NN′, M′C], can also be supported when off-the-equilibrium beliefs satisfy $q \leq \frac{2}{5}$.

Exercise 3—Cheap Talk Game[c]

Figure 9.22 depicts a cheap talk game. In particular, the sender's payoff coincides when he sends message m_1 or m_2, and only depends on the receiver's response (either a, b or c) and the nature's type. You can interpret this strategic setting as a lobbyist (Sender) informing a Congressman (Receiver) about the situation of the industry he represents: messages "good situation" or "bad situation" are equally costly for him, but the reaction of the politician to these message (and the actual state of the industry) determine the lobbyist payoff. A similar argument applies for the payoffs of the receiver (Congressman), which do not depend on the particular message he receives in his conversation with the lobbyist, but are only a function of the specific state of the industry (something he cannot observe) and the action he chooses (e.g., the policy that he designs for the industry after his conversation with the lobbyist). For instance, when the sender is type t_1, in the left-hand side of the tree, payoff pairs only depend on the receiver's response (a, b or c) but do not depend on the sender's message, e.g., when the receiver responds with a players obtain (4,3), both when the original message was m_1 and m_2. This cheap talk game can be alternatively represented with Fig. 9.23.

Part (a) Check if a separating strategy profile in which only the sender with type t_1 sends message m_1 can be sustained as a PBE.

Part (b) Check if a pooling strategy profile in which both types of the sender choose message m_1 can be supported as a PBE.

Part (c) Assume now that the probability that the sender is type t_1 is p and, hence, the probability that the sender is type t_2 is $1 - p$. Find the values of p such that the

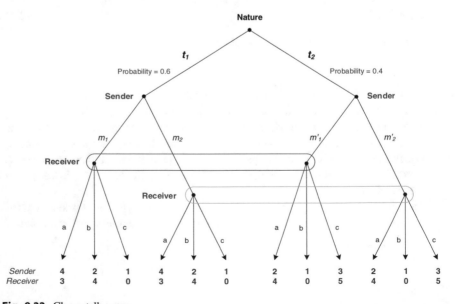

Fig. 9.22 Cheap talk game

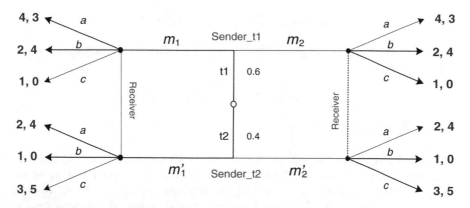

Fig. 9.23 An alternative representation of the cheap talk game

pooling PBE you found in part (b) of the exercise can be sustained and predict that the receiver responds using action *b*.

Answer

Part (a) *Separating strategy profile (m_1, m_2'):*

Figure 9.24 depicts the separating strategy profile (m_1, m_2'), where message m_1 only originates from a sender of type t_1 (see shaded branch in the upper part of the tree) while message m_2 only stems from a sender with type t_2 (see lower part of the tree).

Receiver's beliefs:

Let $\mu(t_i|m_j)$ represent the conditional probability that the receiver assigns to the sender being of type t_i given that message m_j is observed, where $i,j \in \{1,2\}$. Thus, after observing message m_1, the receiver assigns full probability to such a message originating only from t_1-type of sender,

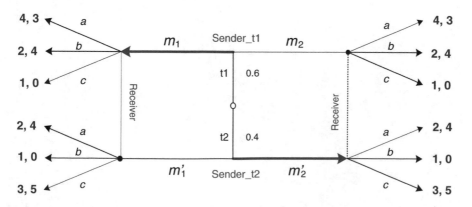

Fig. 9.24 Separating strategy profile

$$\mu(t_1|m_1) = 1, \quad \text{and} \quad \mu(t_2|m_1) = 0$$

while after observing message m_2, the receiver infers that it must originate from t_2-type of sender,

$$\mu(t_1|m_2) = 0, \quad \text{and} \quad \mu(t_2|m_2) = 1$$

Receiver's optimal response:

- After observing m_1, the receiver believes that such a message can only originate from a t_1-type of sender. Graphically, the receiver is convinced to be in the upper left-hand corner of Fig. 9.24. In this setting, the receiver's optimal response is b given that it yields a payoff of 4 (higher than what he gets from a, 3, and from c, 0.)
- After observing message m_2, the receiver believes that such a message can only originate from a t_2-type of sender. That is, he is convinced to be located in the lower right-hand corner of Fig. 9.24. Hence, his optimal response is c, which yields a payoff of 5, which exceeds his payoff from a, 4; or b, which entails a zero payoff.

Figure 9.25 summarizes these optimal responses of the receiver, by shading branch b upon observing message m_1 in the left-hand side of the tree, and c upon observing m_2 in the right-hand side of the tree.

Sender's optimal messages:

- If his type is t_1, by sending m_1 he obtains a payoff of 2 (since m_1 is responded with b), but a lower payoff of 1 if he deviates towards message m_2 (since such a message is responded with c). Hence, the sender doesn't want to deviate from m_1.

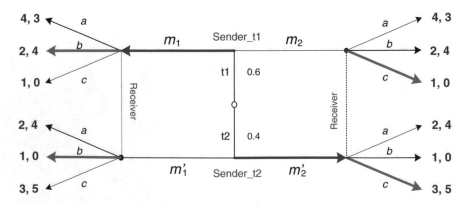

Fig. 9.25 Optimal responses in the separating strategy profile

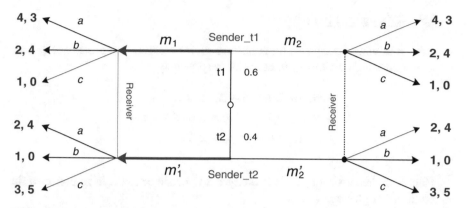

Fig. 9.26 Pooling strategy profile

- If his type is t_2, he obtains a payoff of 3 by sending message m_2 (which is responded with c) and a payoff of 1 if he deviates to message m_1 (which is responded with b). Hence, he doesn't have incentives to deviate from m_2.

Hence, the initially prescribed separating strategy profile *can* be supported as a PBE. As a curiosity, note that the opposite separating strategy profile (m_2, m_1') can also sustained as a PBE (you can check that as a practice).

Part (b) *Pooling strategy profile* (m_1, m_1').

Figure 9.26 depicts the pooling strategy profile (m_1, m_1') in which both types of senders choose the same message m_1.

Receiver's beliefs:

After observing message m_1 (which occurs in equilibrium), beliefs coincide with the prior probability distribution over types. Indeed, applying Bayes' rule we find that

$$\mu(t_1|m_1) = \frac{p(t_1) * p(m_1|t_1)}{p(m_1)} = \frac{0.6 * 1}{0.6 * 1 + 0.4 * 1} = 0.6$$

$$\mu(t_2|m_1) = \frac{p(t_2) * p(m_1|t_2)}{p(m_1)} = \frac{0.4 * 1}{0.6 * 1 + 0.4 * 1} = 0.4$$

After receiving message m_2 (what happens off-the-equilibrium path), beliefs cannot be updated using Bayes' rule since

$$\mu(t_1|m_2) = \frac{p(t_1) * p(m_2|t_1)}{p(m_2)} = \frac{0.6 * 0}{0.6 * 0 + 0.4 * 0} = \frac{0}{0}$$

and hence off-the-equilibrium beliefs must be arbitrarily specified, i.e. $\mu = \mu(t_1|m_2) \in [0,1]$.

Receiver's optimal response:

- After receiving message m_1 (in equilibrium), the receiver's expected utility from responding with actions a, b, and c is, respectively

$$\text{Action } a: 0.6 \times 3 + 0.4 \times 4 = 3.4$$
$$\text{Action } b: 0.6 \times 4 + 0.4 \times 0 = 2.4, \text{ and}$$
$$\text{Action } c: 0.6 \times 0 + 0.4 \times 5 = 2.0$$

Hence, the receiver's optimal strategy is to choose action a in response to the equilibrium message of m_1.
- After receiving message m_2 (off-the-equilibrium), the receiver's expected utility from each of his three possible responses are

$$EU_{Receiver}(a|m_2) = \mu * 3 + (1 - \mu) * 4 = 4 - \mu,$$
$$EU_{Receiver}(b|m_2) = \mu * 4 + (1 - \mu) * 0 = 4\mu, \text{ and}$$
$$EU_{Receiver}(c|m_2) = \mu * 0 + (1 - \mu) * 5 = 5 - 5\mu$$

where the receiver's response critically depends on the particular value of his off-the-equilibrium belief μ, i.e., $4\mu > 4 - \mu$ if and only if $\mu > \frac{4}{5}$, $4 - \mu > 5 - 5\mu$ if and only if $\mu > \frac{1}{4}$, and $4\mu > 5 - 5\mu$ if and only if $\mu > \frac{5}{9}$. Hence, if off-the-equilibrium beliefs lie within the interval $\mu \in \left[0, \frac{1}{4}\right]$, $5 - 5\mu$ is the highest expected payoff, thus inducing the receiver to respond with c; if they lie on the interval[4] $\mu \in \left(\frac{1}{4}, \frac{4}{5}\right]$, $4 - \mu$ becomes the highest expected payoff and the receiver chooses a; finally, if off-the-equilibrium beliefs lie on the interval $\mu \in \left(\frac{4}{5}, 1\right]$, 4μ is the highest expected payoff and the receiver responds with b. For simplicity, we only focus on the case in which off-the-equilibrium beliefs satisfy $\mu = 1$ (and thus the responder chooses b upon observing message m_2).

Figure 9.27 summarizes the optimal responses of the receiver (a after m_1, and b after m_2) in this pooling strategy profile.

Sender's optimal message:

- If his type is t_1, the sender obtains a payoff of 4 from sending m_1 (since it is responded with a), but a payoff of only 2 when deviating towards m_2 (since it is responded with b). Hence, he doesn't have incentives to deviate from m_1.

---

[4]To graphically understand this result, you can plot lines $4-\mu$, 4μ, and $5-5\mu$ on the same figure, with μ lying between 0 and 1 on the horizontal axis. You will see that the line corresponding to payoff $5-5\mu$ is the highest when μ is sufficiently low (left part of the figure). If we increase μ (moving rightward), line $4-\mu$ becomes the highest, and if we further increase μ, line 4μ corresponds to the highest expected payoff.

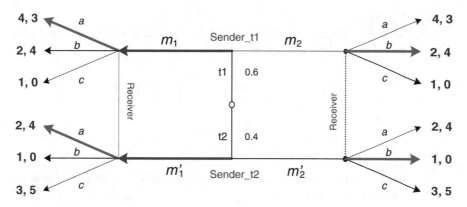

Fig. 9.27 Optimal responses in the pooling strategy profile

- If his type is t_2, the sender obtains a payoff of 2 by sending m_1 (since it is responded with a), but a payoff of only 1 by deviating towards m_2 (since it is responded with b). Hence, he doesn't have incentives to deviate from m_1.

Therefore, the initially prescribed pooling strategy profile where both types of sender select m_1 can be sustained as a PBE of the game. As a practice, check that this equilibrium can be supported for any off-the-equilibrium beliefs, $\mu(t_1|m_2)$, the receiver sustains upon observing message m_2.

You can easily check that the opposite pooling strategy profile (m_2, m_2') can also be sustained as a PBE. However, this PBE is only supported for a precise set of off-the-equilibrium beliefs, and the receiver does not respond with action b in equilibrium.

Part (c) Fig. 9.28 depicts the pooling strategy profile (m_1, m_1'), where note that the prior probability of type t_1 is now, for generality, p, rather than 0.6 in the previous parts of the exercise.

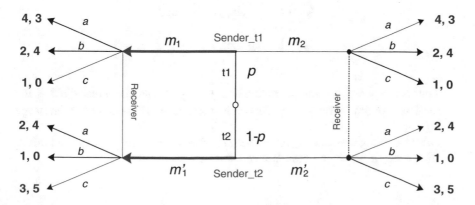

Fig. 9.28 Pooling strategy profile

Receiver's beliefs:

After observing message m_1 (which occurs in equilibrium), his beliefs coincide with the prior probabilities over types,

$$\mu(t_1|m_1) = \frac{p*1}{p*1+(1-p)*1} = p, \text{ and}$$

$$\mu(t_2|m_1) = \frac{(1-p)*1}{p*1+(1-p)*1} = 1-p$$

After receiving message m_2 (off-the-equilibrium path), his beliefs cannot be updated using Bayes' rule since

$$\mu(t_1|m_2) = \frac{p*0}{p*0+(1-p)*0} = \frac{0}{0}$$

Hence, off-the-equilibrium beliefs must be left arbitrarily specified, i.e., $\mu = \mu(t_1|m_2) \in [0,1]$.

Receiver's optimal response:

- After receiving a message m_1 (in equilibrium), the receiver's expected utility from responding with actions a, b, and c is, respectively,

  Action a: $p \times 3 + (1-p)*4 = 4-p$
  Action b: $p \times 4 + (1-p)*0 = 4p$, and
  Action c: $p \times 0 + (1-p)*5 = 5-5p$

  For the receiver to be optimal to respond choosing action b (as required in the exercise) it must be the case that

  $$4p > 4 - p, \text{ i.e., } p > \frac{4}{5}, \text{ and}$$

  $$4p > 5 - 5p, \text{ i.e., } p > \frac{5}{9}$$

  Thus, since $p > \frac{4}{5}$ is more restrictive than $p > \frac{5}{9}$, we can claim that if $p > \frac{4}{5}$ the receiver's optimal strategy is to choose b in response to the equilibrium message m_1.

- After receiving message m_2 (off-the-equilibrium), the receiver's expected utility from each of his three possible responses is, respectively,

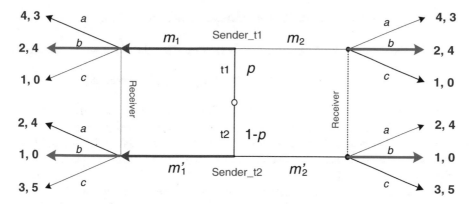

Fig. 9.29 Optimal responses in the pooling strategy profile

$$EU_{Receiver}(a|m_2) = \mu * 3 + (1 - \mu) * 4 = 4 - \mu,$$
$$EU_{Receiver}(b|m_2) = \mu * 4 + (1 - \mu) * 0 = 4\mu, \text{ and}$$
$$EU_{Receiver}(c|m_2) = \mu * 0 + (1 - \mu) * 5 = 5 - 5\mu$$

which coincides with the receiver's expected payoffs in part (b). From our discussion in part (b), the receiver's response depends on his off-the-equilibrium belief μ. However, for simplicity, we consider that $\mu = 1$, which implies that the receiver responds with b after observing the off-the-equilibrium message m_2.

Figure 9.29 summarizes the optimal responses of the receiver in this pooling strategy profile.

Sender's optimal message:

- If his type is t_1 (in the upper part of Fig. 9.29), the sender obtains a payoff of 2 from sending m_1 (since it is responded with b), and the same exact payoff when deviating towards m_2 (since it is responded with b). Hence, he does not have strict incentives to deviate from m_1.
- If his type is t_2 (in the lower part of Fig. 9.29), the sender obtains a payoff of 1 by sending m_1 (since it is responded with b), and the same payoff by deviating towards m_2 (since it is responded with b). Hence, he does not have strict incentives to deviate from m_1.

Therefore, the initially prescribed pooling strategy profile, where both types of sender select m_1 and the receiver responds with b, can be sustained as a PBE of the game when $p > \frac{4}{5}$.

Exercise 4—Firm Competition Under Cost Uncertainty[C]

Consider two firms competing in prices. Each firm faces a direct demand function of:

$$q_i = 1 - p_i + p_j \quad \text{for every firm } i,j = 1,2, \text{ where } i \neq j$$

The firms compete as Bertrand oligopolists over two periods of production. Assume that, while the unit cost of firm 2 is deterministically set at zero, i.e., $c_2 = 0$, the unit cost of firm 1, c_1, is stochastically determined at the beginning of the first period. The distribution function of c_1 is defined over the support $[-c_0, c_0]$ with expected value $E(c_1) = 0$.

At the beginning of period 1, firm 1 can observe the realization of its unit cost, c_1, while its rival cannot. Firms simultaneously and independently select first-period prices after firm 1's unit cost has been realized. At the beginning of period 2, both firms can observe the prices set in the first period of production. In particular, firm 2 observes the price selected by firm 1 in period 1 and draw some conclusions about firm 1's unit cost, c_1. After observing first-period prices, the firms simultaneously select second-period prices.

Find equilibrium prices in each of the two periods. [*Hint*: Assume that firm 1's first period price is linear in its private information, c_1, i.e., $p_1^1 = f(c_1) = A_0 + A_1 C_1$, where A_0 and A_1 are positive constants.]

Answer
Let us start with a summary of the time structure of the game:

$t = 0$: Firm 1 privately observes c_0
$t = 1$: Both firms i and j simultaneously select their own price (p_i^1, p_j^1), where the superscript indicates the first period.
$t = 2$: Observing first-period prices (p_i^1, p_j^1), firms i and j simultaneously choose second-period prices $\left(p_i^2, p_j^2\right)$.

We start analyzing the second-period game
Firm 1. Let us first examine the informed Firm 1. This, firm chooses the second-period price, p_1^2, that solves

$$\max_{p_1^2}(1 - p_1^2 + p_2^2)(p_1^2 - c_1)$$

Taking first order conditions with respect to p_1^2, yields

$$1 - 2p_1^2 + p_2^2 + c_1 = 0$$

and solving for p_1^2 we obtain firm 1's best response function in the second-period game (which is positively sloped in its rival's price, p_2^2),

$$p_1^2 = \frac{1 + c_1 + p_2^2}{2} \leftarrow BRF_1^2$$

Firm 2. Let us now analyze the uninformed Firm 2. This firm chooses price p_2^2 that solves

$$\max_{p_2^2} (1 + p_1^2 - p_2^2)(p_2^2)$$

We recall that $c_2 = 0$, i.e., this firm faces no unit costs. Taking first order conditions with respect to p_2^2, we obtain

$$1 - 2p_2^2 + p_1^2 = 0$$

and solving for p_2^2 we find firm 2's best response function in the second-period game

$$p_2^2 = \frac{1 + p_1^2}{2} \leftarrow BRF_2^2$$

(which is also positively sloped in its rival's price). Plugging one best response function into another, we can find the price that the informed Firm 1 sets:

$$p_1^2 = \frac{1 + c_1 + \frac{1 + p_1^2}{2}}{2}$$

which solving for p_1^2 yields a price of

$$p_1^2 = 1 + \frac{2}{3}c_1 \qquad (A)$$

Therefore, the uninformed Firm 2 sets a price

$$p_2^2 = 1 + \frac{1}{2}f^{-1}(p_1^1)$$

which, rearranging, yields

$$p_2^2 = \frac{2(3 + 1 - f^{-1}(p_1^1))}{9} \qquad (B)$$

Explanation: The price that firm 1 sets is a function of its privately observed cost, $p_1^1 = f(c_1)$. Then, upon observing a price from firm 1 in the first period, the uninformed firm 2 can infer firm 1's costs by inverting the above function, as follows: $f^{-1}(p_1^1) = c_1$. This helps firm 2 set a second-period price p_2^2.

Let's now analyze the first period:

Firm 2. The uninformed Firm 2 chooses a price p_2^1 that maximizes

$$\max_{p_2^1} \left(1 - p_2^1 + p_1^1\right)p_2^1$$

Taking first order conditions with respect to p_2^1 yields:

$$1 - 2p_2^1 + p_1^1 = 0$$

and solving for p_2^1, we obtain firm 2's best response function in the first-period game

$$p_2^1 = \frac{1 + p_1^1}{2} \leftarrow BRF_2^1$$

Firm 1. The informed Firm 1 solves

$$\max_{p_1^1} \left(p_1^1 - c_1\right) \cdot \left(1 - p_1^1 + p_2^1\right) + \left(p_1^2 - c_1\right) \cdot \left(1 - p_1^2 + p_2^2\right)$$

where the first term indicates first-period profits, while the second term reflects second-period profits. Note that this firm takes into account the effect of its first-period price on its second-period profits because a lower price today could lead firm 2 to believe that firm 1's costs are low (that is, firm 1 would be seen as a "tougher" competitor); ultimately improving firm 1's competitiveness in the second-period game. Firm 2's profit-maximization problem, in contrast, does not consider the effect of first-period prices on second-period profits since firm 1 is perfectly informed about firm 2's costs.

Firm 1 anticipates second-period prices, $p_1^2 = 1 + \frac{2}{3}c_1$ for firm 1 and $p_2^2 = \frac{2(3 + 1 - f^{-1}(p_1^1))}{9}$ for firm 2. Hence, his maximization problem can be rewritten as

$$\max_{p_1^1} \left(p_1^1 - c_1\right) \cdot \left(1 - p_1^1 + p_2^1\right) + \left(1 + \frac{2}{3}c_1 - c_1\right) \cdot \left(1 - 1 + \frac{2}{3}c_1 + \frac{2(3 + 1 - f^{-1}(p_1^1))}{9}\right)$$

Taking first order condition with respect to p_1^1, we obtain

$$1 - 2p_1^1 + p_2^1 + c_1 + 2\left(1 - \frac{1}{3}c_1\right)\frac{1}{3}f'\left(f^{-1}(p_1^1)\right) = 0$$

Simplifying and solving for p_1^1, we find firm 1's best response function in the first period game

$$p_1^1(p_2^1) = \frac{1 + c_1}{2} + \frac{p_2^1}{2} + \frac{(3 - c_1)f'\left(f^{-1}(p_1^1)\right)}{9}$$

Inserting this expression of p_1^1 into $p_2^1 = \frac{1+p_1^1}{2}$, we obtain the optimal second-period price for Firm 1

$$p_2^1 = \frac{1+c_1}{2} + \frac{2}{3} + \frac{2(3-c_1)f'\left(f^{-1}\left(p_1^1\right)\right)}{27}.$$

Plugging this result into the best response function $p_1^1(p_2^1)$, yields the optimal first-period price for Firm 1

$$p_1^1 = \frac{3+2c_1}{3} + \frac{4(3-c_1)f'\left(f^{-1}\left(p_1^1\right)\right)}{27}$$

As suggested in the exercise, let us now assume that there exists a linear function $p_1^1 = f(c_1)$ that generates the previous strategy profile. That is,

$$p_1^1 = A_0 + A_1 \cdot c_1 = f(c_1)$$

where A_0 and A_1 are positive constants. Intuitively, a firm with zero unit costs, $c_1 = 0$, would charge a first-period price of A_0, while a marginal increase in its unit costs, c_1, would entail a corresponding increase in prices of A_1. Figure 9.30 illustrates this pricing function for firm 1. (Note that this is a separating strategy profile, as firm 1 charges a different first-period price depending on its unit cost as long as $A_1 > 0$. A pooling strategy profile would exist if $A_1 = 0$.)

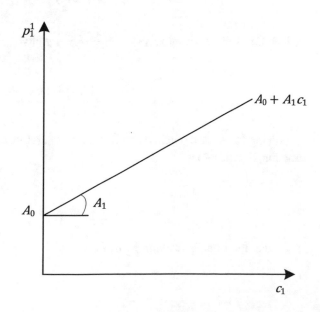

Fig. 9.30 Firm 1's pricing function

Setting it equal to the expression for p_1^1 found above, we obtain

$$A_0 + A_1 c_1 = \frac{3 + 2c_1}{3} + A_1 \frac{4(3 - c_1)}{27}$$

since A_1 measures the slope of the pricing function (see Fig. 9.30), thus implying $A_1 = f'\left(f^{-1}\left(p_1^1\right)\right)$. Rearranging the above expression, we find

$$A_1(31c_1 - 12) = 27 + 18c_1 - 27A_0$$

which, solving for A_1, yields

$$A_1 = \frac{27(1 - A_0) + 18c_1}{31c_1 - 12}$$

In addition, when firm 1's costs are nil, $c_1 = 0$, the above expression becomes

$$27A_0 = 27 + 12A_1$$

or, after solving for A_0,

$$A_0 = 1 + \frac{4}{9}A_1$$

Inserting this result into $A_1 = \frac{27(1-A_0) + 18c_1}{31c_1 - 12}$, yields

$$A_1 = \frac{27\left(1 - \left(1 + \frac{4}{9}A_1\right)\right) + 18c_1}{31c_1 - 12}.$$

Solving for A_1, we obtain $A_1 = \frac{18}{31} \approx 0.58$. Therefore, the intercept of the pricing function, A_0, becomes

$$A_0 = 1 + \frac{4}{9}\frac{18}{31} = \frac{39}{31} \approx 1.26.$$

Hence, the pricing function p_1^1 of firm 1, $p_1^1 = A_0 + A_1 c_1$, becomes

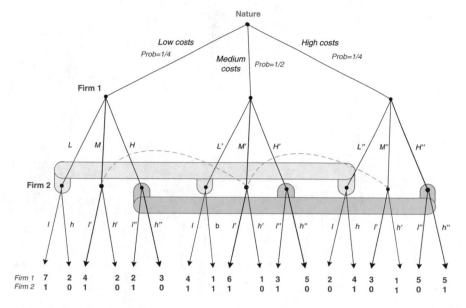

Fig. 9.31 Signaling game with 3 types, each with 3 possible messages

$$p_1^1 = \frac{18}{31} + \frac{39}{31}c_1.$$

Exercise 5—Signaling Game with Three Possible Types and Three Messages[c]

Consider the signaling game depicted in Fig. 9.31. Nature first determines the production costs of firm 1. Observing this information, firm 1 decides whether to set low (L), medium (M) or High (H) prices. Afterwards, observing firm 1's prices, but not observing its costs, firm 2 responds with either high (h) or low (l) prices. Find a PBE in which firm 1 chooses intermediate prices both when its costs are low and medium, but selects high prices when its costs are high.

Answer

Figure 9.32 shades the branches in which firm 1 chooses intermediate prices both when its costs are low and medium, M and M' respectively, but a high price otherwise (H''). This type of strategy profile is often referred to as "semi-separating" since it has two types of sender pooling into the same message (medium prices in this example), but has another type of sender choosing a different ("separating") message (i.e., H'' by the high-cost firm in our example).

Receiver beliefs:
After observing medium prices, the probability that such action originates from a firm 1 with low costs can be computed using Bayes' rule, as follows

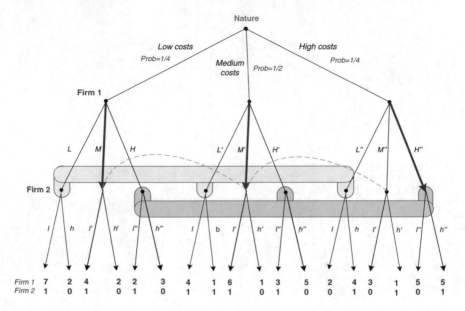

Fig. 9.32 Semi-separating strategy profile (M, M′, H″)

$$prob(low\,costs|M) = \frac{prob(low\,costs) * prob(M|low\,costs)}{prob(M)}$$

$$= \frac{\frac{1}{4} * 1}{\frac{1}{4} * 1 + \frac{1}{2} * 1 + \frac{1}{4} * 0} = \frac{\frac{1}{4}}{\frac{3}{4}} = \frac{1}{3}$$

Intuitively, since low- and medium-cost firms choose medium prices in this strategy profile, the conditional probability that, upon observing medium prices, firm 1's costs are low is just the relative frequency of low cost firms within the subsample of low- and medium-cost firms. A similar argument applies to medium-cost firms. Specifically, upon observing a medium price, the probability that such price originates from a firm with medium costs is

$$prob(medium\,costs|M) = \frac{\frac{1}{2} * 1}{\frac{1}{4} * 1 + \frac{1}{2} * 1 + \frac{1}{4} * 0} = \frac{\frac{1}{2}}{\frac{3}{4}} = \frac{2}{3}$$

which is the relative frequency of medium-cost firms within the subsample of low- and medium-costs firms. Finally, high-cost firms should never select a medium price in this strategy profile, implying that firm 2 assigns no probability to the fact that the medium price originates from this type of firm, that is

$$prob(high\,costs|M) = 0$$

Similarly, after observing a high price, the probability that such a price originates from a firm 1 with low costs can be computed using Bayes' rule, as follows

$$prob(low\,costs|H) = \frac{\frac{1}{4}*0}{\frac{1}{4}*0+\frac{1}{2}*0+\frac{1}{4}*1} = 0$$

since a high price should never originate from a low-cost firm in this strategy profile,

$$prob(medium\,costs|H) = \frac{\frac{1}{2}*0}{\frac{1}{4}*0+\frac{1}{2}*0+\frac{1}{4}*1} = 0$$

since a high price should not originate from a medium-cost firm either, and

$$prob(high\,costs|H) = \frac{\frac{1}{4}*1}{\frac{1}{4}*0+\frac{1}{2}*0+\frac{1}{4}*1} = \frac{\frac{1}{4}}{\frac{1}{4}} = 1$$

given that only high-cost firms choose high prices in this strategy profile.

Finally regarding low prices, we know that these can only occur off-the-equilibrium path, since no type of firm 1 selects this price level in the strategy profile we are testing. Hence, firm 2's off-the-equilibrium beliefs are

$$prob(low\,costs|L) = \gamma_1 \in [0, 1]$$

(Recall that, as shown in previous exercises, the use of Bayes' rule doesn't provide a precise value for off-the-equilibrium beliefs such as γ_1, and we must leave the receiver's beliefs unrestricted, i.e., $\gamma_1 \in [0, 1]$). Similarly, the conditional probability that such low prices originate from a medium-cost firm is

$$prob(medium\,costs|L) = \gamma_2 \in [0, 1],$$

And, as a consequence, the conditional probability that such low price originates from a high-cost firm is

$$prob(High|L) = 1 - \gamma_1 - \gamma_2$$

Receiver (firm 2):

Given the above beliefs, after observing a medium price (M), firm 2 responds with either low (l) or high (h) prices depending on which action yields the highest expected utility. In particular, these expected utilities are

$$EU_2(l|M) = \frac{1}{3}*1 + \frac{2}{3}*1 = 1, \text{ and}$$

$$EU_2(h|M) = \frac{1}{3} * 0 + \frac{2}{3} * 0 = 0$$

Therefore, after observing a medium price, firm 2 responds selecting a low price, l.

After observing high prices (H), firm 2 similarly compares its payoff from selecting l and h. (Note that in this case, the receiver does not need to compute expected utilities, since it is convinced to be dealing with a high-cost firm, i.e., firm 2 is convinced of being at the right-most node of the game tree.) In particular, responding with l'' at this node yields a payoff of zero, while responding with h'' entails a payoff of 1.Thus, firm 2 responds with high prices (h) after observing that firm 1 sets high prices.

Finally, after observing low prices (L) (off-the-equilibrium path), firm 2 compares its expected payoff from responding with l and h, as follows

$$EU_2(l|L) = \gamma_1 * 1 + \gamma_2 * 1 + (1 + \gamma_1 - \gamma_2) * 0 = \gamma_1 + \gamma_2, \text{ and}$$

$$EU_2(h|L) = \gamma_1 * 0 + \gamma_2 * 1 + (1 + \gamma_1 - \gamma_2) * 1 = 1 + \gamma_1$$

Hence, after observing low prices (L), firm 2 responds choosing low prices (l) if and only if $\gamma_1 + \gamma_2 > 1 + \gamma_1$, or $\gamma_2 > 1 - 2\gamma_1$. In our following discussion of the sender's optimal messages, we will have to consider two cases: one in which $\gamma_2 > 1 - 2\gamma_1$ holds, thus implying that firm 2 responds with low prices after observing the off-the-equilibrium message of low prices from firm 1; and that in

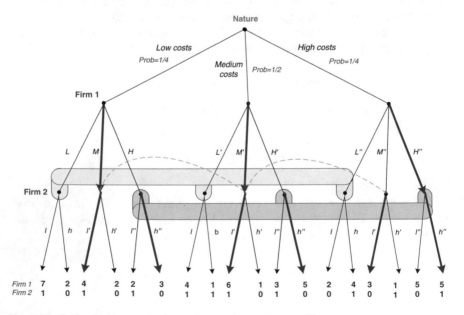

Fig. 9.33 Optimal responses in the semi-separating strategy profile

which $\gamma_2 \leq 1 - 2\gamma_1$, whereby firm 2 responds with high prices to such an off-the-equilibrium message.

Figure 9.33 summarizes the optimal responses of firm 2 by shading l' and h''.

<u>Sender</u>

Low costs. If its costs are low (left-hand side of the game tree) and firm 1 selects medium prices (as prescribed in this strategy profile), it obtains a payoff of 4, since firm 1 can anticipate that its message will be responded with l' (see shaded branches in Fig. 9.33). If instead, it deviates towards a high price, it is responded with h'' yielding a payoff of only 3. If instead, it deviates towards a low price, firm 1 obtains a payoff of 7 if firm 2's off-the-equilibrium beliefs satisfy $\gamma_2 > 1 - 2\gamma_1$ (which lead firm 2 to respond with l), but a payoff of only 2 if $\gamma_2 \leq 1 - 2\gamma_1$ (case in which firm 2 responds with h). Hence, in order for firm 1 to prefer medium prices (obtaining a payoff of 4) we need that off-the-equilibrium beliefs satisfy $\gamma_2 \leq 1 - 2\gamma_1$ (otherwise firm 1 would have incentives to deviate towards a low price).

Medium costs. If firm 1's costs are medium and it sets a medium price (as prescribed), firm 1 obtains a payoff of 6 (see shaded arrows at the center of Fig. 9.33). Deviating towards a high price would only provide a payoff of 5. Similarly, deviating towards a low price gives firm 1 a payoff of 4 (when $\gamma_2 > 1 - 2\gamma_1$, and firm 2 responds with *l*) or a payoff of 1 (when $\gamma_2 \leq 1 - 2\gamma_1$, and firm 2 responds with *h*). Hence, regardless of firm 2's off-the-equilibrium beliefs, firm 1 does not have incentives to deviate from setting a medium price, as prescribed in this semi-separating strategy profile.

High costs. A similar argument applies to firm 1 when its costs are high (see right-hand side of Fig. 9.33). If it sets a high price, as prescribed, its profit is 5. Deviating towards a medium price would reduce its profits to 3, and so would a deviation to low prices (where firm 1 obtains a payoff of 2 if firm 2 responds with *l* or

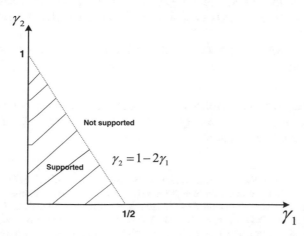

Fig. 9.34 Off-the-equilibrium beliefs of firm 2

a payoff of 4 if firm 2 responds with h). Thus, independently on firm 2's off-the-equilibrium beliefs, firm 1 does not have incentives to deviate from high prices.

Summarizing, the only condition that we found for this separating strategy profile to be sustained as a PBE is that off-the-equilibrium beliefs must satisfy $\gamma_2 \leq 1 - 2\gamma_1$. Figure 9.34 represents all combinations of γ_1 and γ_2 for which the above strategy profile can be sustained as a PBE of this game.[5]

Exercise 6—Second-Degree Price Discrimination[B]

Assume that Alaska Airlines is the only airline flying between the Seattle and Pullman. There are two types of passengers, tourist and businessmen, with the latter being willing to pay more than tourists. The airline, however, cannot directly tell whether a ticket purchaser is a tourist or a business traveler. Nevertheless, the two types differ in how much they are willing to pay to avoid having to purchase their tickets in advance. (Passengers do not like to commit themselves in advance to traveling at a particular time.)

More specifically, the utility functions of each traveler net of the price of the ticket, p, are:

$$\text{Business traveler:} \ v - \theta_B p - w$$
$$\text{Tourist traveler:} \ v - \theta_T p - w$$

where w is the disutility of buying the ticket w days prior to the flight and the price coefficients are such that $0 < \theta_B < \theta_T$. The proportion of travelers who are tourists is λ. Assume that the cost that the airline incurs when transporting a passenger is $c > 0$, regardless of the passenger's type.

Part (a) Describe the profit-maximizing price discrimination scheme (second degree price discrimination) under the assumption that Alaska Airlines sells tickets to both types of travelers. How does it depend on the underlying parameters λ, θ_B, θ_T, and c.

Part (b) Under what circumstances does Alaska Airlines choose to serve only business travelers?

Answer

Part (a) In order to find the profit-maximizing discrimination scheme, we must find pairs of prices and waiting time, (p_B, w_B) and (p_T, w_T), that solve the following constrained maximization problem

---

[5]This condition allows for different combinations of off-the-equilibrium beliefs. For instance, when $\gamma_1 < 1/2$ but $\gamma_2 > 1/2$, the uninformed firm is essentially placing a large probability weight on the off-the-equilibrium message of low price originating from a medium-cost firm, and not from a low- or high-cost firm. If, instead, $\gamma_1 = 1/2$ but $\gamma_2 = 0$, the uninformed firm believes that the off-the-equilibrium message of low price is equally likely to originate from a low- or a high-cost firm.

$$\max \lambda p_T + (1 - \lambda)p_B - 2c$$

subject to

$$
\begin{aligned}
v - \theta_B p_B - w_B &\geq v - \theta_B p_T - w_T & (\text{IC}_B) \\
v - \theta_T p_T - w_T &\geq v - \theta_T p_B - w_B & (\text{IC}_T) \\
v - \theta_B p_B - w_B &\geq 0 & (\text{IR}_B) \\
v - \theta_T p_T - w_T &\geq 0 & (\text{IR}_T) \\
p_B, w_B, p_T, w_T, &\geq 0
\end{aligned}
$$

Intuitively, incentive compatibility condition IC_B (IC_T) represents that business (tourist, respectively) travelers prefer the pair of price and waiting time offered to their type then that of the other type. Individual rationality constraint IR_B (IR_T) describe that business (tourist) travelers obtain a positive utility level from accepting the pair of ticket price and waiting time from Alaska Airlines (we normalize the utility from not traveling, in the right-hand side of the individual rationality constraints, to be zero).

In addition, constraints IC_B and IR_T must be binding. Otherwise, the firm could extract more surplus from either type of traveler, thus increasing profits. Therefore, IR_T can be expressed as

$$v - \theta_T p_T - w_T = 0 \Leftrightarrow w_T = v - \theta_T p_T$$

while IC_B can be written as

$$v - \theta_B p_B - w_B = v - \theta_B p_T - v + \theta_T p_T \Leftrightarrow w_B = v - \theta_B p_B - (\theta_T - \theta_B)p_T$$

Hence, the firm's maximization problem becomes selecting ticket prices p_B and p_T (which reduces our number of choice variables from four in the previous problem to only two now) as follows

$$\max_{p_T, p_B} \lambda p_T + (1 - \lambda)p_B - 2c$$

subject to
$$
\begin{aligned}
(p_T, w_T) &= (p_T, v - \theta_T p_T) \\
(p_B, w_B) &= (p_B, v - \theta_B p_B - (\theta_T - \theta_B)p_T)
\end{aligned}
$$

Before continuing any further, let us show that the waiting time for business travelers, w_B, does not affect Alaska Airlines profits.

Proof Assume, by contradiction, that $w_B > 0$. Then w_B can be reduced by $\varepsilon > 0$ units, and increase the price charged to this type of passengers, p_B, by $\frac{\varepsilon}{\theta_B}$, so that business travelers' utility level does not change, and the firm earns higher profits.

We now need to check that tourists will not choose this new package offered to business travelers. That is,

$$v - \theta_T p_T - w_T \geq v - \theta_B p_B - w_B$$

rearranging, $\theta_T p_T + w_T \leq \theta_B p_B + w_B$, which implies,

$$\theta_T \left(p_B + \frac{\varepsilon}{\theta_T} \right) + (w_B - \varepsilon) < \theta_T \left(p_B + \frac{\varepsilon}{\theta_B} \right) + (w_B - \varepsilon)$$

This result, however, contradicts that $w_B > 0$ can be part of an optimal contract. Hence, only $w_B = 0$ can be part of an optimal incentive contract. (Q.E.D.)

Hence, from the above claim $w_B = 0$, which yields

$$v - \theta_B p_B - (\theta_T - \theta_B)p_T = 0$$

and solving for price p_B, we obtain

$$p_B = \frac{1}{\theta_B}[v - (\theta_T - \theta_B)p_T]$$

Plugging this price into the firm's maximization problem, we further reduce the set of choice variables to only one price, p_T, as follows:

$$\max_{p_T} \lambda p_T + (1 - \lambda) \frac{1}{\theta_B}[v - (\theta_T - \theta_B)p_T] - 2c$$

which can be alternatively expressed as

$$\max_{p_T} p_T \left[\lambda - (1 - \lambda) \frac{\theta_T - \theta_B}{\theta_B} \right] + (1 - \lambda) \frac{1}{\theta_B} v - 2c$$

Taking first-order conditions with respect to p_T, we find

$$\lambda - (1 - \lambda) \frac{\theta_T - \theta_B}{\theta_B} = 0$$

and rearranging

$$\frac{\lambda}{1 - \lambda} = \frac{\theta_T - \theta_B}{\theta_B}$$

At this point we can identify three solutions: one interior and two corners, as we next separately describe.

(1) If $\frac{\lambda}{1-\lambda} = \frac{\theta_T - \theta_B}{\theta_B}$, any contract (p_B, w_B) and (p_T, w_T) satisfying the following conditions is optimal:

$$\text{Prices: } p_T \geq 0 \text{ and } p_B = \frac{1}{\theta_B}[v - (\theta_T - \theta_B)p_T] \geq 0$$

Waiting times: $w_B = 0$ and $w_T = 0$

(2) If $\frac{\lambda}{1-\lambda} > \frac{\theta_T - \theta_B}{\theta_B}$, the willingness to pay of tourist travelers is relatively high (recall that a low value of parameter θ_T indicates that a higher price does not reduce the traveler's utility significantly. Since θ_T is relatively low, a marginal increase in prices does not substantially reduce the tourists' utility.) Then, it is optimal to raise price p_T (which reduces the waiting time of this type of traveler w_T). In fact, it is optimal to do it until w_T is reduced to $w_T = 0$. Hence,

$$w_T = v - \theta_T p_T = 0,$$

which yields a price of $p_T = \frac{v}{\theta_T}$ and

$$w_B = v - \theta_B p_B - (\theta_T - \theta_B)\frac{v}{\theta_T} = 0,$$

which entails a price of $p_B = \frac{v}{\theta_T}$. Therefore, the optimal contract in this case coincides for both type of travelers, $(p_T, w_T) = (p_B, w_B) = \left(\frac{v}{\theta_T}, 0\right)$, i.e., a pooling scheme.

(3) $\frac{\lambda}{1-\lambda} < \frac{\theta_T - \theta_B}{\theta_B}$. In this case, tourists' willingness to pay is relatively low, i.e., the disutility they experience from a higher price, θ_T, is very high. It is, hence, optimal to reduce the price p_T as much as possible ($p_T = 0$), which implies that its corresponding waiting time w_T becomes $w_T = v - \theta_t p_T = v$ (because $\theta_t p_T = 0$). For business travelers, this implies a waiting time of

$$w_B = v - \theta_B p_B - (\theta_T - \theta_B)p_T = 0,$$

which entails a price of $p_B = \frac{v}{\theta_B}$. Therefore, in case 3 optimal contracts satisfy $(p_T, w_T) = (0, v)$ and $(p_B, w_B) = \left(\frac{v}{\theta_B}, 0\right)$

Intuitively, case 3 (in which only business travelers are serviced) is more likely to emerge when a) the proportion of business travelers is large enough (small λ); and/or b) business travelers suffer less from higher prices (small θ_B). Case 2 (serving both types of travelers) is more likely to arise when: a) the proportion of business travelers is small (large λ); and/or b) tourists suffer less from higher prices (small θ_T). Figure 9.35 summarizes under which conditions of the unit cost, c, the discriminating monopolist chooses to serve all types of travelers, only business travelers, or no traveler at all.

Fig. 9.35 Summary of price discrimination results

Serving everyone: The (pooling) price of serving every type of traveler described in case 2 is $p_T = p_B = \frac{v}{\theta_T}$. Hence, when costs are below this price, the firm serves all types of travelers.

Only Business travelers: When $c > \frac{v}{\theta_B}$ the above pooling pricing scheme would imply a sure loss of money for the firm. In this case, the firm chooses to serve only business travelers since $p_B = \frac{v}{\theta_B} > c$ (as in case 3 above).

No operation: When $p_B = \frac{v}{\theta_B} < c$, costs are so high that it is not profitable to serve business travelers. Hence the firm decides to not operate at all.

Exercise 7—Applying the Cho and Kreps' (1987) Intuitive Criterion in the Far West[B]

Consider the signaling game depicted in Fig. 9.36, usually referred as "Breakfast in the Far West" or "Beer-Quiche game." The game describes a saloon in the Far West, in which a newcomer (player 1) enters. His shooting ability (he is either a wimpy or a surely type) is his private information. Observing his own type, he orders breakfast: either quiche or beer (no more options today; this is a saloon in the Far West!). The troublemaker in this town (player 2) shows up who, observing the choice for breakfast of the newcomer but not observing the newcomer's type, decides whether to duel or not duel him. Despite being a troublemaker, he would

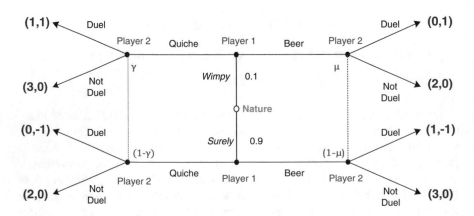

Fig. 9.36 Breakfast in the Far West ("Beer-Quicke" game)

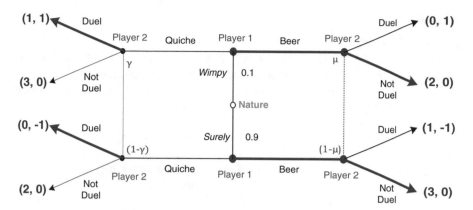

Fig. 9.37 Pooling equilibrium (Beer, Beer)

prefer a duel with a wimpy type and avoid the duel otherwise. This signaling game has several PBEs but, importantly, it has two pooling equilibria: one in which both types of player 1 choose to have quiche for breakfast, and another in which both types have beer for breakfast. (At this point of the chapter you should be able to confirm this result. You can do it as practice.)

Part (a) Check if the pooling equilibrium in which both types of player 1 have *beer* for breakfast survives the Cho and Kreps' (1987) Intuitive Criterion. Figure 9.37 depicts the pooling PBE where both types of player 1 have beer for breakfast, shading the branches that each player selects.

Part (b) Check if the pooling equilibrium in which both types of player 1 have *quiche* for breakfast *survives* the Cho and Kreps' (1987) Intuitive Criterion.

Figure 9.38 illustrates the pooling PBE in which both types of player 1 have quiche for breakfast.

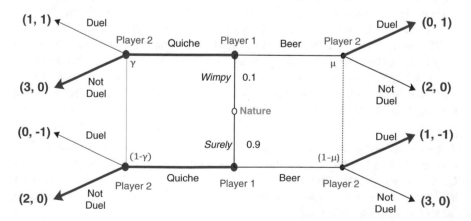

Fig. 9.38 Pooling equilibrium (Quiche, Quiche)

Answer

Part (a) First step:

In the first step of Cho and Kreps' (1987) Intuitive Criterion, we want to eliminate those off-the-equilibrium messages that are equilibrium dominated. For the case of a wimpy player 1, we need to check if having quiche can improve his equilibrium utility level (from having beer). That is, if

$$\underbrace{u_1^*(Beer|Wimpy)}_{Equil.\,Payoff,\,2} < \underbrace{\underset{a_2}{\text{Max}}\, u_1(Quiche|Wimpy)}_{\substack{Highest\,payoff\,from\,deviating \\ towards\,Quiche,\,3}}$$

This condition can be indeed satisfied if player 2 responds to the off-the-equilibrium message of quiche not dueling: the wimpy player 1 would obtain a payoff of 3 instead of 2 in this pooling equilibrium. Hence, the wimpy player 1 has incentives to deviate from this separating PBE. Intuitively, this would occur if, by deviating towards quiche, the wimpy type of player 1 can still convince player 2 not to duel, i.e., player 1 would be having his preferred breakfast while still avoiding the duel.

Let us now check the surely type of player 1. We need to check if an (off-the-equilibrium) message of quiche could ever be convenient for the surely type. That is, if

$$\underbrace{u_1^*(Beer|Surely)}_{Equil.\,Payoff,\,3} < \underbrace{\underset{a_2}{\text{Max}}\, u_1(Quiche|Surely)}_{\substack{Highest\,payoff\,from\,deviating \\ towards\,Quiche,\,2}}$$

But this condition is not satisfied: the surely player 1 obtains in equilibrium a payoff of 3 (from having beer for breakfast), and the highest payoff he could obtain from deviating to quiche is only 2 (as a surely type, he dislikes quiche but loves to have a beer for breakfast). Hence, the surely type would never deviate from beer.

Therefore, Player 2's beliefs, after observing the (off-the-equilibrium) message of quiche, can be restricted to $\Theta^{**}(Quiche) = \{Wimpy\}$. That is, if player 2 were to observe player 1 choosing quiche, he would believe that player 1 must be wimpy, since this is the only type of player 1 who could benefit from deviating from the pooling equilibrium of beer.

Second step:

After restricting the receiver's beliefs to $\Theta^{**}(Quiche) = \{Wimpy\}$, player 2 plays duel every time that he observes the (off-the-equilibrium) message of quiche, since he is sure that player 1 must be wimpy. The second step of the Intuitive Criterion analyzes if there is any type of sender (wimpy or surely) and any type of message he would send that satisfies:

$$\min_{a\in A^*[\Theta^{**}(m),m]} u_i(m, a, \theta) > u_i^*(\theta)$$

In this context, this condition does not hold for the wimpy type: the minimal payoff he can obtain by deviating (having quiche for breakfast) once the receiver's beliefs have been restricted to $\Theta^{**}(Quiche) = \{Wimpy\}$, is 1 (given that when observing that player 1 had quiche for breakfast, player 2 infers that he must be wimpy, and responds dueling). In contrast, his equilibrium payoff from having beer for breakfast in this pooling equilibrium is 2. A similar argument is applicable to the surely type: in equilibrium he obtains a payoff of 3, and by deviating towards quiche he will be dueled by player 2, obtaining a payoff of 0 (since he does not like quiche, and in addition, he has to fight). As a consequence, no type of player deviates towards quiche, and the pooling equilibrium in which both types of player 1 have beer for breakfast *survives* the Cho and Kreps' (1987) Intuitive Criterion.

Part (b) First step:

Let us now analyze the pooling PBE in which both types of player 1 choose quiche for breakfast. In the first step of the Intuitive Criterion, we seek to eliminate those off-the-equilibrium messages that are equilibrium dominated. For the case of a wimpy player 1, we need to check if having beer can improve his equilibrium utility level (of having quiche in this pooling PBE). That is, if

$$\underbrace{u_1^*(Quiche|Wimpy)}_{Equil.\,Payoff,\,3} < \underbrace{\max_{a_2} u_1(Beer|Wimpy)}_{\substack{Highest\,payoff\,from\,deviating \\ towards\,Beer,\,2}}$$

This condition is not satisfied: the wimpy player 1 obtains a payoff of 3 in this equilibrium (look at the lower left-hand corner of Fig. 9.38: the wimpy type loves to have quiche for breakfast and avoid the duel!). By deviating towards beer, the highest payoff that he could obtain is only 2 (if he were to still avoid the duel). Let us now check if the equivalent condition holds for the surely type of player 1,

$$\underbrace{u_1^*(Quiche|Surely)}_{Equil.\,Payoff,\,2} < \underbrace{\max_{a_2} u_1(Beer|Surely)}_{\substack{Highest\,payoff\,from\,deviating \\ towards\,Beer,\,3}}$$

This condition is satisfied in this case: the surely player 1 obtains in equilibrium a payoff of 2 (from having quiche for breakfast, which he dislikes), and the highest payoff he could obtain from deviating towards beer is 3. Notice that this would happen if, upon observing the off-the-equilibrium message of beer, player 2 infers that such a message can only originate from a surely type, and thus refrains from dueling. In summary, the surely type is the only type of player 1 who has incentives to deviate towards beer.

Player 2's beliefs, after observing the (off-the-equilibrium) message of beer, can then be restricted to $\Theta^{**}(Beer) = \{Surely\}$.

Second step:
After restricting the receiver's beliefs to $\Theta^{**}(Beer) = \{Surely\}$, player 2 responds by not dueling player 1 as he assigns full probability to player 1 being a surely type. The second step of the Intuitive Criterion analyzes if there is any type of sender (wimpy or surely) and any type of message he would send that satisfies:

$$\min_{a \in A^*[\Theta^{**}(m),m]} u_i(m, a, \theta) > u_i^*(\theta)$$

This condition holds for the surely type: the minimal payoff he can obtain by deviating (having beer for breakfast) after the receiver's beliefs have been restricted to $\Theta^{**}(Beer) = \{Surely\}$, is 3 (given that player 2 responds by not dueling any player who drinks beer). In contrast, his equilibrium payoff from having quiche for breakfast in this pooling equilibrium is only 2. As a consequence, the surely type has incentives to deviate to beer, and the pooling equilibrium in which both types of player 1 have quiche for breakfast *does not survive* the Cho and Kreps' (1987) Intuitive Criterion.[5]

---

[5] For applications of the Cho and Kreps' (1987) Intuitive Criterion to games in which the sender has a continuum of possible messages and the receiver a continuum of available responses, see the Exercise 9.4 about the labor market signaling game in this Chapter.

More Advanced Signaling Games

10

Introduction

The final chapter presents extensions and variations of signaling games, thus providing more practice about how to find the set of PBEs in incomplete information settings. We first study a poker game where, rather than having only one player being privately informed about his cards (as in Chap. 8), both players are privately informed. In this context, the first mover's actions can reveal information about his cards to the second mover, thus affecting the latter's incentives to bet or fold relative to a context of complete information.

We then explore a game in which one player is privately informed about his benefits from participating in a project, and evaluate his incentives to acquire a costly certificate that publicly discloses such benefit to other players. In the same vein, we then examine an entry game where firms are privately informed about their own type (e.g., production costs) before choosing to simultaneously and independently enter a certain industry. In this setting, firms can spend on advertising in order to convey their benefits from entering the industry to other potential entrants, where we identify symmetric and asymmetric PBEs.

We extend the Spence's (1974) job market signaling game examined in Chap. 9 to a context in which workers can choose an education level from a continuum (rather than only two education levels), and the firm can respond with a continuum of salaries. We first study under which conditions separating and pooling PBEs can be sustained in this version of the game when the worker has only two types, and then extend it to contexts in which the worker might have three or an infinite number of types. In order to provide further practice of the application of Cho and Kreps' (1987) Intuitive Criterion in games where players have a continuum of available strategies, we offer a step-by-step explanation of this criterion.

The original version of the chapter was revised: The erratum to the chapter is available at: 10.1007/978-3-319-32963-5_11

Afterwards we consider two types of entry-deterrence games. In the first game, the incumbent might be of a "sane" type (if he prefers to avoid price wars when entry occurs) or a "crazy" type (if he enjoys such price wars); following an incomplete information setting similar to that in Kreps et al. (1982). After simplifying the game, we systematically describe how to test for the presence of separating and pooling equilibria. In the second game, which follows Kreps and Wilson (1982), an incumbent faces entry threats from a sequence of potential entrants, and must choose whether to start a price war with the first entrant in order to build a reputation that would protect him from further entry. While this game seems at first glance more complicated than standard signaling games with a single potential entrant, we show that it can be simplified to a point where we easily apply a similar methodology to that developed in most games of Chap. 9.

Exercise 1—Poker Game with Two Uninformed Players[B]

Consider a simple poker game in which each player receives a card which only he can observe.

1. There is an ante of $\$A$ in the pot. Player 1 has either a good hand (G) or a bad hand (B) chosen by Nature with probability p and $1 - p$, respectively. The hand is observed by player 1 but not by player 2. Player 1 then chooses whether to bet or resign: If player 1 resigns, player 2 wins the ante. If player 1 bets, he puts $\$B$ in the pot.

2. If player 1 bets then player 2, observing his own card (either good or bad), must decide whether to call or fold: if player 2 responds folding, player 1 gets the pot, winning the ante. If, instead, player 2 responds calling, he puts $\$1$ in the pot. Player 1 wins the pot if he has a high hand; player 2 wins the pot if player 1's hand is low.

Part (a) Draw the extensive form game.

Part (b) Fully characterize the set of equilibria for all sets of parameter values p, B, and A.

Answer

Part (a) Figure 10.1 depicts the game tree of this poker game:

Nature determines whether both players receive a good hand (GG), only player 1 does (GB), etc. Only observing his good hand (left-hand side of the tree), player 1 decides whether to bet (B) or fold (F). Note that player 1's lack of information about player 2's hand is graphically represented by the fact that player 1 chooses to bet or fold on an information set. For instance, when player 1 receives a good hand, he does not observe player 2's hand, i.e., is uncertain about being located in node GG or GB. If player 1 bets, player 2 observes such an action and must respond betting or folding when his own hand is good (but without knowing whether player 1's hand is good, in μ_1, or bad, in $1 - \mu_1$), and similarly when his own hand is bad (and he does not know whether his rival's hand is good, in μ_3, or bad, in $1 - \mu_3$). A similar argument applies to the case in which player 2 observes player 1 folding, leaving player 2 in information set (μ_2, $1 - \mu_2$) when his own hand is good, and (μ_4, $1 - \mu_4$) when his hand is bad.

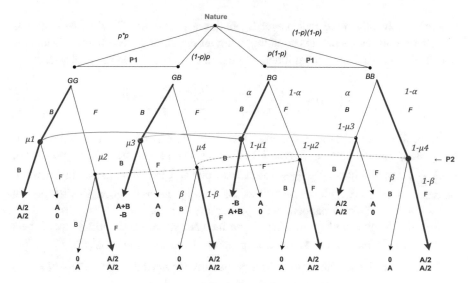

Fig. 10.1 A poker game with two uninformed players

First, note that betting is a strictly dominant strategy for every player who receives a good hand, irrespective of the strategy his opponent selects. You can see this result by starting, for instance, to examine player 1's decision to fold or bet when he receives a good card (in the GG and GB nodes). In particular, betting would yield an expected payoff of $p(A/2) + (1 - p)(A + B)$, but a lower expected payoff of $p(A/2) + (1 - p)(A/2)$ from folding. A similar argument applies to player 2 when his cards are good. However, when players receive a bad hand, they do not have a strictly dominant strategy. For compactness, let α denote the probability that player 1 bets when he receives a bad hand, and β the probability that player 2 bets when he receives a bad hand, as indicated in Fig. 10.1.

Part (b) Fully characterize the set of equilibrium for all parameter values p, B, and A.

As a roadmap, note that we need to first examine player 2's optimal responses in all four information sets in which he is called on to move: those in which he observes player 1 betting, $(\mu_1, 1 - \mu_1)$ and $(\mu_3, 1 - \mu_3)$, and those in which he responds to player 1 folding, $(\mu_2, 1 - \mu_2)$ and $(\mu_4, 1 - \mu_4)$.

Player 1 bets. After observing that player 1 bets, player 2 will respond betting when he receives a bad hand, as depicted in information set $(\mu_3, 1 - \mu_3)$, if

$$EU_2(\text{Bet}|\text{Bad hand}, P_1\text{Bets}) \geq EU_2(\text{Fold}|\text{Bad hand}, P_1\text{Bets}),$$

that is

$$(-B) \cdot \text{prob}(p_1 \text{ Bets}|p_1 \text{ has a good hand}) + \frac{A}{2} \cdot \text{prob}(p_1 \text{ Bets}|p_1 \text{ has a bad hand}) \geq 0$$

where, graphically, note that player 2 obtains a payoff of $-B$ when he responds betting and player 1's hand is good, but a higher payoff of $\frac{A}{2}$ when player 1's hand is bad. Using conditional probabilities, we can rewrite the above inequality as

$$(-B)\frac{p}{p+(1-p)\alpha} + \frac{A}{2}\frac{\alpha(1-p)}{p+(1-p)\alpha} \geq 0$$

Note that the two large ratios use Bayes' rule in order to determine the probability that, upon observing player 1 betting, such a player received a good hand, $\frac{p}{p+(1-p)\alpha}$, or a bad hand, $\frac{\alpha(1-p)}{p+(1-p)\alpha}$. Intuitively, player 1 bets with probability 1 when he has a good had (which occurs with probability p), but with probability α when he has a bad hand (which happens with probability $1-p$).

Solving for probability α in the above inequality, we obtain that player 2 bets when having a bad hand and observing player 1 betting if and only if $\alpha \geq \frac{2Bp}{A(1-p)}$. Note that, in order for this ratio to be a probability we need that: $0 \leq \frac{2Bp}{A(1-p)} \leq 1$, i.e., the prior probability p must be relatively low, $p < \frac{A}{2B+A}$.

In contrast, when player 2 observes that player 1 bets but player 2's hand is good, he responds betting (as indicated in information set $(\mu_1, 1 - \mu_1)$), since

$$EU_2(\text{Bet}|\text{Good hand, Bet}) \geq EU_2(\text{Fold}|\text{Good hand, Bet})$$

or

$$\frac{A}{2}\mu_1 + (A+B)(1-\mu_1) \geq 0\mu_1 + 0(1-\mu_1)$$

which reduces to

$$(A+B) \geq \left(\frac{A}{2} + B\right)\mu_1$$

which holds since $A, B > 0$, and $\mu_1 \leq 1$. Thus, player 2 responds to player 1's bet with a bet when his hand is good, under all parameter values.

Player 1 folds. Let us now move to the case in which, upon receiving a good hand, player 2 observes that player 1 folded; as depicted in information set $(\mu_2, 1 - \mu_2)$. In this case, player 2 responds betting if

$$EU_2(\text{Bet}|\text{Good hand, Fold}) \geq EU_2(\text{Fold}|\text{Good hand, Fold})$$

or

$$A\mu_2 + A(1-\mu_2) \geq \frac{A}{2}\mu_2 + \frac{A}{2}(1-\mu_2)$$

which simplifies to $A \geq \frac{A}{2}$, a condition which is always true. Hence, player 2 bets when having a good hand, both after observing that player 1 bets and that he folds.

Let us finally analyze the case in which player 2 receives a bad hand and observes that player 1 folds; as depicted in information set $(\mu_4, 1 - \mu_4)$. In this setting, player 2 responds betting if

$$EU_2(\text{Bet}|\text{Bad hand, Folds}) \geq EU_2(\text{Fold}|\text{Bad hand, Folds})$$

which implies

$$A\frac{0p}{0p + (1 - \alpha)(1 - p)} + A\frac{(1 - \alpha)(1 - p)}{0p + (1 - \alpha)(1 - p)}$$
$$\geq \frac{A}{2}\left(\frac{0p}{0p + (1 - \alpha)(1 - p)}\right) + \frac{A}{2}\left(\frac{(1 - \alpha)(1 - p)}{0p + (1 - \alpha)(1 - p)}\right)$$

which simplifies to $2A \geq A$. Hence, player 2 will respond betting if, despite having a bad hand, he observes that player 1 folded.

Player 1. Let us now move to the first mover, player 1. When having a good hand, betting is strictly dominant, as discussed above. However, when his hand is bad (as indicated in the two branches in the right-hand side of the tree), he finds profitable to bet if

$$EU_1(\text{Bet}|\text{Bad hand}) \geq EU_1(\text{Fold}|\text{Bad hand})$$

that is,

$$\underbrace{p[1 \cdot (-B) + 0 \cdot A]}_{\substack{\text{if } P_2\text{'s hand is good} \\ \text{he responds betting}}} + \underbrace{(1 - p)\left[\beta\frac{A}{2} + (1 - \beta)A\right]}_{\substack{\text{if } P_2\text{'s hand is bad} \\ \text{he bets with prob } \beta}} \geq \underbrace{p\left[1 \cdot 0 + 0 \cdot \frac{A}{2}\right]}_{\substack{\text{if } P_2\text{'s hand is good} \\ \text{he responds betting}}} + \underbrace{(1 - p)\left[\beta \cdot 0 + (1 - \beta)\frac{A}{2}\right]}_{\substack{\text{if } P_2\text{'s hand is bad} \\ \text{he bets with prob } \beta}}$$

and solving for probability p we find that player 1 bets after receiving a bad hand if and only if

$$p < \frac{A}{2B + A} \equiv \bar{p}$$

(Note that this condition only depends on the prior probability of nature, p, and on the size of the payoffs A and B). Therefore, player 1 bets, despite receiving a bad hand, when the chances that player 2 has a better hand are relatively low, i.e., $p < \bar{p}$.[1] (For a numerical example, note that if the ante contains \$1 and the amount of money

---

[1]Importantly, this condition holds both when α satisfies $\alpha \geq \frac{2Bp}{A(1-p)}$, which implies that player 2 responds betting when his hand is bad, thus implying $\beta = 1$, and when $\alpha \geq \frac{2Bp}{(A(1-p))}$, implying that player 2 responds folding when his hand is bad, i.e., $\beta = 0$, since the cutoff strategy $p < \bar{p}$ is independent on alpha.

that each player must add to the ante when betting is also \$1, i.e., $A = B = \$1$, then we obtain that the cutoff probability is $\bar{p} = 1/3$.)

Hence, we can summarize the PBE of this game as follows:

Player 1: Bet after receiving a good hand for all p, but bet after receiving a bad hand if and only if $p < \bar{p}$. Player 1's posterior beliefs coincide with his prior p.

Player 2: Bet after receiving a good hand for all p. After receiving a bad hand, and observing that player 1 bets, player 2 responds betting if and only if $\alpha \geq \frac{2Bp}{A(1-p)}$; while after observing that player 1 folds, player 2 responds betting under all parameter values. Player 2's beliefs are given by Bayesian updating, as follows:

$$\mu_1 = \frac{p}{p + (1-p)\alpha}, \; \mu_2 = \frac{0 \cdot p}{1 \cdot p + (1-p)(1-\alpha)} = 0,$$

$$\mu_3 = \frac{p}{p + (1-p)\alpha}, \; \text{and } \mu_4 = \frac{0 \cdot p}{0 \cdot p + (1-p)(1-\alpha)} = 0$$

As a remark, note that $\mu_3 = 0$ ($\mu_4 = 0$) means that player 2, in equilibrium, anticipates that the bet (fold, respectively) decision from player 1 he observes must indicate that player 1's cards are bad, both when player 2's hand is good (in μ_3) and bad (in μ_4).

Exercise 2—Incomplete Information and Certificates[B]

Consider the following game with complete information. Two agents, A and B, simultaneously and independently decide whether to join, J, or not join, NJ, a common enterprise. The following matrix represents the payoffs accruing to agents A and B under each strategy profile:

		Player B	
		J	NJ
Player A	J	1, 1	-1, 0
	NJ	0, -1	0, 0

Note that every agent $i = \{A, B\}$ obtains a payoff of 1 if the other agent $j \neq i$ joins, but a payoff of -1 if the other agent does not join (while the agent who did not join obtains a payoff of zero). Finally, if no agent joins, both agents obtain a payoff of zero.

Part (a) Find the set of pure and mixed Nash equilibria.

Part (b) Let us now consider the following extension of the previous game in which we introduce incomplete information: before players are called to move (deciding whether they join or not join), nature selects a type $t \in [0, 1]$, uniformly distributed between

0 and 1, i.e., $t \sim U[0, 1]$, which only affects player B's payoff when both players join. Intuitively, the benefit of player B from having both players joining the common enterprise is larger than that of player A, $1, but such additional benefit t is player B's private information. Thus, after such choice by nature, the payoff matrix becomes

Player B

		J	NJ
Player A	J	1, 1+t	-1, 0
	NJ	0, -1	0, 0

Find the set of Bayesian Nash equilibria (BNE) of this game.

Part (c) Consider now a variation of the above incomplete information game: First, nature determines the specific value of t, either $t = 0$ or $t = 1$, which only player B observes. Player B proposes to the uninformed player A that the latter will make a transfer p to the former. If player A accepts the offer, the transfer is automatically implemented. Upon observing the value of t, player B chooses whether or not to acquire a certificate at a cost $c > 0$. Player A can only observe whether player B acquired a certificate, and knows that both types ($t = 0$ and $t = 1$) are equally likely. Find under which conditions can a pooling equilibrium be sustained in which player B acquires a certificate, both when his type is $t = 0$ and when it is $t = 1$.

Part (d) In the version of the game described in part (c), identify under which conditions can a separating PBE be sustained, in which player B acquires a certificate only when his type is $t = 1$.

Answer

Part (a) Complete information version of the game:

Player B

		J	NJ
Player A	J	1, 1	-1, 0
	NJ	0, -1	0, 0

PSNE. Let us first examine each player's best response function. If player A joins (top row), then player B's best response, BR_B, is to join; while if player A does not join (bottom row), then BR_B is not join. By symmetry, if player B joins, then player A's best response, BR_A is to join; while if player B does not join, then BR_A is not join. Intuitively, both players have incentives to mimic the action selected by the other player, as in similar coordination games we analyzed in previous chapters. Thus, the game has the following two pure (and symmetric) strategy Nash equilibrium:

$$\text{PSNE}: \ \{(J,J), \ (NJ,NJ)\}$$

Hence, the game reflects the same strategic incentives as in a Pareto coordination game with two PSNE, where one equilibrium yields unambiguously larger payoffs to both players than the other equilibrium, i.e., (J, J) Pareto dominates (NJ, NJ).

MSNE. Let us now find the MSNE of this game, denoting p (q) the probability with which player A (player B, respectively) joins. Making player A indifferent between joining and not joining, we obtain

$$EU^A(J) = EU^A(NJ), \text{ that is}$$

$$q \cdot 1 + (1-q)(-1) = 0 \leftrightarrow q = \frac{1}{2}$$

Similarly $p = \frac{1}{2}$ given that agents are symmetric. Hence, the mixed strategy Nash equilibrium of the game is

$$\text{MSNE}: \ \left\{ \left(\frac{1}{2}J, \frac{1}{2}NJ\right), \left(\frac{1}{2}J, \frac{1}{2}NJ\right) \right\}$$

Part (b) Let us now consider the incomplete information extension of the previous game, where $t \sim U[0,1]$, as depicted in the next payoff matrix.

Player B

		J	NJ
	J	1, 1+t	-1, 0
Player A			
	NJ	0, -1	0, 0

Let us first examine the informed player (player B). Denoting by μ the probability with which player A joins, then player B's expected utility from joining and not joining are

$$EU^B(J) = \mu(1+t) + (1-\mu)(-1) = \mu + \mu t - 1 + \mu$$
$$= \mu(2+t) - 1, \text{ and } EU^B(NJ) = 0$$

where μ denotes the probability with which his opponent joins the enterprise. Hence, player B will decide to join if and only if:

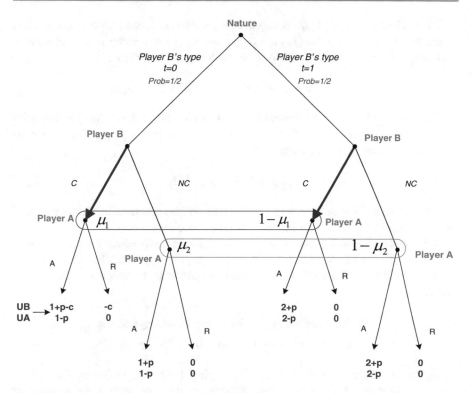

Fig. 10.2 Pooling strategy profile—certificates

$$EU^B(J) \geq EU^B(NJ) \leftrightarrow \mu(2+t) - 1 \geq 0 \leftrightarrow \mu \geq \frac{1}{2+t}$$

This implies the existence of two BNEs, as we next describe:

<u>1st BNE</u>

If $\mu \geq \frac{1}{2+t}$ then both players join, and the BNE is $(\sigma^A, \sigma^B, \mu) = \left\{ J, J; \mu > \frac{1}{2+t} \right\}$

<u>2nd BNE</u>

If $\mu < \frac{1}{2+t}$ then no player joins, and the BNE is $(\sigma^A, \sigma^B, \mu) = \left\{ NJ, NJ; \mu < \frac{1}{2+t} \right\}$

Note that both of these BNEs involve structurally consistent beliefs: first, when both players join, $\mu = 1$, which satisfies $\mu > \frac{1}{2+t}$ since $t \sim U[0,1]$; second, when no player joins, $\mu = 0$, which satisfies $\mu < \frac{1}{2+t}$. Intuitively, if every player believes that his opponent will likely join, he responds joining; thus giving rise to the first type of BNE where both players join the enterprise. Otherwise, neither of them joins, and the second type of BNE emerges.

Part (c) Pooling PBE

Figure 10.2 depicts the pooling strategy profile in which the privately informed player, player B, acquires a certificate, both when his type is $t = 0$ and $t = 1$.

After observing that player B acquired a certificate. When player A is called on
to move on information set $(\mu_1, 1 - \mu_1)$, he accepts the transfer p after observing
that player B acquired a certificate (in equilibrium) if and only if[2]:

$$\mu_1(1 - p) + (1 - \mu_1)(2 - p) \geq 0 \leftrightarrow 2 - p \geq \mu_1$$

Since we are in a pooling strategy profile in which both types of player B acquire
a certificate, then $\mu_1 = \mu(t = 0|cert) = 1/2$ (i.e., posterior = prior). This allows us
to further reduce the above inequality to

$$2 - p \geq 1/2 \leftrightarrow p \leq 3/2.$$

that is, any transfer p in the range $p \in [0, 3/2]$ will be accepted by the uninformed
player A after observing that player B acquired a certificate.

After observing that player B does not acquire a certificate. When player A is
called on to move on information set $(\mu_2, 1 - \mu_2)$, i.e., he observes the
off-the-equilibrium action of no certificate from player B, he rejects the transfer p if
and only if

$$EU^A(Accept|NC, \mu_2) \leq EU^A(Reject|NC, \mu_2), \text{that is}$$
$$\mu_2(1 - p) + (1 - \mu_2)(2 - p) \leq 0 \leftrightarrow 2 - p \leq \mu_2$$

At this point, note that any transfer p satisfying $2 - p \leq \mu_2$, or $2 - \mu_2 \leq p$, will be
rejected. Several off-the-equilibrium beliefs μ_2 support such a behavior. For
instance, if $\mu_2 = 1$, we obtain that $2 - 1 \leq p$, or $1 \leq p$, leading player A to reject the
transfer proposed by player B if no certificate is offered and the transfer A needs to
give to B is weakly larger than \$1. Intuitively, note that $\mu_2 = 1$ implies that the
uninformed player A assigns the worst possible beliefs about player B's type upon
observing no certificate, i.e., player B's type must be low ($t = 0$). If, instead, his
beliefs assign full probability to agent B's type being good, i.e., $\mu_2 = 0$, then the
above incentive compatibility condition becomes $2 - 0 \leq p$, or $2 \leq p$, thus implying
that player A is willing to make transfers to player B as long as the transfer does not
exceed \$2).

Informed player B. Let us now move to the informed player B (the first mover in
this sequential move game with incomplete information). In order to keep track of
player A's optimal responses in this game, i.e., accepting after observing that player
B acquired a certificate but rejecting otherwise, Fig. 10.3 shades the acceptance
(rejection) branch of player A after observing that player B acquires (does not
acquire, respectively) a certificate.

- *Player B's type is $t = 0$*. In this case, player B anticipates that a certificate will be
  responded with acceptance, yielding for him a payoff of $1 + p - c$; but not

---

[2]Notice that here p represents a monetary transfer or payment not necessarily contained in the
unitary range, different to the probability p used in part (a) of this exercise.

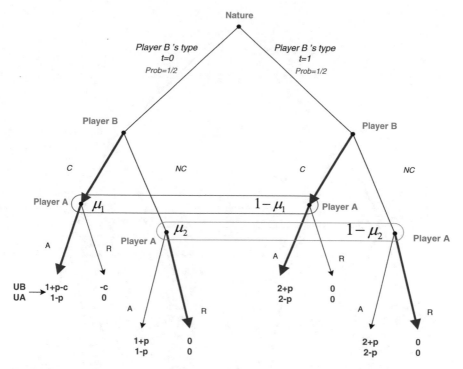

Fig. 10.3 Pooling strategy profile (including responses)

acquiring a certificate will be rejected, entailing a payoff of 0. Hence, he offers the certificate if $1 + p - c \geq 0$ and since the transfer enters positively, he will choose the highest offer p that guarantees acceptance, i.e., $p = 3/2$, implying that he will acquire a certificate as long as its cost is sufficiently low, i.e., $1 + 3/2 - c \geq 0 \leftrightarrow c \leq 5/2$.

- *Player B's type is $t = 1$.* Similarly, when his type is $t = 1$, player B offers the certificate if $2 + p > 0$. By the same argument as above, any transfer $p \leq 1$ is accepted upon offering no certificate (note that this occurs off-the-equilibrium path), leading player B to select the highest offer that guarantees rejection, i.e., $p = 1$. This inequality thus becomes $2 + 1 > 0$, which holds for all parameter conditions.

Hence, a <u>Pooling strategy profile</u> can be sustained as a PBE, and it can be summarized as follows:

- Player B acquires a certificate, both when his type is $t = 0$ and when $t = 1$, as long as the certificate costs are sufficiently low, i.e., $c < 5/2$, and asks for a transfer of $p = \frac{3}{2}$ after acquiring a certificate (in equilibrium), and a transfer of $p = 1$ after not acquiring a certificate (off-the-equilibrium).

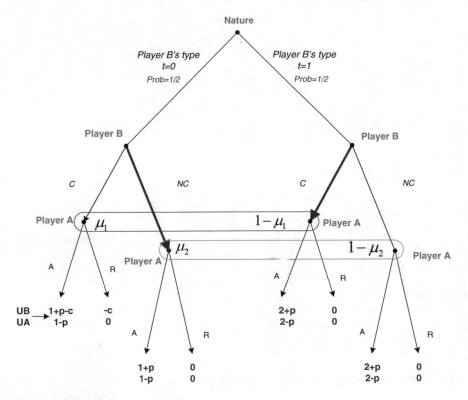

Fig. 10.4 Separating strategy profile

- Player A responds accepting to make any transfer $p \leq \frac{3}{2}$ to player B after observing a certificate, and any transfer $p \leq 1$ after observing no certificate. His equilibrium beliefs are $\mu_1 = \frac{1}{2}$ (after observing a certificate, since both types of player B acquire such a certificate in equilibrium and both types are equally likely), while his off-the-equilibrium beliefs are $\mu_2 = 1$.[3]

Part (d) <u>Separating PBE</u>

Let us now check if the separating strategy profile in which only $t = 1$ acquires certificate (while $t = 0$ does not) can be sustained as a PBE. In order to facilitate our visual presentation, Fig. 10.4 shades the arrow corresponding to certificate by player B when his type is $t = 1$, and that corresponding to no certificate when $t = 0$.

Beliefs. In a separating equilibrium, we must have that, upon observing the certificate, player A's posterior beliefs are,

---

[3]Note that if, instead, off-the-equilibrium beliefs are $\mu_2 = 0$, then the highest transfer that player B can request from player A increases from $p = 1$ to $p = 2$ when he does not acquire a certificate. All other results of the pooling PBE would remain unaffected.

$$\mu_1 = \mu(t = 0|C) = 0, \text{and}$$
$$\mu_2 = \mu(t = 0|NC) = 1,$$

since player B does not acquire a certificate in this strategy profile when his type is $t = 0$. Let us next examine player A's optimal responses after observing a certificate or no certificate from player B.

Player A after observing that player B acquired a certificate. Given the above beliefs, if player A observes a certificate (in μ_1) he rejects if and only if

$$EU^A(Accept|NC, \mu_1) = 1 - p \le EU^A(Reject|NC, \mu_1) = 0,$$

that is, if $p \ge 1$. Intuitively, player A is convinced to be in the leftmost node (marked with μ_1 in Fig. 10.4), thus making his optimal response straightforward.

Player A after observing that player B did not acquire a certificate. If, instead, player A observes that player B did not acquire a certificate (in μ_2), he responds accepting if and only if

$$EU^A(Accept|NC, \mu_2) = 2 - p \ge EU^A(Reject|Nc, \mu_2) = 0$$

that is, if $p \le 2$.

Player B. Let us now analyze player B, who anticipates under which conditions a certificate will be accepted and rejected by player A as described in our above decisions.

- *Player B's type is $t = 1$.* In particular, when his type is $t = 1$, player B offers a certificate (as prescribed in this separating equilibrium) if and only if

$$EU^B(Cert|t = 1) = 2 + p - c \ge EU^B(NC|t = 1) = 0$$

Hence, since the transfer p enters positively into player B's utility, he will acquire a certificate and select the highest value of p that still guarantees acceptance, i.e., $p = 2$. In this case, the cost of acquiring the certificate must satisfy $2 + 2 - c \ge 0$, i.e., $c \le 4$.

- *Player B's type is $t = 0$.* In contrast, when player B's type is $t = 0$, he doesn't offer a certificate (as prescribed) if and only if

$$EU^B(Cert, p \ge 1|t = 0) = 1 + p - c \le EU^B(NC, p \ge 1|t = 0) = 0$$

where for a certificate to be accepted (entailing a payoff of $1 + p - c$ for player B), the transfer p must satisfy $p \ge 1$, as described in our analysis of player A. Therefore, when player B's type is $t = 0$, his best response is to not acquire a

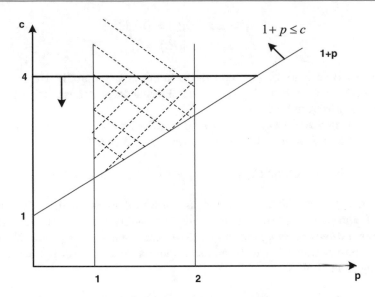

Fig. 10.5 (p, c)-pairs for which a separating PBE can be sustained

certificate (as prescribed in this separating strategy profile) if and only if $1 + p - c \leq 0$, or $1 + p \leq c$.

Given that p must satisfy $p \geq 1$, and $p \leq 2$, i.e., $1 \leq p \leq 2$, we can graphically represent the region of (p, c)-pairs for which the separating equilibrium can be sustained in the shaded area of Fig. 10.5, where $1 + p \leq c$, $1 \leq p \leq 2$, and $c \leq 4$. Intuitively, the certificate must be sufficiently costly for player B to not have incentives to acquire it when his type is $t = 0$, i.e., $1 + p \leq c$, but sufficiently inexpensive for player B to have incentives to acquire it when $t = 1$, i.e., $c \leq 4$. In addition, the transfer that player B requests from player A, p, cannot be too high, since otherwise player A would not be induced to accept the proposal (and, as a consequence, the common enterprise will not be started), i.e., $1 \leq p \leq 2$. We can, hence, summarize this separating PBE as follows:

- Player B acquires a certificate when his type is $t = 1$ if and only if the certificate is not too costly $c \leq 4$, but does not acquire the certificate when his type is $t = 0$ if and only if it is sufficiently costly, i.e., $c \geq 1 + p$. Regarding the transfers that he requests from the uninformed player A, player B asks a transfer $p = 2$ when he acquires a certificate (which guarantees acceptance), but asks for a transfer $p > 1$ when he does not acquire a certificate (which entails a rejection).
- Player A accepts player B's offer if and only if the transfer he must give to player B is sufficiently low, i.e., $p \leq 2$ after observing a certificate, but $p \leq 1$ after observing no certificate.
- Player A's beliefs are $\mu_1 = \mu(t = 0|C) = 0$ after observing a certificate and $\mu_2 = \mu(t = 0|NC) = 1$ after observing no certificate.

Exercise 3—Entry Game with Two Uninformed Firms[B]

Suppose there are two symmetric firms, A and B, each privately observing its type
(t), either g (good) or b (bad) with probabilities p and $1 - p$, respectively, where
$0 < p < 1$. For instance, "good" can represent that a firm is cost efficient, while
"bad" can indicate that the firm is cost inefficient. They play an entry game in which
the two firms simultaneous choose to enter or stay out of the industry. Denote the
outcome of the entry game by the set of types that *entered*, i.e., (g, g) means two
good firms entered, (b) means only one bad firm entered, and so on. If a firm does
not enter, its payoff is 0. When it enters, its payoff is determined by its own type and
the set of types that entered, with the following properties:

$$u^g(g) > u^g(g,b) > u^g(g,g) > u^b(b) > 0 > u^b(b,b) > u^b(g,b)$$

where the superscript is the firm's type and the argument is the outcome (profile of
firms that entered the industry).

Find a symmetric separating Perfect Bayesian equilibrium (PBE) in which every
firm i enters only when its type is good assuming that the following condition holds:
$pu^b(g,b) + (1 - p)u^b(b) < 0$.

Answer

Since this is a simultaneous-move game of incomplete information (both firms are
uninformed about each other's types), and firms are ex-ante symmetric (before each
observes its type), we next analyze the entry decision of every firm i. First for the
case in which its own type is good, and then when its type is bad.

   Firm i's type is good. When firm i's type is good, $t_i = g$, the expected utility that
firm i obtains from entering is

$$EU_i(Enter|t_i = g) = \underbrace{p[r'u^g(g,g) + (1 - r')u^g(g)]}_{\text{if firm } j \text{ is good}, j \text{ can decide either to enter or not}}$$

$$+ \underbrace{(1 - p)[ru^g(g,b) + (1 - r)u^g(g)]}_{\text{if firm } j \text{ is bad}, j \text{ can decide either to enter or not}}$$

where r' denotes the probability that firm j enters when its type is good, $t_i = g$;
while r represents the probability that firm j enters when its type is bad, $t_i = b$.

   But we know that when a firm's type is good it always enters given that its
payoff is strictly positive by definition, i.e., entering is a strictly dominant strategy
for a good firm since staying out yields a payoff of zero. That is, a good firm i is
better off entering when firm j does not enter, $u^g(g) > 0$, and when firm j enters,
both when firm j's type is good, $u^g(g,g) > 0$, and bad $u^g(g,b) > 0$. This implies

that firm j when its type is good, $t_j = g$, will decide to enter, i.e., $r' = 1$. Therefore the above EU_i can be simplified to:

$$EU_i(Enter|t_i = g) = pu^g(g,g) + (1-p)\left[ru^g(g,b) + (1-r)u^g(g)\right]$$

On the other hand, firm i obtains a payoff of zero from staying out regardless of firm j's type and irrespective of whether firm j enters or not. Indeed,

$$EU_i(Not\,Enter|t_i = g) = p[r' \cdot 0 + (1-r') \cdot 0] + (1-p)[r \cdot 0 + (1-r) \cdot 0] = 0$$

Therefore, when deciding whether to join or not, a good firm i will compare these two expected utilities, inducing it to enter if and only if

$$EU_i(Enter|t_i = g) > EU_i(Not\,Enter|t_i = g)$$

$$p \cdot \underbrace{u^g(g,g)}_{+} + (1-p)\left[r \cdot \underbrace{u^g(g,b)}_{+} + (1-r)\underbrace{u^g(g)}_{+}\right] > 0$$

which holds for all values of p and r, since $u^g(g,g), u^g(g,b)$, and $u^g(g)$ are all positive by definition. Hence, a good firm i will enter, i.e., we don't have to impose additional conditions in order to guarantee that this type of firm enters.

Firm i's type is bad. Let us now consider the case in which firm i's type is bad, i.e., $t_i = b$. The expected utility that this firm obtains from entering is

$$EU_i(Enter|t_i = b)$$

$$= p\overbrace{\left[r'u^b(g,b) + (1-r')u^b(b)\right]}^{if\,firm\,j\,is\,good} + \overbrace{(1-p)\left[ru^b(b,b) + (1-r)u^b(b)\right]}^{if\,firm\,j\,is\,bad}$$

But given that firm j enters when its type is good, then $r' = 1$, and the previous expression reduces to:

$$EU_i(Enter|t_i = b) = pu^b(g,b) + (1-p)\left[ru^b(b,b) + (1-r)u^b(b)\right]$$

This firm's expected utility from not entering is zero, i.e., $EU_i(Not\,Enter|t_i = b) = 0$, since firm i attains a zero payoff from staying out regardless of firm j's decision and type. Hence, for firm i to stay out when its type is bad (as prescribed in the separating strategy profile) we need that

$$pu^b(g,b) + (1-p)\left[ru^b(b,b) + (1-r)u^b(b)\right] \leq 0$$

But, since the strategy profile we are testing requires firms to enter when their type is good, we have that $r = 1$, which further simplifies the above inequality to

$$p \underbrace{u^b(g,b)}_{-} + (1-p) \underbrace{u^b(b,b)}_{-} \leq 0$$

which holds since $u^b(g,b) < 0$ and $u^b(b,b) < 0$ by definition. Hence, firm i will not enter when being of bad type.

Therefore, we have found a symmetric PBE in which, every firm $i = \{A, B\}$ enters (stays out) when its type is good (bad, respectively), and each firm believes the other firm will only enter when its type is good.

Exercise 4—Labor Market Signaling Game and Equilibrium Refinements[B]

Consider the following labor market signaling model. A worker with privately known productivity θ chooses an education level e. Upon observing e, each firm responds with a wage offer that corresponds to the worker's expected productivity. (That is, we assume that the labor market is perfectly competitive, so firms pay the value of the expected marginal productivity to every worker. Otherwise, competing firms would have incentives to increase their salaries in order to attract workers with higher productivities.) The worker's payoff from receiving a wage w and acquiring an education level e, given an innate productivity θ, is $u(w, e|\theta) = w - \frac{e^3}{\theta}$, i.e., the cost of acquiring education, $\frac{e^3}{\theta}$, is convex, and decreases in the worker's innate productivity, θ. (Note that, for simplicity, we assume that the worker's innate productivity, θ, is unaffected by the education level he acquires.)

Part (a) Assuming that there are two equally likely types $\theta_1 = 1$ and $\theta_2 = 2$, characterize the set of all separating equilibria.

Part (b) Which of these equilibria survive the Cho and Kreps' (1987) Intuitive Criterion?

Answer

Part (a) *Separating PBE when* $\theta_1 = 1$ *and* $\theta_2 = 2$. Figure 10.6 depicts (w, e)-pairs, and two indifference curves: one for the high-productivity worker, IC_H, i.e., $u = w - \frac{e^3}{2}$ since $\theta_2 = 2$, in which this type of worker reaches the same utility level at all points of IC_H; and other for the low-productivity worker, IC_L, whereby this worker reaches the same utility level among all points satisfying $w = u + e^3$. Intuitively, note that an increase in education must be associated with an increase in wages in order to guarantee that the worker's utility level is unaffected, which explains why the worker's indifference curve is increasing in education. Furthermore, an increase in education must be accompanied be a more-than-proportional increase in wages in order to guarantee that the worker's utility level remains unaltered, which explains why indifference curves are convex in e. In addition, note that indifference curves closer to the northeast of the figure represent a higher utility level, since they are associated to a higher wage.

Fig. 10.6 Separating PBE in
the labor market signaling
game

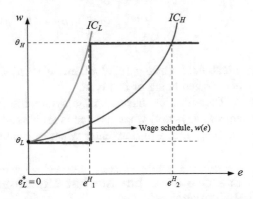

The set of all separating PBE satisfies:

- low productivity worker chooses no education, i.e., $e_L^*(\theta_L) = 0$
- high productivity worker chooses an education level in the range $e_H^*(\theta_H) \in \left[e_1^H, e_2^H\right]$, where the lower bound e_1^H solves $u_L = w_H - e^3$, i.e., $w_H = u_L + e^3$ as represented in the figure. A similar argument applies for the upper bound e_2^H but, for compactness, its exact expression is presented below.
- Firms, after observing equilibrium education levels, $e_L^*(\theta_L) = 0$ and $e_H^*(\theta_H) \in \left[e_1^H, e_2^H\right]$, offer wages:

$$w^*\left(e_L^*(\theta_L)\right) = \theta_L = 1, \text{ and}$$
$$w^*\left(e_H^*(\theta_H)\right) = \theta_H = 2$$

while for any off-the-equilibrium education level $e \neq e_L^*(\theta_L) \neq e_H^*(\theta_H)$ firms offer a wage schedule $w^*(e)$ which lies weakly below the indifference curve of both types of workers (as represented in the thick black line in the Fig. 10.6).[4]

In order to determine the exact value of e_1^H and e_2^H, we need that the following incentive compatibility conditions are satisfied:

$$EU\left(e_L^*, w^*(e)|\theta_L\right) \geq EU\left(e_H^*, w^*(e)|\theta_L\right)$$

so the low-productivity worker does not have incentive to deviate from his education level, e_L^*, to that of the high-productivity worker, e_H^*. Similarly,

---

[4]The wage schedule then lies below IC_H for all off-the-equilibrium education levels, i.e., for all $e < e_1^H$ and $e > e_2^H$. When the firm observes, instead, an equilibrium education level of zero, the firm manager offers a wage of $w(0) = 0$, as depicted in the vertical intercept of wage schedule $w(e)$. Similarly, upon receiving an equilibrium education level of $e \in \left[e_1^H, e_2^H\right]$, the firm manager offers salary $w(e) = 2$, as depicted in Fig. 10.6 for all $e \in \left[e_1^H, e_2^H\right]$.

$$EU\left(e_H^*, w^*(e)|\theta_H\right) \geq EU\left(e_L^*, w^*(e)|\theta_H\right)$$

implying that the high-productivity worker does not have incentives to reduce his education level from e_H^* to e_L^*. For the utility functions and parameters given in the exercise, the above two incentive compatibility conditions reduce to,

$$1 - \frac{0^3}{1} \geq 2 - \frac{(e^H)^3}{1} \quad \text{which implies} \quad e^H \geq 1$$

$$2 - \frac{(e^H)^3}{2} \geq 1 - \frac{0}{2} \quad \text{which implies} \quad \sqrt[3]{2} \geq e^H$$

Therefore, the education level that the high-productivity worker selects in equilibrium, $e_H^*(\theta_H) \in \left[e_1^H, e_2^H\right]$ must lie in the interval $e_H^*(\theta_H) \in \left[1, \sqrt[3]{2}\right]$, where $\sqrt[3]{2} \simeq 1.26$

Note that the above strategy profile is indeed a PBE:

- firms are optimizing given their benefits about workers' types and given the equilibrium strategies for the workers.
- workers are optimizing given firms' equilibrium strategies (wages) and given firms' beliefs about the workers' types.
- firms' beliefs are computed by Bayes' rule when possible.

Note that the θ_L-worker will never acquire more education than $e_L^*(\theta_L) = 0$: by acquiring more education, such as any e_1^H satisfying $e_1^H < 1$, he is still identified as a low-productivity worker and offered a wage of θ_L, and he is incurring a large costs in acquiring this additional education. Furthermore, even if he acquires $e_1^H = 1$ (or higher) and he is identified as a high-productivity worker (fooling the firm), the increase in his wage will not offset the increase in his education costs (which are especially high for him), ultimately indicating that he does not have incentives to deviate from his equilibrium education level $e_L^* = 0$. A similar argument applies to the high-productivity worker: he is already identified as such by the firm manager, who offers him a salary of \$2. Acquiring an even larger education level would only reduce his utility (since acquiring education is costly) without increasing his salary. Lowering his education level is not profitable either, since any education level below e_1^H would imply that he is identified as a low-productivity worker, and thus receives a wage of \$1. Since the savings in education costs he experiences from lowering his education level do not offset the reduction in wages, the high-productivity worker does not have incentives to deviate from his equilibrium education level.

Part (b) *Intuitive criterion*

Figure 10.7 depicts indifference curve IC_L which crosses the (w, e) equilibrium pair the low-productivity worker, i.e., an education level of $e_L*(\theta_2) = 0$ responded with a wage of $w_L = 1$. Similarly, IC_H crosses the (w, e) equilibrium pair for the

Fig. 10.7 Applying the intuitive criterion

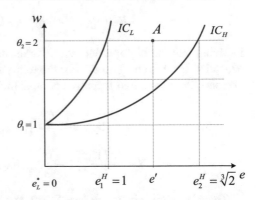

high-productivity worker, namely, an education level of $e_H^*(\theta_H) = \sqrt[3]{2}$ responded with a wage of w_H=\$2. This education level is the upper bound of the equilibrium interval $e_H^*(\theta_H) \in [1, \sqrt[3]{2}]$ of admissible education levels in equilibrium. We next seek to show that all education levels in this interval, other than the least-costly education level in lower bound of the interval, $e_H^*(\theta_H) = 1$, violate the Cho and Kreps' (1987) Intuitive Criterion.

First Step
Firms' beliefs after observing off-the-equilibrium messages, such as e' in Fig. 10.7 need to be restricted to $\Theta^{**}(e') \in \Theta$, which represents all those types of workers for whom sending e' is never equilibrium dominated:

$$\Theta^{**}(e') = \left\{ \theta \in \Theta | U^*(\theta_i) \leq \max_s U(s, e', \theta_i) \right\}, \quad \text{where } s \in S^*(\Theta, e')$$

In words, upon observing effort e', the firm restricts its beliefs to those types of workers $\theta \in \Theta$ for whom their maximal deviating payoff exceeds their equilibrium payoff we can see this inequality is only satisfied for the θ_H-type, but is never satisfied for the low type. That is, even if firms believe that θ_L is in fact a θ_H-worker and pay $w(e') = 2$, it is not convenient for a θ_L-worker to send such a message because of the high education costs that he would need to incur, i.e., $1 - \frac{0^2}{1} \geq 2 - \frac{(e')^3}{1} \leftrightarrow e' \geq 1$. Hence, we say that message e' is equilibrium dominated for the θ_L-worker. Graphically, this argument can be checked by noticing that the indifference curve of the low-productivity worker that passes through point A (guaranteeing a salary of 2 after acquiring an education level of e') would lie to the southeast of the indifference curve that this worker reaches in equilibrium (crossing the vertical axis when he acquires a zero education and is offered a wage of 1). Hence, this type of worker does not have incentives to deviate from his equilibrium education level of zero, even if by doing so he could alter the firm's equilibrium beliefs, i.e., even if he could "fool" the firm into believing his type is high.

However, sending message e' is *not* equilibrium dominated for the θ_H-worker. That is, if firms believe that he is a θ_H-worker when he sends an education level e' rather than the (higher) e_2^H, his salary would be unaffected while his education costs would go down. This is graphically illustrated by the fact that the indifference curve of the θ_H-worker passing through point A in the figure (corresponding to wage $\theta_H = 2$ and education level e') would lie to the northwest of the indifference curve that he reaches when acquiring the education level e_2^H, IC_H, thus indicating that this worker would be able to reach a higher utility level by deviating to e'.

Therefore, the only type of worker for whom sending e' is never equilibrium dominated is $\Theta(e') = \{\theta_H\}$, i.e., only the θ_H-worker could benefit by deviating from his equilibrium message.

Second Step

Once we have restricted firms' beliefs to $\Theta^{**}(e') = \{\theta_H\}$ we need to find whether there exists a message that can be sent by θ_H that implies a higher level of utility than sending his equilibrium message. That is, we need to find a pair (θ_H, e) for which the following inequality holds:

$$U^*(\theta_H) < \min_s U^*(s, e, \theta_H), s \in S^*(\Theta^{**}(e), e)$$

As we can see, once we have restricted firms' beliefs to $\Theta^{**}(e') = \{\theta_H\}$ firms will simply offer wages of $w(e) = \theta_H$ for all those education levels that the θ_L-worker finds unprofitable, i.e., any $e \geq e_1^H$.

As a consequence, the θ_H-worker will be better-off by acquiring an education level lower than the one he acquires in any of the separating PBEs where $e > e_1^H$. Indeed,

$$\theta_H - c\left(e_H^*, \theta_H\right) < \theta_H - c(e, \theta_H) \leftrightarrow c\left(e_H^*, \theta_H\right) > c(e, \theta_H)$$

which is true given that $e_H > e$ and that costs are increasing in education.

Therefore, all the inefficient separating PBEs can be eliminated by the Intuitive Criterion. However, the so-called "Riley outcome" (the efficient separating PBE where the high-productivity worker only needs to acquire the lowest possible amount of education that separates him from the low-productivity worker, i.e., $e = e_1^H$) cannot be eliminated by the Intuitive Criterion. Intuitively, lowering his education level below e_1^H would identify him as a low-productivity worker, receiving a wage offer of $\theta_L = 1$, which is not profitable for the high-productivity worker. Hence, $e = e_1^H$ is the so-called least-costly separating equilibrium, since the θ_H-worker acquires the minimum amount of education that helps him convey his type to the firms, separating himself from the low-productivity worker.[5]

---

[5]For more details on the application of the Cho and Kreps' (1987) Intuitive Criterion and other equilibrium requirements, see Espinola-Arredondo and Munoz-Garcia (2010) "The Intuitive and Divinity Criterion: Interpretation and Step-by-Step examples," *Journal of Industrial Organization Education*, vol. 5(1), article 7.

Exercise 5—Entry Deterrence Through Price Wars[A]

Consider two firms competing in an industry, an incumbent and an entrant. The entrant has already entered the market, and in period 1 the incumbent must decide whether to fight or accommodate; if she fights, the entrant loses $f_E > 0$ (i.e., profit $= -f_E$) and if she accommodates the entrant makes a profit of $a_E > 0$. The incumbent may be either "sane", in which case he loses $f_I > 0$ from fighting, and gains $a_I > 0$ from accommodating; or "crazy", in which case his only available action is to fight. In period 2, the entrant must choose whether to stay in the market or exit; if she exits, the incumbent makes a profit of m; if she stays, the incumbent gets to fight or accommodate again and the payoffs are as described before. Assume that there is no discounting, and that the prior probability of a "sane" incumbent is $0 < p < 1$ (the actual type of the incumbent is private knowledge to the incumbent).

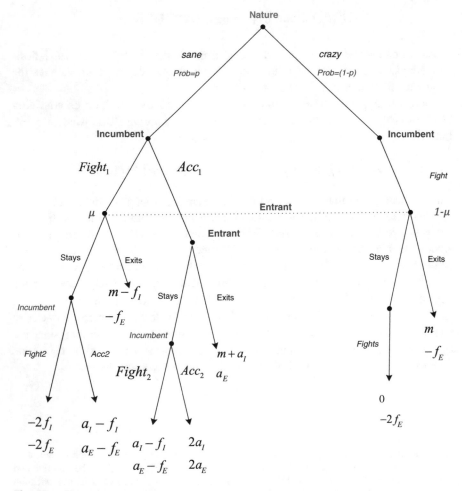

Fig. 10.8 Entry deterrence game—I

Part (a) Draw the game in its extensive form.

Part (b) Show that if payoffs satisfy $m - f_I \leq 2a_I$, then there is a separating equilibrium.

Part (c) Show that if $m - f_I \leq 2a_I$ does not hold there is a pooling equilibrium when $p \leq \dfrac{f_E}{f_E + a_E}$.

Answer

Part (a) Figure 10.8 depicts the game tree. Here is a summary of the entrant's payoff: $-f_E$ are his losses from fighting, while a_E is his benefit from accommodating. For the incumbent we have that $-f_I$ represents the cost of fighting for the sane incumbent, and a_I is the benefit of accommodating for the sane incumbent. This game tree can be alternatively represented as in Fig. 10.9.

We can apply backwards induction in the proper subgames of the above game tree. In particular, after Fight$_1$ (in the right-hand side of the tree) and the entrant responding with Stay, we encounter a proper subgame in which the sane incumbent operates under complete information. In particular, the sane incumbent prefers to choose Acc$_2$, which yields a payoff of $a_I - f_I$, than Fight$_2$, which entails a lower payoff of $-2f_I$. In addition, after Acc$_1$ (in the left-hand side of the tree) and the entrant responding with Stay, we also find a proper subgame in which the sane incumbent is called on to move under complete information. In this case, the sane incumbent prefers Acc$_2$, which yields a payoff of $2a_I$, than Fight$_2$, which only entails $a_I - f_I$. Anticipating this response from the sane incumbent after Acc$_1$, the entrant chooses Stay, since its payoff from doing so, $2a_E$, exceed that from exiting, a_E. These optimal responses help us reduce the game tree as depicted in Fig. 10.10.

Part (b) Once we have reduced the game tree, we can apply our standard approach to search for PBEs, first to check for the existence of separating PBEs, and then for pooling PBEs.

Separating PBE. Figure 10.11 shades the branches corresponding to separating strategy profile in which only the sane incumbent accommodates can be sustained (note that the crazy incumbent can only fight by definition).

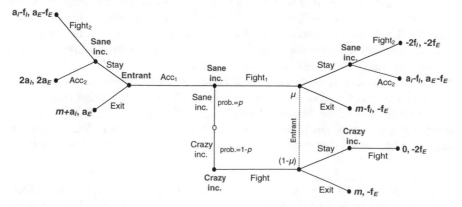

Fig. 10.9 Entry deterrence game—II

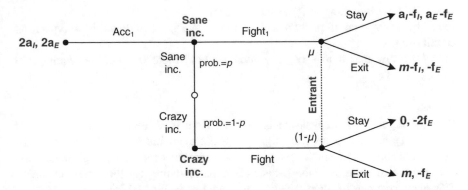

Fig. 10.10 Reduced Entry deterrence game

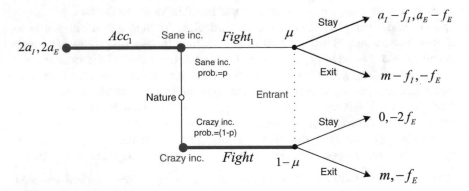

Fig. 10.11 Separating strategy profile

In this strategy profile, the entrant's beliefs are $\mu = 0$, thus considering that a decision to fight can never originate from a sane incumbent. Given these beliefs, the entrant focuses on the node at the lower right-hand corner of the game tree, and responds exiting, since $-f_E < -2f_E$. We depict this response of the entrant in Fig. 10.12, with the blue shaded branch corresponding to exit. (Note that the entrant chooses exit irrespective of the incumbent's type, since the entrant cannot condition its response on the unobserved incumbent's type.)

Once we found such a response from the uninformed entrant, we can show that the sane incumbent prefers Acc_1 (as prescribed), obtaining $2a_I$, rather than deviate towards $Fight_1$, where he would only obtain a payoff of $m - f_I$, where $2a_I > m - f_I$ holds by definition. Hence, the sane incumbent sticks to the prescribed strategy of accommodation, and this separating strategy profile can be sustained as a PBE.[6]

---

[6]Note that this is the only separating equilibrium of the game, since the crazy incumbent cannot choose to accommodate (his only available action is to fight), as depicted in the lower part of Fig. 10.12.

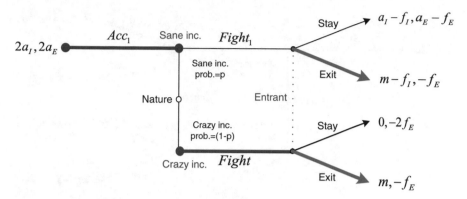

Fig. 10.12 Separating strategy profile (with responses)

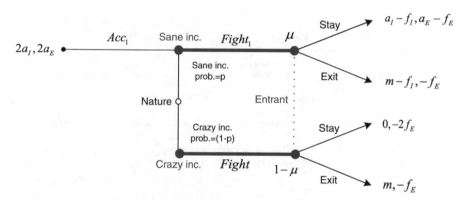

Fig. 10.13 Pooling strategy profile

Part (c) *Pooling PBE*. Let us now examine if a pooling strategy profile, where both types of incumbent fight, can be supported as a PBE. Fig. 10.13 illustrates this strategy profile.

First, in this strategy profile the entrant's beliefs cannot be updated using Bayes' rule, and must coincide with his priors, i.e., $\mu = p$. Intuitively, since both types of incumbent are choosing to fight, the entrant cannot refine its beliefs about the incumbent's type upon observing that the incumbent fought. In this case, the entrant chooses to stay if and only if

$$p(a_E - f_E) + (1 - p)(-2f_E) \geq p(-f_E) + (1 - p)(-f_E) = -f_E$$

or alternatively, if $p \geq \dfrac{f_E}{f_E + a_E} \equiv \bar{p}$. We separately consider the case in which the entrant responds staying $(p \geq \bar{p})$ and the case in which it responds exiting $(p < \bar{p})$.

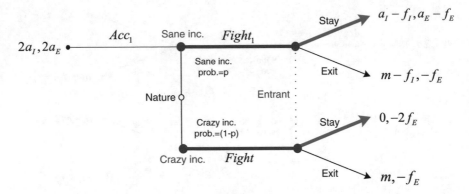

Fig. 10.14 Pooling strategy profile (with responses)—Case 1

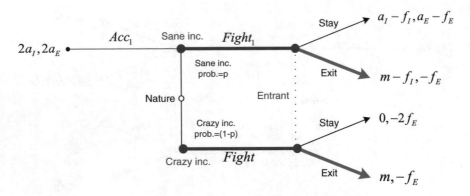

Fig. 10.15 Pooling strategy profile (with responses)—Case 2

Case 1 $p \geq \bar{p}$. Figure 10.14 depicts the case in which the entrant responds staying (blue shaded arrows). In this setting, the sane incumbent would choose to Fight$_1$ (as prescribed) if $a_I - f_I \geq 2a_I$, or alternatively $-f_I \geq a_I$, which cannot hold, since $a_I > 0$ and $-f_I < 0$ by definition. Hence, the pooling strategy profile in which both types of incumbent fight cannot be sustained as a PBE when $p \geq \bar{p}$.

Case 2, $p < \bar{p}$. Let us next check the case in which $p < \bar{p}$, which indicates that the uninformed entrant responds exiting, as Fig. 10.15 depicts (see blue shaded branches). In this context, the sane incumbent chooses Fight$_1$ (as prescribed) if and only if $m - f_I \geq 2a_I$. Hence, as long as this condition holds, the pooling strategy profile in which both types of incumbents fight can be sustained as a PBE if the prior probability of the incumbent being sane, p, is sufficiently low, i.e., $p < \bar{p}$.

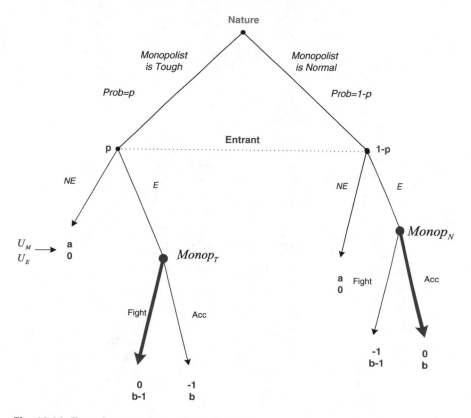

Fig. 10.16 Entry deterrence with only one entrant

Exercise 6—Entry Deterrence with a Sequence of Potential Entrants[C]

The following entry model is inspired on the original paper of Kreps and Wilson (1982)[7]. Consider an incumbent monopolist building a reputation as a tough competitor who does not allow entry without a fight. The entrant first decides whether to enter the market, and, if he does, the monopolist chooses whether to fight or acquiesce. If the entrant stays out, the monopolist obtains a profit of $a > 1$, and the entrant gets 0. If the entrant enters, the monopolist gets 0 from fighting and -1 from acquiescing if he is a "tough" monopolist, and -1 from fighting and 0 from acquiescing if he is a "normal" monopolist. The entrant obtains a profit of b if the monopolist acquiesces and $b - 1$ if he fights, where $0 < b < 1$. Suppose the entrant

---

[7]Kreps, David and Robert Wilson (1982) "Reputation and Imperfect Information,' *Journal of Economic Theory*, 27, pp. 253-279.

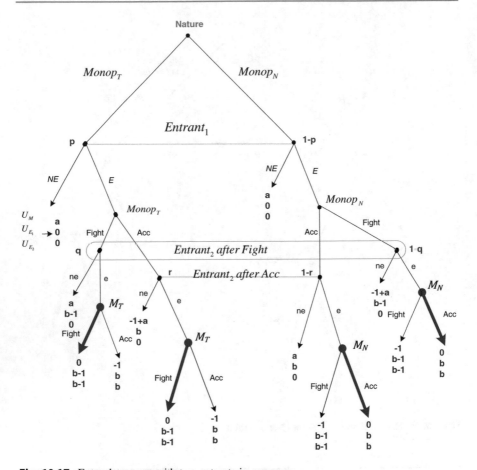

Fig. 10.17 Entry deterrence with two entrants in sequence

believes the monopolist to be tough (normal) with probability p $(1 - p$, respectively), while the monopolist observes his own type.

Part (a) Depict a game tree representing this incomplete information game.

Part (b) Solve for the PBEs game tree, and solve for the PBE of this game.

Part (c) Suppose the monopolist faces two entrants in sequence, and the second entrant observes the outcome of the first game (there is no discounting). Depict the game tree, and solve for the PBE. [*Hint*: You can use backward induction to reduce the game tree as much as possible before checking for the existence of separating or pooling PBEs. For simplicity, focus on the case in which prior beliefs satisfy $p \leq b$.]

Answer

Part (a) See Fig. 10.16.

Part (b) The tough monopolist fights with probability 1, since fight is a dominant strategy for him; while the normal monopolist accommodates with probability 1,

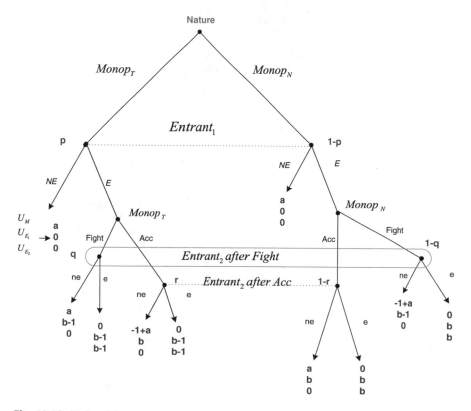

Fig. 10.18 Reduced-form game

since accommodation constitutes a dominant strategy for this type of incumbent. Indeed, at the node labeled with $Monop_T$ on the left-hand side of the game tree in Fig. 10.16, the monopolist's payoff from fighting, 0, is larger than from accommodating, -1. In contrast, at the node labeled with $Monop_N$ for the normal monopolist (see right-hand side of Fig. 10.16), the monopolist's payoff from fighting, -1, is strictly lower than from accommodating, 0. Hence, the entrant's decision on whether or not to enter will be based on:

$$\mathrm{EU}_E(Enter|p) = p(b-1) + (1-p)b = b - p, \text{ and } \mathrm{EU}_E(Not\ Enter|p) = 0$$

Therefore, the entrant enters if and only if $b - p > 0$, or alternately, $b > p$. We can, hence, summarize the equilibrium as follows:

The entrant enters if $b > p$, but doesn't enter if $b \le p$.

The incumbent fights if tough, but accommodates if normal.

Part (c) The following game tree depicts an entry game in which the incumbent faces two entrants in sequence (see Fig. 10.17).

Hence, applying backward induction on the proper subgames (those labeled with M_T and M_N in the last stages of the game tree) we can reduce the previous game tree

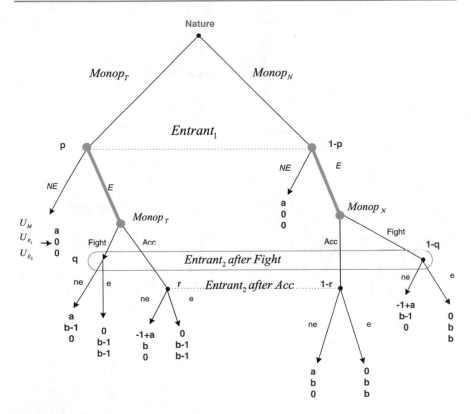

Fig. 10.19 Entry of the first potential entrant

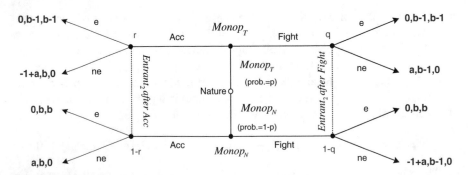

Fig. 10.20 A further reduction of the game

to that in Fig. 10.18. For instance, after the second entrant chooses to enter despite observing a fight with the first entrant (left side of game tree), the tough monopolist chooses between fighting and accommodating in the node labeled M_T. In this case, the tough monopolist prefers to fight, which yields a payoff of zero, rather than

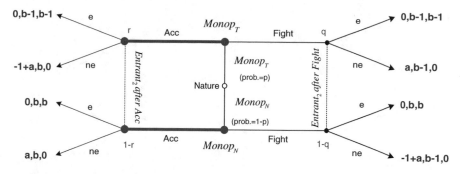

Fig. 10.21 Pooling strategy profile—Acc

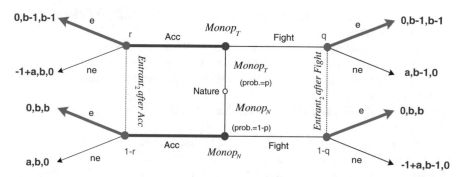

Fig. 10.22 Pooling strategy profile—Acc (with responses)

accommodate, which entails a payoff of -1. A similar analysis applies to the other node labeled with M_T (where the second entrant has entered after observing that the incumbent accommodated the first entrant). However, an opposite argument applies for the nodes marked with M_N in the right-hand side of the tree, where the normal monopolist prefers to accommodate the second entrant, regardless of whether he fought or accommodated the first entrant, since his payoff from accommodating (zero) is larger than from fighting (-1).

In addition, note that the first entrant behaves in exactly the same way as in exercise (a): entering if and only if $b > p$. Hence, when $p \le b$, the first entrant enters, as shown in exercise (a). Figure 10.19 shades this choice of the first entrant (green shaded branches).

Therefore, upon entry, the first entrant gives rise to a beer-quiche type of signaling game, which can be more compactly represented as the game tree in Fig. 10.20. Intuitively, all elements before $Monop_T$ and $Monop_N$ can be predicted (i.e., the first entrant enters as long as $p \le b$), while the subsequent stages characterize a signaling game between the monopolist (privately informed about its type) and the second entrant. In this context, the monopolist uses his decision to fight or

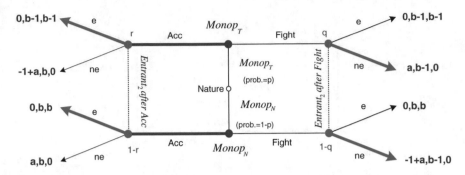

Fig. 10.23 Pooling strategy profile—Acc (with responses)

accommodate the first entrant as a message to the second entrant, in order to convey or conceal his type.

(For every triplet of payoffs, the first corresponds to the monopolist, the second to the first entrant, and the third to the second entrant.) Let us now check if a pooling strategy profile in which both types of monopolists accommodate can be sustained as a PBE.

Pooling PBE with Acc. Figure 10.21 shades the braches corresponding to such a pooling strategy.

In this setting, posterior beliefs cannot be updated using Bayes' rule, which entails $r = p$. As in similar exercises, the observation of accommodation by the uninformed entrant does not allow him to further refine his beliefs about the monopolist's type. Hence, the second entrant responds entering (e) after observing that the monopolist accommodates (in equilibrium), since

$$p(b - 1) + (1 - p)b > p \cdot 0 + (1 - p) \cdot 0 \leftrightarrow b > p$$

which holds in this case.

If, in contrast, the second entrant observes the off-the-equilibrium message of fight, then this player also enters if

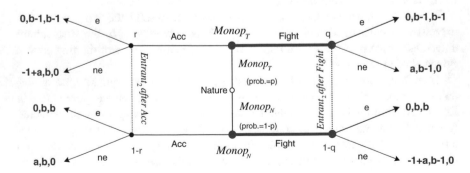

Fig. 10.24 Pooling strategy profile—Fight

$$q(b-1)+(1-q)b > q \cdot 0 + (1-q) \cdot 0 \leftrightarrow b > q$$

Hence, the second entrant enters regardless of the incumbent's action if off-the-equilibrium beliefs, q, satisfy $q < b$, as depicted in Fig. 10.22 (see blue shaded branches in the right-hand side of the figure). Otherwise, the entrant only enters after observing the equilibrium message of Acc.

The tough monopolist, M_T, is hence indifferent between Acc (as prescribed) which yields a payoff of 0, and deviate to Fight, which also yields a payoff of zero. A similar argument applies to the normal monopolist, M_N, in the lower part of the game tree. Hence, the pooling strategy profile where both types of incumbents accommodate can be sustained as a PBE.

A remark on the Intuitive Criterion. Let us next show that the above pooling equilibrium, despite constituting a PBE, violates the Cho and Kreps' (1987) Intuitive Criterion. In particular, the tough monopolist has incentives to deviate towards Fight if, by doing so, he is identified as a tough player, $q = 1$, which induces the entrant to respond not entering. In this case, the tough monopolist obtains a payoff of a, which exceeds his equilibrium payoff of 0. In contrast, the normal monopolist doesn't have incentives to deviate since, even if his deviation to Fight deters entry, his payoff from doing so, $-1 + a$, would still be lower than his equilibrium payoff of 0, given that $-1 + a < 0$ or $a < 1$. Hence, only the tough monopolist has incentives to deviate, and the entrant's off-the-equilibrium beliefs can thus be restricted to $q = 1$ upon observing that the monopolist fights. Intuitively, the entrant infers that the observation of Fight can only originate from the tough monopolist. In this case, the tough incumbent indeed prefers to select Fight, thus implying that the above pooling PBE violates Cho Kreps' (1983) Intuitive Criterion.(Q.E.D)

Let us finally check if this pooling strategy profile can be sustained when off-the-equilibrium beliefs satisfy, instead, $q \geq b$, thus inducing the entrant to respond not entering upon observing the off-the-equilibrium message of Fight, as illustrated in the game tree of Fig. 10.23 (see blue shaded branches in the right-hand side of the tree). In this case, the M_T has incentives to deviate from Acc, and thus the pooling strategy profile where both M_T and M_N select to Acc cannot be sustained as a PBE.

Pooling PBE with Fight. Let us now examine the opposite pooling strategy profile (Fight, Fight), in which both types of monopolists fight, as depicted in Fig. 10.24.

Hence, equilibrium beliefs after observing Fight cannot be updated, and thus satisfy $q = p$; while off-the-equilibrium beliefs are arbitrary $r \in [0, 1]$ after observing the off-the-equilibrium message of accommodation. Given these beliefs, upon observing the equilibrium message of Fight, the entrant responds entering since

$$p(b-1)+p \cdot b > p \cdot 0 + (1-p) \cdot 0 \leftrightarrow b > p$$

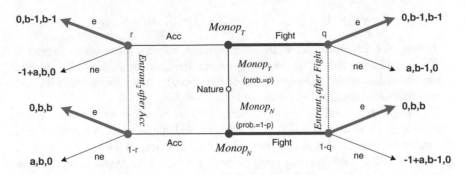

Fig. 10.25 Pooling strategy profile—Fight (with responses)

which holds by definition. If, in contrast, the entrant observes the off-the-equilibrium message of Acc, then it responds entering if

$$r(b-1) + r \cdot b > p \cdot 0 + (1-r) \cdot 0 \leftrightarrow b > r$$

Hence, if $b > r$, the entrant enters both after observing Fight (in equilibrium) and Acc (off-the-equilibrium path). If, instead, $b \leq r$, then the entrant only responds entering after observing Fight, but is deterred from the industry otherwise. We next separately analyze each case.

Case 1. Figure 10.25 illustrates the entrant's responses when $b > r$, and thus the second entrant enters after Acc. In this context, the tough monopolist is indifferent between Fight, obtaining zero profits, and accommodating, which also yields zero profits. A similar argument applies to the normal monopolist in the lower part of the game tree. Hence, in this case the pooling strategy profile (Fight, Fight) can be supported as a PBE if off-the-equilibrium beliefs, r, satisfy $r < b$.

Case 2. If, instead, $r \geq b$, then the entrant is deterred upon observing the off-the-equilibrium message of Acc, as Fig. 10.26 depicts. The pooling strategy profile cannot be sustained in this context, since M_N has incentives to deviate towards Acc, obtaining a payoff of a, which exceeds its payoff of zero when he

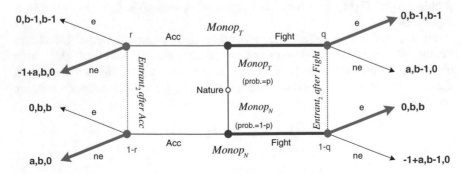

Fig. 10.26 Pooling strategy profile—Fight (with responses)

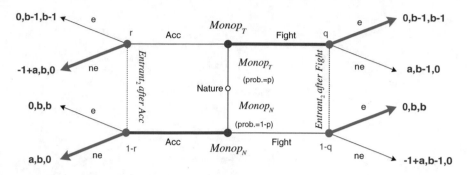

Fig. 10.27 Separating strategy profile (Fight, Acc)

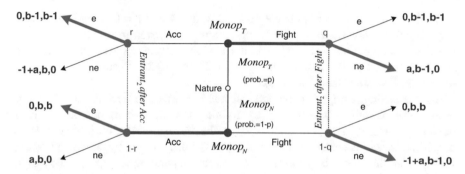

Fig. 10.28 Separating strategy profile (Fight, Acc), with responses

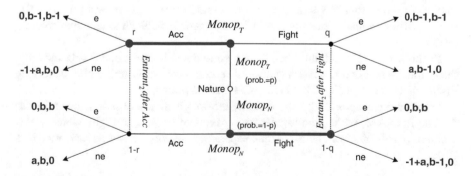

Fig. 10.29 Separating strategy profile (Acc, Fight)

Fights. Hence, the pooling strategy profile (Fight, Fight) cannot be supported as a PBE when off-the-equilibrium beliefs satisfy $r \geq b$.

Separating PBE (Fight, Acc). Let us next examine if the separating strategy profile in which only the tough monopolist fights can be sustained as a PBE. Figure 10.27 depicts this strategy profile.

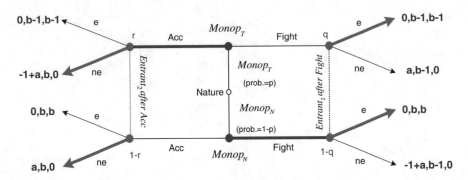

Fig. 10.30 Separating strategy profile (Acc, Fight), with responses

In this case, entrant's beliefs are updated to $q = 1$ and $r = 0$ using Bayes' rule, implying that, upon observing Fight, the entrant is deterred from the market since $b - 1 < 0$, given that $b < 1$ by definition. Upon observing Acc, the entrant is instead attracted to the market since $b > 0$. Figure 10.28 illustrates the entrant's responses (see blue shaded arrows).

In this setting, no type of monopolist has incentives to deviate: (1) the tough monopolist obtains a payoff of a by fighting (as prescribed) but only 0 from deviating towards Acc; and similarly (2) the normal monopolist obtains 0 by accommodating (as prescribed) but a negative payoff, $-1 + a$, by deviating towards Fight, given that $a < 1$ by definition. Hence, this separating strategy profile can be sustained as a PBE.

Separating PBE (Acc, Fight). Let us now check if the alternative separating strategy profile, in which only the normal monopolist fights, can be supported as a PBE (We know that this strategy profile sounds crazy, but we want to formally show that it cannot be sustained as a PBE.) Figure 10.29 illustrates this strategy profile.

In this setting, the entrant's beliefs can be updated to $r = 1$ and $q = 0$, inducing the entrant to respond not entering after observing Acc, but entering after observing Fight, as depicted in Fig. 10.30 (see blue shaded branches).

Given these responses by the entrant, the tough monopolist has incentives to deviate from Acc, which yields a negative payoff of $-1 + a$, to Fight, which yields a higher payoff of zero. Therefore, this separating strategy cannot be supported as a PBE.

Errata to: Strategy and Game Theory

Felix Munoz-Garcia and Daniel Toro-Gonzalez

Erratum to:
F. Munoz-Garcia and D. Toro-Gonzalez, *Strategy and Game*
Theory, **Springer Texts in Business and Economics,**
DOI 10.1007/978-3-319-32963-5

The original version of the book was inadvertently published with incorrect information throughout the book, correction are given as below:

Chapter 1—Dominance Solvable Games

- Page 1, Introduction. At the end of the second paragraph the index "i" must be in italics.
- Page 1, Introduction. At the end of the third (last) paragraph, the sixth line from the final "lover" must be changed by "lower".
- Exercise 7, Page 11, Fig. 1.14 should delete column x rather than z. The figure and subsequent text should be changed as follows:

The updated online version of this book can be found at
http://dx.doi.org/10.1007/978-3-319-32963-5

F. Munoz-Garcia
School of Economic Sciences, Washington State University, Pullman, WA, USA

D. Toro-Gonzalez
School of Economics and Business, Universidad Tecnológica de Bolívar,
Cartagena, Bolivar, Colombia

Player 2

Player 1

	y	z
b	1,3	0,2
c	2,1	1,2
d	0,1	2,4

Fig. 1.14 Reduced normal-form game

"We can now move to player 1 again. For him, strategy c strictly dominates b, since it provides an unambiguously larger payoff than b regardless of the strategy selected by player 2 (regardless of the column). In particular, when player 2 chooses y (left-hand column), player 1 obtains a payoff of 2 from selecting strategy c but only one from strategy b. Similarly, if player 2 chooses z (in the right-hand column), player 1 obtains a payoff of one from strategy c but a payoff of zero from strategy b. As a consequence, strategy b is strictly dominated, which allows us to delete strategy b from the above matrix, obtaining the reduced matrix in Fig. 1.15.

Player 2

Player 1

	y	z
c	2,1	1,2
d	0,1	2,4

Fig. 1.15 Reduced normal-form game

At this point, note that returning to player 2 we note that z strictly dominates y, so we can delete strategy y for player 2 and finally, considering player 2 always chooses z, for player 1 strategy d strictly dominates c, since the payoff of 2 is higher than one unit derived from playing c. Therefore, our most precise equilibrium prediction after using IDSDS are the solely remaining strategy profile (d,z), indicating that player 1 will always choose d, while player 2 will always select z."

- Exercise 8, Page 14, In the third paragraph after $x_1 = 0$ change "yields" by "does not yield". Similarly, after $x_2 = 0$ change "but the same" by "but lower".
- Exercise 10, Page 17, The reference to "Fig. 1.21" should be changed for "Fig. 1.23."
- Exercise 11, Page 20,

  - In the first displayed equation at the top of the page, the multiplicative sign should have $i = 1$ in its subscript below and I in the superscript above.
  - In the second displayed equation, its second line should have subscript i everywhere instead of j.
  - In the third displayed equation, the subscript of the multiplicative sign should be $j \neq i$ instead of $j = 1$.

- Exercise 12, Page 21, Starting the answer change the word "above" for "below".

Chapter 2—Nash Equilibrium And Simultaneous-Move Games with Complete Information

- Page 25, Introduction. Nash equilibrium: Change the word "people" for "profile".
- Exercise 4,

  - Page 31, After $c > 0$, add "with no fixed costs"
  - Page 32, At the end of the second paragraph of the answer key, "i" should be in italics.
  - Page 33, In the figures the best response functions are represented by the acronym *BRF*. To be consistent with the acronym used in the accompanying text, all acronyms in the figures should be *BR*.

- Exercise 8,

  - Page 44, At the end, before the last paragraph starting with "Therefore" change the payoff of $3 for a payoff of $2.
  - Page 46, The solution to the quadratic equation, at the beginning of the second paragraph, should be $m_x = \frac{1 \pm \sqrt{1-4(1-n)}}{2}$. The results of the rest of the exercise do not change.

- Exercise 9, Page 48,

  - Add at the end of the first case at the top of the page (immediately after "as depicted in Fig. 2.29"): Similarly, when the location of the three candidates satisfies $x_D^* = x_R^* = x_I^* > 1/2$ each candidate has incentives to deviate towards the left, while if $x_D^* = x_R^* = x_I^* < 1/2$ each candidate has incentives to deviate to the right.
  - Second case. End of the first line. Replace "two candidates chooses" for "two candidates choose".

- Exercise 10, Page 51, The second paragraph in the answer key, the best response function is referred as *BRF*. For consistency, it should be referred to as *BR*.
- Exercise 12,

  - Page 54, At the end of paragraph 2 instead of "...while the probability of being caught is $\frac{1}{1+xy}$" should be "...while the probability of not being caught is $\frac{1}{1+xy}$."
  - Page 55, At the end of part (a) it should read "$BR_G, x(y) = \frac{y}{c^2}$" instead of "$BR_G, x(y) = \frac{y}{c}$". In addition, the next sentence should read "it is convenient

to solve for y which yields $y = c^2x$" instead of "it is convenient to solve for y which yields $y = cx$".

- Page 55, Middle of page, makes a reference to "Fig. 2.29" which should be changed to "Fig. 2.32."
- Page 55, In part (b), please replace the sentence "...you can plug in the first expression into the second expression" for "...you can plug in the second expression into the first expression".
- Page 55, The title of Fig. 2.32 should be "Incentives and Punishment" instead of "Lobbying-Best response functions and Nash equilibrium".
- Page 56, The title of Fig. 2.33 should be "Incentives and Punishment-Comparative Statics" instead of "Lobbying-Comparative Statics".

- Exercise 13,

  - Page 57, The last sentence at the end of part (a) should read "Hence, equilibrium prices are $p_C = 1 - \frac{1-c}{4} - \frac{1-c}{4} - \frac{1-c}{4} = 1 - 3\left(\frac{1-c}{4}\right) = \frac{1+3c}{4}$, and every firm's equilibrium profits are $\pi_C = \left(\frac{1+3c}{4} - c\right)\frac{1-c}{4} = \frac{(1-c)^2}{16}$."
  - Page 58, Last paragraph of the page, fifth line from the bottom, "above this cutoff indicate parameters indicate" must be changed for "above this cutoff indicate".

Chapter 3—Mixed Strategies, Strictly Competitive Games, and Correlated Equilibria

- Exercise 2, Page 69, At the msNE listed immediately above part (c) of the exercise, the second parentheses should have 1/2 for both sets of probabilities.
- Exercise 5,

  - Page 76, The first paragraph should say "Similarly, let p_1 represent the probability that player 1 chooses T, p_2 the probability ..."
  - Page 76, The second displayed equation in the triplet at the center of the page should read $EU_1(C) = 1q + 2(1 - q) = 2 - q$.
  - Page 77, The end of the first paragraph, it should read "showing that only some of them can be sustained in equilibrium."
  - Page 77, In the section "*Mixing between T and C alone*" player 1's indifference condition is $EU_1(T) = EU_1(C)$, which in the next line becomes $1 + 2q = 2 + q$. This displayed equation should actually read $1 + 2q = 2 - q$. The text to the right-hand side of the displayed equation should be then changed for "which yields $q = \frac{1}{3}$. The following paragraph (after the displayed equation) should read "Hence, player 1 randomizes between T and C, assigning a probability of $q = \frac{1}{3}$ to T and $1 - q = \frac{2}{3}$ to C.

Last, the last displayed equation of the page should read "$2 - q = 2$, which yields $q = 0$."

– Page 78, The displayed equation at the top of the page should read $1 + 2q = 2 - q = 2$. The line immediately after should then read "Providing us with two equations, $1 + 2q = 2$ and $2 - q = 2$, which cannot simultaneously hold, i.e., $1 + 2q = 2$ entails $q = 1/2$ while $2 - q = 2$ yields $q = 0$.

– Page 78, Delete the last paragraph of Exercise 5, starting at "Hence, the unique msNE ..." and ending at "...probability on each as well."

• Exercise 7, Page 82, In the last paragraph of the page, replace "this result in Fig. 3.19 by noticing that" with "this result in Fig. 3.22 by noticing that".

• Exercise 8,

– Page 85, The first paragraph of part (c) should end with a parenthesis, "...as illustrated in Fig. 3.27."

– Page 86, The last paragraph of Exercise 8 should include the following explanation at the end of the paragraph: "The equation in of the line connecting points $(2, 7)$ and $(7, 2)$ is $u_2 = 9 - u_1$. To see this, recall that the slope of a line can be found in this context with $m = \frac{2-7}{7-2} = -1$, while the vertical intercept is found by inserting either of the two points on the equation. For instance, using $(2, 7)$ we find that $7 = b - 2$ which, solving for b, yields the vertical intercept $b = 9$. It is then easy to check that point $(4.5, 4.5)$ lies on this line since $4.5 = 9 - 4.5$ holds with equality."

• Exercise 9,

– Page 88, The msNE displayed at the center of the page should read $\left\{ \left(\frac{2}{3} U, \frac{1}{3} D \right), \left(\frac{1}{3} L, \frac{2}{3} R \right) \right\}$.

– Page 89, Figure 3.31 should have the labels changed to "(1,2), psNE (D,R)" in the upper left-hand side of the figure, and to "(2,1), psNE (U,L)" in the lower right-hand side of the figure.

– Page 91, Fig. 3.33 should have the labels changed to "(1,5), psNE (D,R)" in the upper left-hand side of the figure, and to "(5,1), psNE (U,L)" in the lower right-hand side of the figure.

• Exercise 11,

– Page 99, The displayed equation in part (b) should be changed to: "we find that player 1 prefers the latter, i.e.

$$u_1(C, NC) = -10 < 0 = u_1(NC, C)$$

while player 2 prefers the former i.e.,

$$u_2(C, NC) = 0 > -10 = u_2(NC, C).$$

- Page 100, The third paragraph starting at "Hence, this game is…" should have $s = (NC, NC)$ rather than $s = (NC; NC)$.
- Exercise 12,

  - Page 102, The last displayed equation of the page should read $3q + 3 = 6 - 6p \Leftrightarrow p = 1/3$.
  - Page 105, The displayed equation at the middle of the page should read $6q = 3q + 6(1 - q) \Leftrightarrow q = 2/3$. (Only the part after the arrow needs to be fixed.)
  - Page 106, The displayed equation at the top of the page should read $\left\{ \left(\frac{1}{4} Top, \frac{3}{4} Bottom \right), \left(\frac{2}{3} Left, \frac{1}{3} Right \right) \right\}$. In part (d) of the exercise, the second line of expression EU_1 should read $6\left(\frac{1}{4}\right)\left(\frac{2}{3}\right) + 3\left(\frac{2}{3}\right)\left(\frac{3}{4}\right) + 6\left(\frac{1}{3}\right)\left(\frac{3}{4}\right) = 1 + \frac{3}{2} + \frac{1}{2} = 4$. Similarly, the second line of expression EU_2 should read $6\left(\frac{1}{4}\right)\left(\frac{1}{3}\right) + 2\left(\frac{2}{3}\right)\left(\frac{3}{4}\right) = 1.5$. Last, part (e) of the exercise should read "Player 1's expected utility form playing the msNE of the game, 4, coincides with that from playing his maxmin strategy, 4. A similar argument applies to Player 2, who obtains an expected utility of 1.5 under both strategies."

Chapter 4—Sequential-Move Games with Complete Information

- Exercise 1,

  - Page 108, Two lines before the displayed equation, "his monetary payoff is lower than" should be replaced for "his monetary payoff is higher than".
  - Page 109, In the 4th line of Part (a) it should read: "…the **utility** of player 2 (responder) is…" instead of "…the **payoff** of player 2 (responder) is…".
  - Page 109, The equation $1 - m$, immediately after the first displayed equation, should have a minus sign (not a dash) between the 1 and the m.
  - Page 109, At the equilibrium payoffs at the center of the page (displayed equation), the payoff from player 1 should read $\frac{1+\alpha}{1+2\alpha}$. That is, the numerator should be $1 + \alpha$ rather than $1 - \alpha$.
  - Page 109, As a consequence, the derivative two lines below should read $\frac{\partial(1-m^*)}{\partial\alpha} = -\frac{1}{(1+2\alpha)^2}$. This derivate is still negative, so all the subsequent intuition at the end of page 109 remains unaffected.
  - Page 110, Fig. 4.2, should be changed according to the correction on Page 109 in the equilibrium payoffs: $(1 - m^*, m^*) = \left(\frac{1+\alpha}{1+2\alpha}, \frac{\alpha}{1+2\alpha} \right)$. The figure below should replace Fig. 4.2.

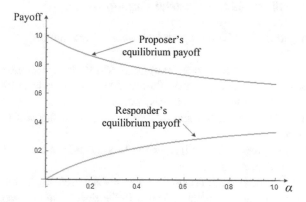

In Part(c), the last sentence of the paragraph, starting at "When $\alpha = 0.5$, the equilibrium split is...", should be replaced for "When $\alpha = 0.5$ the equilibrium split becomes $(1 - m^*, m^*) = \left(\frac{1+0.5}{1+2x0.5}, \frac{0.5}{1+2x0.5} \right) = \left(\frac{1.5}{2}, \frac{0.5}{2} \right) = (0.75, 0.25)$, thus indicating that the proposer offers more to the responder as he cares more about the payoff difference, and when the proposer cares the most about the payoff difference, $\alpha = 1$, the equilibrium split becomes $(2/3, 1/3)$. However, for all the values of α between zero and one, the proposer's payoff is higher than that of the responder.

- Exercise 4, Page 115, In the sixth line, it should read "player 2 has the opportunity to give any, all, or none".
- Exercise 5,

  - Page 117, Instead of $Q = q_1 + q_2$ it should be $Q = q_L + q_F$.
  - Page 118, At the bottom of the line, after "which simplifies into" the displayed equation should be $\frac{1}{2}[(1 + c_F) - q_L]q_L - c_L q_L$. Hence, the last displayed equation of page 118 should read $\frac{1}{2}(1 + c_F) - q_L - c_L = 0$. All subsequent calculations in this exercise are correct.
  - Page 121, In the third line of 'Bertrand competition', "equliibrium" should be replaced by "equilibrium".
  - Page 122, In the figure, the Cournot output should be $\frac{2(1-c)}{3}$.

- Exercise 6, Page 125, Both profits at the top of the page should be corrected to $\pi_1 = 5027.7$ and $\pi_2 = 5181.62$, respectively.
- Exercise 12, Page 134, In the seventh line, it should read: "If, instead, the new iPhone is introduced".
- Exercise 12, Page 136, On the 4th line of the text, instead of '...payoff is only 2...', should be '...payoff is only 0...'.
- Exercise 12, Page 136, In the 8th line of the discussion on Company I- case II, instead of '...its payoff is 2...' should be '...its payoff is 4...'.

Chapter 5—Applications to Industrial Organization

- Exercise 5 (Strategic advertising and product differentiation).

  - Page 161, After the sentence "Plugging firm j's best response function into firm i's, we find" the expression should have a d_j in the last numerator on the right-hand side (end of parenthesis) rather than a d_i.
  - Page 161, Expression 5.3 should not have a 2 in the second term of the denominator.
  - Page 162, Expression q_j^* in the second line should not have a 2 in the second term of the denominator.
  - Page 162, The derivative of output q_j^* with respect to d_i should not have a 2 in the second term of the denominator; neither in the first nor in the second expression.
  - Page 162, The derivative of profit π_j with respect to A_i should not have a 2 in the second term of the denominator; neither in the first nor in the second expression.
  - Page 162, The derivative of d_i with respect to A_i should not have a 2 in the second term of the numerator.

- Exercise 7, Page 165, The output $q_i = q_j$ in the second paragraph should not have a 2 in the second term of the denominator.
- Exercise 9, Page 173, The price p_1 should go in italics.
- Exercise 11, Pages 180–181, Instead of $m - \sqrt{m} - 1$ it should be $m + \sqrt{m} - 1$.

Chapter 6—Repeated Games and Correlated Equilibria

- Exercise 8 (Collusion and Imperfect Monitoring).

  - Page 209, The plus sign in $\frac{d(a+dq_i)}{4b^2}$ (firm i's best response function at the center of Page 209) should have a minus sign "-" so it reads $\frac{d(a-dq_i)}{4b^2}$.
  - Page 210, The first two "4"s in the denominator of the profits from the collusive agreement, π_i^m (at the center of the page), should be 2 s, so the expression reads

$$\pi_i^m = \left(a - b\frac{a}{2(b+d)} - d\frac{a}{2(b+d)} \right) \frac{a}{4(b+d)} = \frac{a^2}{8(b+d)}$$

  - Page 211, The $\delta\pi_i^D$ in the right-hand side at the bottom of the page should read $(1+\delta)\pi_i^D$.

- Page 212, The second displayed equation should have an 8 in the denominator rather than a discount factor δ; reading $\frac{a^2}{8(b+d)}$ rather than $\frac{a^2}{\delta(b+d)}$.
- Page 215, The previous to the last displayed equation should have a "greater than or equal" $\geq$ sign rather than $=$; reading $\frac{a^2}{2(b+d)} \geq \ldots$

Chapter 7—Simultaneous-Move Games with Incomplete Information

- Exercise 1,

    - Page 220, Table 7.2 should have a payoff pair $(1,-1)$ in the cell corresponding to (Bf, Bet), located in the second row, left column.
    - Page 220, In the first bullet point, it should read "For player 2, his best response when player 2 bets (in the left-hand column) is to play Bf since it yields a higher payoff, i.e., 1, than any other strategy, i.e., $BR_1(Bet) = Bf$.
    - Page 220, In the second bullet point, the second sentence should read "If, instead, player 1 chooses Bf (in the second ow), player 2's best response is to fold, $BR_2(Bf) = Fold$, since his payoff from folding, $-1/3$, is larger than from betting, -1.
    - Page 221, Table 7.3 should have a payoff pair $(1,-1)$ in the cell corresponding to (Bf, Bet), located in the second row, left column. In addition, player 2's underlined payoff in the second row should be $-1/3$ (that corresponding to Fold in the right-hand column) rather than that of betting.

- Exercise 2, Page 223, In the paragraph with title "*Player 2's best responses*", the sixth line should read "i.e., $4 + p > 1$, $4 + p > 4$-$3p$, and $4 + p > 1 + 4p$, which hold for all values of p."
- Exercise 5, Page 234, The displayed equation after "rearranging yields" in the middle of the page should read $B(p - \beta p) > L(p - \beta p)$. All subsequent results are correct.

Chapter 8—Auctions

- Exercise 3, Page 248, The expressions in the section with title "Direct approach" should not have $N - 1$ in the exponents. This applies to the expression of the *prob* (*win*), the $EU_i(v_i)$, and its rearranged representation at the bottom of the page.

Chapter 9—Perfect Bayesian Equilibrium and Signaling Games

- Exercise 1,

    - Page 261, The last sentence of the page should read "with U, which provides him a payoff…".

- – Page 262, Part (b) of the exercise should be preceded by space to separate it from the answer key of part (a).
- – Page 265, Footnote 2. Strategy R' should read R', for consistency with the primes in previous parts of the exercise.

- • Exercise 2, Page 274, The paragraph labelled *Case 1* should start as follows "When $q > \frac{2}{5}$, the…" The end of this case (immediately before Case 2) should read "beliefs satisfy $q > \frac{2}{5}$." The paragraph labelled *Case 2* should start as follows "When $q \le \frac{2}{5}$, the firm…:"
- • Exercise 3,

  - – Page 277, The first paragraph of part (a), in the last line, should read "only stems from a sender".
  - – Page 280, The second line after the three expected utility expressions should read "if and only if $\mu > \frac{4}{5}$, $4 - \mu > 5 - 5\mu$" (Now one of the minus signs looks like a hyphen.).

- • Exercise 4,

  - – Page 285, The expression of p_2^2 immediately above expression (B) should have a 2 in the denominator, rather than a 3. As a consequence, expression (B) should become $p_2^2 = 1 + \frac{1}{3}f^{-1}(p_1^1)$.
  - – Page 286, The expression of p_2^2 in the paragraph starting with "Firm 1 anticipates…" should be corrected for $p_2^2 = 1 + \frac{1}{3}f^{-1}(p_1^1)$. A similar argument applies to the expression plugged at the end of the big parenthesis in the profit maximization problem immediately after this paragraph. As a consequence, the subsequent calculations should be replaced for the following:
  - – "Taking first order condition with respect to p_1^1, we obtain $1 - 2p_1^1 + p_2^1 + c_1 + 2\left(1 - \frac{1}{3}c_1\right)\frac{1}{3}f'\left(f^{-1}(p_1^1)\right) = 0$.

Simplifying and solving for p_1^1, we find firm 1's best response function in the first period game

$$p_1^1(p_2^1) = \frac{1+c_1}{2} + \frac{p_2^1}{2} + \frac{(3-c_1)f'\left(f^{-1}(p_1^1)\right)}{9}$$

Inserting this expression of p_1^1 into $p_2^1 = \frac{1+p_1^1}{2}$, we obtain the optimal second-period price for Firm 1

$$p_2^1 = \frac{1+c_1}{2} + \frac{2}{3} + \frac{2(3-c_1)f'\left(f^{-1}(p_1^1)\right)}{27}.$$

Plugging this result into the best response function $p_1^1(p_2^1)$, yields the optimal first-period price for Firm 1

$$p_1^1 = \frac{3 + 2c_1}{3} + \frac{4(3 - c_1)f'\left(f^{-1}(p_1^1)\right)}{27}$$

As suggested in the exercise, let us know...."

- Page 288, This Page should be replaced for the following, starting in the expression at the top of the page:

$$A_0 + A_1 c_1 = \frac{3 + 2c_1}{3} + A_1 \frac{4(3 - c_1)}{27}$$

since A_1 measures the slope of the pricing function (see Fig. 9.30), thus implying $A_1 = f'\left(f^{-1}(p_1^1)\right)$. Rearranging the above expression, we find

$$A_1(31c_1 - 12) = 27 + 18c_1 - 27A_0$$

which, solving for A_1, yields

$$A_1 = \frac{27(1 - A_0) + 18c_1}{31c_1 - 12}$$

In addition, when firm 1's costs are nil, $c_1 = 0$, the above expression becomes

$$27A_0 = 27 + 12A_1$$

or, after solving for A_0,

$$A_0 = 1 + \frac{4}{9}A_1$$

Inserting this result into $A_1 = \frac{27(1-A_0) + 18c_1}{31c_1 - 12}$, yields

$$A_1 = \frac{27\left(1 - \left(1 + \frac{4}{9}A_1\right)\right) + 18c_1}{31c_1 - 12}.$$

Solving for A_1, we obtain $A_1 = \frac{18}{31} \approx 0.58$. Therefore, the intercept of the pricing function, A_0, becomes

$$A_0 = 1 + \frac{4}{9}\frac{18}{31} = \frac{39}{31} \approx 1.26.$$

Hence, the pricing function p_1^1 of firm 1, $p_1^1 = A_0 + A_1 c_1$, becomes

$$p_1^1 = \frac{18}{31} + \frac{39}{31} c_1.$$

- Exercise 6, Page 296, The two maximization problems in the middle of the page should have $2c$ in the last term, rather than c. In addition, the first-order condition with respect to p_T should have the parenthesis corrected, as follows

$$\lambda - (1 - \lambda) \frac{\theta_T - \theta_B}{\theta_B} = 0$$

Chapter 10—More Advanced Signaling Games

- Exercise 1,

  - Page 307, The labels below each term at the center of the page should read P_2's, with the apostrophe after the P_2.
  - Footnote 1, Page 307, It reads "alpha" when it should read α.

- Exercise 3, Delete part (b), both in the question (Page 317), and in the answer key (Pages 319–322).
- Exercise 4, Page 322, last line should read "they are associated with a higher wage".
- Exercise 5,

  - Page 327, The second line of the answer key of part (a) should read "while a_E is the benefit".
  - Page 328, Figure 10.8 should have all subscripts in capital letters, such as f_I, f_E, a_I, and a_E. In addition, all notations in this exercise should go in italics.

- Exercise 6, Page 340, Figure 10.27 should have branch *Acc* shaded in the lower part of the game tree, since the figure represents the separating strategy profile (*Fight, Acc*).

Index

A

All-pay auction, 237, 246, 251, 252

Auctions, 237–239, 244–247, 251, 252, 254, 255

Anti-coordination game, 17, 63, 68, 95, 97

B

Backward induction, 107–109, 125, 129, 130, 132, 136, 171, 332, 333

Bargaining game, 80, 82, 108, 109, 127, 129, 202, 203

Battle of the Sexes game, 16, 26, 29–31, 62, 100, 101

Bayesian Nash Equilibria (BNE), 217

Bayes' rule, 92, 96, 257, 263, 265, 279, 282, 290, 306, 330, 339

Belief updating, 257, 265, 267–271, 274, 339, 341

Beliefs, 257, 258, 260–262, 264, 265, 267, 269–271, 273, 277, 280–282, 293, 300, 302, 312, 324, 329, 335, 341

Bertrand game, 47, 118, 145, 148, 150, 153, 154, 195, 233

Best response function, 20, 32, 34, 36, 38, 39, 51, 52, 55, 57, 66, 68, 70, 85, 117, 123, 124, 126, 130, 148, 154, 155, 158, 161, 169, 175, 209, 214, 285, 309

Best responses, 25, 26, 29, 41, 44, 45, 78, 83, 87, 115, 135, 197, 204, 220, 223, 225

Blume, 154

Branch (of a game tree), 7, 133, 270, 272, 274, 275, 307, 338

C

Cartel, 183, 188–194

Cheating (in repeated game), 185, 186

Cho and Kreps' (1987) intuitive criterion, 258, 298–303, 322, 325, 337

Collusion, 188, 192, 194, 196, 208, 211, 215

Competition in prices, 47, 121, 124, 145, 146, 151, 154, 195

Competition in quantities, 26, 31–34

Complete information, 25, 26, 145, 218, 230, 231, 303, 308, 309, 327

Conditional probability, 92, 93, 277, 291

Concavity, 131

Continuous action, 151

Convex hull of equilibrium payoffs, 85, 86, 91

Convexity, 206

Cooperation, 183–187, 191, 195, 211, 212

Coordination games, 17, 30, 53, 78, 309

Correlated equilibrium, 62, 83, 85, 86, 91, 93, 95–97

Cournot game, 26, 31, 32, 118, 145, 160, 188, 192, 218

Credible punishment, 192

D

Defect (in repeated games), 183, 185–187, 207

Deviation, 38, 49, 146, 147, 150–154, 183, 184, 190, 192, 193, 195, 196, 207–209, 212, 215, 293

Direct approach, 38, 239, 241–243, 247–249

Direct demand, 122, 124, 213, 283

Discrete action, 151

Dominance solvable games, 1, 9

Duopolists, 56

Duopoly, 145–147, 152, 156, 157, 171, 230

E

Efficiency gains, 26, 56, 59

Entry deterrence, 327–329, 332–334

Printed in the United States
By Bookmasters